Under An Ionized Sky

I wanted to illuminate the whole earth.
There is enough electricity to become a second sun.
Light would appear around the equator,
as a ring around Saturn.
— NIKOLA TESLA

Under An Ionized Sky

From Chemtrails to Space Fence Lockdown

ELANA FREELAND

SEQUEL TO

Chemtrails, HAARP, and the
Full Spectrum Dominance of Planet Earth

 FERAL HOUSE

Feral House
1240 W. Sims Way
Suite 124
Port Townsend WA 98368
www.feralhouse.com

Book design by Unflown Studio

Dedicated to those who have spent thousands of hours researching matters this book touches upon, and to those who have died or are now in pain for speaking out.

ACKNOWLEDGEMENTS

Thanks first go to the Internet chemtrails activists networking in a horizontal and democratic fashion on social media and by email, daily studying the skies, satellite and cell phone photos and film footage, patents, scientific papers and essays in order to decipher this historical moment and what is unfolding around and above us. These activists continue attempting to educate the public even in the face of being laughed at or ostracized by friends and family duped by media spin on "conspiracy theories."

In particular, I thank Billy Hayes "The HAARP Man," Jim Lee, Harold Saive, Rose and Greg at The CON Trail, Clifford and Carol Carnicom, Michael J. Murphy, Ann Fillmore, Suzanne Maher, Anthony Patch, Harald Kautz-Vella, Leuren Moret, Christina "Rad Chick" Consolo, Sofia Smallstorm and WeatherWar101, Sean Gautreaux, Peter A. Kirby, DJ Walsh, J. Marvin Herndon, Christopher Fontenot, Pete Ramon, David Masem, and the many radio show hosts who have invited other activists and me to share this clarion wakeup call with their radio and Internet audiences.

No one ever writes a book like this alone.

Table of Contents

Foreword

There is a world that most of us know of and that we are comfortable with. There is another world that we should know of but do not, and that should, by all rights, make us uncomfortable. We are responsible to learn of its existence, as it surrounds and affects our every move regardless of the level of bliss or ignorance that we adopt to ease our discomfort.

Elana Freeland, in the second of a trilogy series, thrusts upon us the realities of that more complex world in her most recent work, *Under an Ionized Sky: From Chemtrails to Space Fence Lockdown.* We are destined within it to discover that the technological throes that dominate the modern world are much more developed, enmeshed, and controlling of our lives than most of us would like to either admit or confront.

I have known Elana for some time now, and it was not known in our earlier café chats that we would someday share deeply in a common pursuit of truth. The focus of our work may differ in certain respects, but it is the same battle. Elana is providing a map of what is and what may well lie ahead in a world now gone berserk in the idols, greed, and powers of technocracy. It is our job to maintain our spiritual integrity in the clash that is before us, and this book will ask us to trade our ignorance and complacency for the awareness and strength that is needed in battle.

The book is meticulously researched and documented, and she has done a great deal of your homework for you. The morass of controlling systems spoken of here does indeed exist, and they are of purpose and agenda. There is much for all of us to learn here and it is to our benefit to get up to speed quickly on the state of affairs. I hope that you will immerse yourself and gain the quick education that we all need, as proclaiming ignorance or innocence will be at our own peril. As Elana says, "It's a brave new world." Let us be brave together in its future.

—CLIFFORD E. CARNICOM

Preface

▼

We in this country, in this generation, are—by destiny rather than choice—the watchmen on the walls of world freedom.
— President John F. Kennedy, undelivered speech
to the Dallas Citizens Council, Nov. 22, 1963

Many people have criticized my research into the destruction of the WTC complex because I have not named the exact technology that was used in the destruction, including its make, model, and serial number. But it is erroneous to blindly discard evidence that does not conveniently describe a known weapon or blindly discard evidence that contradicts one's pet theory. Remember, empirical evidence is the truth that theory must mimic, not the other way around. The pages of this book include a very great amount of evidence—evidence that must be explained. . .
— Judy Wood, Ph.D., *Where Did the Towers Go? Evidence of Directed Free-Energy Technology on 9/11*, 2005

What occurred in New York City on September 11, 2001 and remains to haunt thousands, if not millions, was a blatant demonstration of the destructive power of what two interfering beams (possibly three, the third for holograms) from directed energy weapons (DEWs) can do. Everything subsequently framed as the "war on terror" sabotaged the direction of the entire nation, and the Nazi neocons and Deep State Democrats responsible are still at large.

The 9/11 display of power was not the first time that Americans witnessed a DEW in action without realizing what they were looking at, though it was, to date, the most devastating. The downing of Paris-bound TWA Flight 800 with 230 people on board on July 17, 1996 was also a scalar DEW. The National Transportation Safety Board (NTSB) report documented dozens of witnesses who saw "something resembling a flare or firework ascend and culminate in an explosion." As with 9/11, the initial hours after the crash were carefully controlled.[1]

---

[1] Dylan Stableford, "TWA Flight 800 crash not due to gas tank explosion, former investigators say." *Yahoo! News*, June 18, 2013. Later that year, American Airlines bought out TWA.

Then there was the plane crash of Senator Paul Wellstone (D-MN), a member of the Democratic Farmer-Labor Party known to be critical of the War on Drugs preceding the War on Terror. When Wellstone visited Bogotá, Colombia in December 2000, he was drenched in chemical spray while witnessing aerial fumigation of a coca field.[2] Two years later on October 25, 2002, while investigating 9/11, he and his wife were killed in a plane crash.

> The NTSB accident report is disputed in the [2004] book *American Assassination: The Strange Death of Senator Paul Wellstone* by Don "Four Arrows" Jacobs and James H. Fetzer. The authors allege there are problems with "the official story" and that there is evidence of an official cover-up. One of their principal claims is that FBI personnel left the field office in St. Paul for northern Minnesota before the plane crashed. They propose that the plane was shot down with a directed-energy weapon. Snowshoe Documentary Films produced *Wellstone: They Killed Him*, a two-hour film that makes similar allegations.[3]

Then there was the space shuttle *Columbia*, STS-107, that blew up over Texas upon re-entry on February 1, 2003 after its twenty-eighth mission. A digital photograph snapped by an amateur San Francisco astronomer caught

> a purplish electrical bolt striking the craft as it streaked across the California sky...In the critical shot, a glowing purple rope of light corkscrews down toward the plasma tail, appears to pass behind it, then cuts sharply toward it from below. As it merges with the plasma trail, the streak itself brightens for a distance, then fades.[4]

In 2005, *Where Did the Towers Go? Evidence of Directed Free-Energy Technology on 9/11* by Judy Wood, Ph.D., closely examined how five-hundred-thousand-ton buildings a half-mile tall, steel girders, concrete, and people's bodies could disintegrate to dust in eight to ten seconds:

> ...a weapon that could produce all of the effects we saw on 9/11 would certainly not be in the public domain, no matter whose weapon it was ...9/11 was the

---

2 Jenny O'Connor, "Colombia's Agent Orange?" *CounterPunch*, October 31, 2012. Plan Colombia, begun in 2000, has cost the U.S. billions in the name of the Drug War. Its purpose, however—according to Peter Dale Scott's Foreword to the 2011 book *Cocaine, Death Squads and the War on Terror: U.S. Imperialism and Class Struggle in Colombia* by Oliver Villar and Drew Cottle (Monthly Review Press)—is "to target specific enemies and thus ensure that the drug traffic remains under the control of those traffickers who are allies of the Colombian state security apparatus and/or the CIA."

3 Wikipedia.

4 Sabin Russell, "S.F. man's astounding photo / Mysterious purple streak is shown hitting Columbia 7 before it disintegrated." *San Francisco Chronicle*, February 5, 2003. Also see Michael E. Salla, Ph.D., "Photographic Analysis Confirms that Space Shuttle Columbia was Destroyed by a Plasma Beam Weapon," *Exopolitics.org*, October 19, 2007.

demonstration of a technology of horrendous magnitude.[5]

By means of evidentiary elimination, Wood discounts kerosene-fueled fire, thermite, and mini-nukes. No high heat meant it was not a laser. Next, she deduced a Hutchison Effect[6]—interfering microwave energies in an electrostatic field remotely introduced into a rotating electric field to produce a magnetic field to make molecular dissociation occur:

> By beaming a specially tuned microwave energy pulse rather than laser or infrared beams, that energy can be targeted within a storm's interior . . .[7]

The Hutchison Effect follows from ELF (extremely low frequency) steering of radio waves, high voltage, and electrostatic fields that can then produce (1) levitation; (2) fusion of dissimilar materials, e.g., metals "melting" without burning adjacent materials like paper and trees, the mythologized but very real 1943 Philadelphia Experiment, the 2x4s that skewered palm trees during Hurricane Andrew in 1992; (3) spontaneous fracturing of metals; and (4) changes in crystalline structure and physical properties.

The backup system included eleven Stealth fighters possibly loaded with DEWs and computer programs, and explosive charges pre-placed in the towers for a boost,[8] while off the coast Hurricane Erin—whipped up by chemtrails and ionospheric heaters—waited to provide yet more energy if needed. Thunder (evidence of an electrical field) was heard at JFK, LaGuardia, and Newark Airports. As each tower collapsed in seconds, the Geophysical Institute Magnetometer Array (GIMA) at the University of Alaska tracked the anomalous changes in the Earth's magnetic field, *the timing, magnitude, and relationships of fluctuations uncannily and clearly related to the molecular dissociation of the towers.*

Cars were on fire but not adjacent trees or paper. Thousands of pieces of paper from ground zero were found in Brooklyn seven miles away. People described being "picked up and moved a block" by black clouds that "had substance," as if they were inside an electrical storm or volcano. The St. Elmo's Fire flames were impossible to put out with water. Like lightning, St. Elmo's Fire is plasma: ionized air that emits a glow like a corona discharge, *luminescence without heat.* Bluish ball-like formations were seen, the same bluish ball lightning or fire columns that accompany electrical tornado systems. Were artificial "tornadoes"

---

5 Judy Wood, Ph.D. *Where Did the Towers Go? Evidence of Directed Free-Energy Technology on 9/11.* 2005.

6 Physicist savant John Hutchison (1945–) studied Tesla's electrogravitic and longitudinal wave work and that of Thomas Townsend Brown (1905–1985), George Piggott ("Overcoming Gravitation"), Francis E. Nipper (1847–1926), and Martin L. Perl (1927–).

7 Leonard David, "Taking the twist out of a twister." *Space.com,* March 3, 2000.

8 Carolyn Williams Palit, "The Direct Energy Age." *No Exotic Warfare Zone,* October 2005: "The smaller beams from psychotronic scalar weapons being used on Americans are very related to the big holographic scalar weapons used on 9/11."

(rotating fields) disintegrating the towers while plasma orbs kept an eye on the holocaust?

Photographer David Handschuh described what he witnessed when the South Tower disintegrated:

> And then this noise filled the air that sounded like a high-pressure gas line had been ruptured. And it seemed to come from all over, not from any one direction . . .And then all of a sudden the second tower explodes. . .I was just walking by myself and heard. . .another loud noise that kind of *echoed the first noise* . . .Looked up and OK it's another aircraft. . .and the whole building was just disintegrating.[9]

Lieutenant Richard Smiouskas described a "huge volume like ten stories high billowing, pushing black smoke and *like a glitter*."[10]

Disintegrating glass? Barium and quartz? Plasma is highly conductive and fluoresces *light*.

Targeted individual Carolyn Williams Palit (whose brilliant insights I quote over and over but who apparently has not outlived her persecutors) believed that the two airplanes hitting the North and South towers were holograms projected onto plasma to cloak the missiles used. Two or three seconds before the missiles hit the towers, the plasmoid holograms caused a Faraday cabin effect (like a car hit by lightning). Palit explained the hologram process this way: Two plasma beams twist around each other like pearl necklaces, one beam transmitting 2D visuals of an airliner (like a Birkeland current), and one beam heating the chemtrails to make metal-induced, ionized gas plasma (like a splay-beam weapon). A third beam is added from another ionospheric heater installation, possibly from Brookhaven National Laboratory (BNL) out on Long Island, to complete the 3D hologram. David Masem, who worked as a Unix Systems Administrator at BNL 2000–2003:

> Not only does BNL have a mini-HAARP, but it also has a collider / accelerator buried in the Earth that is one mile in diameter with four collision detect points every quarter of a mile. I worked at the collision detect point for Project AEROSOL where collisions were done to determine the effects of nanoparticulates on the environment. Immediately after 9/11, the U.S. Navy came in and took control of BNL operations and since then has ramped up its nanotechnology department. The collider / accelerator at BNL or RHIC (Relativistic Heavy Ion Collider) was at that time networked to two other collider / accelerators: CERN and another in Japan. With its close proximity to Manhattan, it would have been perfectly located to use its mini-HAARP

9 Wood, *Where Did the Towers Go?*
10 Ibid.

steering beam as a directed energy weapon to further propagate 9/11. Oh, the stories I could tell you about Brookhaven National Labs would make your head spin.[11]

Thanks to post-9/11 billions[12] and the USA PATRIOT Act of 2001 (Uniting and Strengthening America by Providing Appropriate Tools Required to Intercept and Obstruct Terrorism) that expanded military / intelligence surveillance and communications, Project Cloverleaf aerosol spraying and ionospheric heater manipulation of the atmosphere was kicked into high gear.

When the electrostatically charged Ames strain of anthrax was sent through the mail to critics of the Bush II administration, the trail led the FBI to Battelle Memorial Institute, a CIA front company specializing in weapons,[13] at Wright-Patterson Air Force Base. Under Project Jefferson and the CIA's Project Clear Vision, Battelle had engineered a static charge on the anthrax so that the spores would float out of the opened envelope and into the air.[14] Battelle's corporate partner Bioport—founded by a financier of the bin Laden family through the infamous and shadowy (Bush family) Carlyle Group—made multi-millions from the rush to stockpile anthrax vaccine. And what happened to the five Raytheon executives covering UAV [unmanned aerial vehicle] projects with Mode 4 or 5 upgrades? They were on AA Flights 77 and 11 and UA Flight 175[15] and probably thought they were on a UAV test flight.

Raytheon owned HAARP patents and figured deeply in the government terrorist event of September 11, 2001:

- Raytheon oversaw the Florida flight school that trained the "hijackers"

- Raytheon had acquired Hughes Aircraft and its fleet of A-3 Skywarriors for testing air-to-air and air-to-ground missile systems, guidance and EW systems for the U.S. Air Force, Navy, and Marines, meaning that Raytheon could have provided the AGM A-3 Skywarrior with Mode 4 or 5 IFF transponders that gouged the sixteen-foot hole through three rings of the Pentagon in Washington, D.C. (No 757 could have done it.)

---

11 Email, October 4, 2016.

12 Ned Resnikoff, "'Homeland Security' has received $791 billion since 9/11." *msnbc*, March 1, 2013.

13 Bob Fitrakis, "Battelle Exposed In Anthrax Biochemical Conspiracy." FreePress.org, January 14, 2002.

14 Battelle had hired William C. Patrick III, a bioweapons expert, to study how anthrax behaves when sent through the mail. On August 1, 2008, Dr. Bruce E. Ivins, another anthrax suspect, was "suicided"; two months later on October 1, 2010, Patrick died. See Fitrakis' "Ghoulish And Secretive Biowarfare Expert William C. Patrick III Dead," at *fraudsterbob.com*, October 17, 2010.

15 A director of research at ECOlogic Corporation, a software engineering firm, was also on AA77. ECOlogic worked on data systems for NASA.

- Raytheon oversaw the flight that spirited the bin Laden family out of the country while all other flights were grounded.

To accommodate the directed energy beams, a particular chemical mix had been laid in the atmosphere over New York City. Billy Hayes "The HAARP Man" told me that "cooking" a heavy chemical spray of barium nitrate, aluminum oxide, and ferrous oxide $[Ba(NO_3)_2 + AlO_2 + Fe_3O_2 + O]$ can result in compounds that become highly RF (radio frequency) reactive in a suspension of RR-144 and RR-129 chaffs.[16] When "cooked" with specific amino acids and nano-silicon particulates, mini-synthetic ruby crystals are formed, which, with specific aminoglycosides, can cause multiple myeloma, a very rare blood plasma cancer. Quantities of such synthetic ruby crystals were found in the lungs of those who jumped from the Towers, and multiple myelomas were found among first responders:

Many types of cancers have been reported, including leukemia and the rare disease called multiple myeloma. . .[R]esearchers associated with the World Trade Center Medical Monitoring and Treatment Program at the Mount Sinai School of Medicine examined many sick first responders. One result was that they found eight times the expected level of multiple myeloma in people below the age of 45.[17]

Eight times. Draw a line from the symptoms and illnesses listed in this one *Foreign Policy Journal* article to the typical biological fallout from the technology in my 2014 book *Chemtrails, HAARP, and the Full Spectrum Dominance of Planet Earth* (2014), then follow the chemicals conjured from jet engines and ionospheric heaters and you arrive at the resurrection of the SSS Space Fence here to finalize planetary lockdown.

My journey into "national security" technology began with this one millennial crime perpetrated against three thousand souls, their families, our nation, and the future of humanity itself.

—ELANA FREELAND
Olympia, Washington
May 20, 2016

---

16 "Cooking" = zapping with radio frequency or microwaves.

17 Kevin Ryan, "Energetic Materials as a Potential Cause of the 9/11 First Responder Illnesses." *Foreign Policy Journal*, February 4, 2011; reference to JM Moline et al., "Multiple Myeloma in World Trade Center Responders: A Case Series." *Journal of Occupational & Environmental Medicine* 51(8): 896–902, 2009.

INTRODUCTION

Planetary Lockdown, Geoengineering and "The Deep State"

▼

Just as, at the dawning of a new geological era, the whole world collapses in a gigantic crack, new mountains rise up while gaping abysses open up, and new plains and seas take shape, so will the present structure of Europe be capsized in an immense cataclysm. . .The only chance for Germany to resist this pressure will be to seize the initiative and take control of the inevitable upheaval from which will come a new dawning of history.
— Hermann Rauschning, *Hitler Speaks / Voice of Destruction*, 1939

We've arranged a society on science and technology in which nobody understands anything about science and technology, and this combustible mixture of ignorance and power sooner or later is going to blow up in our faces. I mean, who is running the science and technology in a democracy if the people don't know anything about it?
— Carl Sagan to Charlie Rose, May 27, 1996

Before leaving office, the 44th President issued a shot over the bow of the incoming President by striking the word "limited" from the National Defense Authorization Act (NDAA), thus publicly reinstating the "Star Wars" program of thirty-three years ago—and on Christmas Eve, no less.

> Republican Congressman Trent Franks, who introduced and shepherded the policy changes in the House, said he drew inspiration from former president Ronald Reagan's Strategic Defense Initiative of the 1980s, which was intended to use lasers and other space-based weaponry to render nuclear weapons

"impotent and obsolete." Known as "Star Wars," the initiative cost taxpayers US$30 billion, but no system was ever deployed.[1]

Eighteen days later, mainstream media began normalizing the geoengineering we've been undergoing for two decades in the name of "easing climate change."[2] Since then, space news has been hot and heavy: in February, increased ozone over 3.5 million square miles of the Pacific Ocean and the Western U.S. ("the most unusual meteorological event we've had in decades");[3] in March, "magnetized Rossby waves on the Sun" making it easier to "predict space weather much further in advance" (National Center for Atmospheric Research):

> On Earth, Rossby waves are associated with the path of the jet stream and the formation of low- and high-pressure systems, which in turn influence local weather events.[4]

In April, "Anthropogenic Weather" was finally admitted in *Space Science Reviews* —

> Anthropogenic effects on the space environment started in the late 19th century and reached their peak in the 1960s when high-altitude nuclear explosions were carried out by the USA and the Soviet Union. These explosions created artificial radiation belts near Earth that resulted in major damages to several satellites. . .*Other anthropogenic impacts on the space environment include chemical release experiments, high-frequency wave heating of the ionosphere and the interaction of VLF waves with the radiation belts . . .*[5] (Emphasis added.)

In May, NASA brazenly announced "a massive, human-made 'barrier' surrounding Earth," a "humungous bubble we created out in space" deserving of "calling for a whole new geological epoch to be named after us."[6]

Then on June 17, the AMC-9 satellite in geostationary orbit 36,000 kilometers above the Earth's surface since 2003 lost contact with its Luxembourg-based SES telecommunications operator and began drifting and breaking up. Radar film

---

1 Tyler Durden, "While Blaming Trump for 'Arms Race,' Obama Signs Momentous 'Star Wars II' Defense Bill." *Zero Hedge,* December 24, 2016.

2 Jamie Condliffe, "Geoengineering Gets Green Light from Federal Scientists." *MIT Technology Review,* January 11, 2017.

3 "A Massive 'Blob' of Abnormal Conditions in the Pacific Has Increased Ozone Levels." *ScienceAlert,* 17 February 2017.

4 "Planetary waves, first found on Earth, are discovered on Sun." *PhysOrg,* March 27, 2017.

5 T.I. Gombosi et al., "Anthropogenic Space Weather." *Space Science Reviews,* 13 April 2017.

6 "NASA Space Probes Have Detected a Human-Made Barrier Surrounding Earth." *ScienceAlert,* 18 May 2017.

footage of the moment revealed three orb-like objects flying near the AMC-9 in triangular formation with another orb following aft.[7] Space situational awareness concern disseminated to the public was about the breakup debris, not about a possible space attack.[8]

Thirteen days later, President Trump revived the National Space Council (to be led by Vice President Mike Pence),[9] after which the massive $696 billion NDAA passed the U.S. House of Representatives:

> Tucked in the bill is a proposal endorsed by House Armed Services Committee leaders to create a Space Corps as a new military branch under the umbrella of the Air Force. Rep. Mike Rogers, the Strategic Forces Subcommittee chairman who proposed the idea, argued that the Air Force was prioritizing its fighter jets over space, and a dedicated service was needed to stay ahead of China and Russia in what many see as the next frontier of warfare.[10]

The race to control space began in 1945 when Operation Paperclip[11] brought committed Nazi engineers, technicians, and scientists to the United States to engineer their wonders during the Hegelian ruse known to history as the Cold War. One example among the 10,000 Nazis who sought refuge in the U.S. was Arthur Rudolph, former colleague of Wernher von Braun, aerospace engineer and NASA director of the Marshall Space Flight Center. Rudolph had been director of the Mittelwerk underground rocket factory nicknamed "Dante's Inferno" where 52,000 prisoners turned out 6,000 V-2 rockets. From 1951 to 1961, Rudolph worked for Martin Marietta in Waterton, Colorado. Initially in charge of R&D for the Pershing missile, Rudolph became an American citizen, headed the Saturn project for NASA, and received the Distinguished Service Award. In 1984, he renounced his U.S. citizenship and returned to Germany, having faithfully served the transfer of the Third Reich to the United States. In 1995, Martin Marietta merged with Lockheed Corporation to form Lockheed Martin.

Thus it was through the military-industrial complex that the Trojan horse of amoral, cryptic Nazism entered the naïve, resource-rich United States. In short order, the National Security Act, formation of the CIA, and Cold War followed. Rockets, satellites, computers, MK-ULTRA brain engineering, and exotic propulsion craft thrust the twentieth century into the twenty-first century of a space age.

---

7 BPEarthWatch, "4 Unidentified Objects Take Out Satellite / On Radar!" July 2, 2017, www.youtube.com/watch?v=NQ6xsqhDTaU.

8 Eric Berger, "A satellite may be falling apart in geostationary orbit." *ArsTechnica*, July 2, 2017.

9 Bob Fredericks, "Trump signs executive order reviving National Space Council." *New York Post*, June 30, 2017.

10 Jeremy Herb, "House passes defense bill that would create 'Space Corps.'" *CNN*, July 14, 2017.

11 The Joint Intelligence Objectives Agency (JIOA) circumvented then-President Truman's anti-Nazi order by scrubbing Nazi affiliations and granting them new identities and security clearances.

The Space Age—*Star Trek*'s "final frontier"—began with the necessity of dominating airspace, the near space around the Earth, and weather. Military research into weather control was kept quiet throughout the Cold War while dire warnings about a "little ice age," "greenhouse gases," "desertification," then "extreme weather," "global warming," and "climate change" due to carbons were released to keep the dollars flowing, along with showcase international conferences packed with Ph.D.s recommending expensive "solutions" under the rubric of *geoengineering*, the intentional human-directed manipulation of the Earth's climatic systems (*Stanford Environmental Law Journal*).

The truth is that weather was to be engineered as a "force multiplier"[12] for military operations discussed in my previous book *Chemtrails, HAARP, and the Full Spectrum Dominance of Planet Earth*: (1) Weather modification, (2) environmental / geophysical modification, (3) electromagnetic manipulation, (4) military full spectrum dominance, (5) biological manipulation, (6) intelligence / surveillance, and (7) detection / obscuration of exotic propulsion technology. This was independent scientist Clifford Carnicom's list in his 2005 film *Cloud Cover*. Now that phase two of the ionospheric heater technology is being instituted—the Space Fence—I have re-configured his list to reflect some of the detail behind each operation. Hopefully, this book will fill in the rest. Remember: weather engineering is the *sine qua non* force multiplier that all the rest depend upon. (This list, too, will no doubt be re-configured as our knowledge base grows.)

WEATHER ENGINEERING / GEOENGINEERING

• Chemical / electromagnetic creation of plasma cirrus cloud cover

PLANETARY / GEOPHYSICAL OPERATIONS

•Manipulate the ionosphere to charge,
build and steer storms over and around regions
• Utilize droughts, floods, hurricanes, tornadoes, earthquakes for
environmental modification and disaster capitalism profits
•Earth harvesting for REITs (real estate investment trusts)
• Sun simulation / solar experiments[13]

---

12 Col. Tamzy J. House et al. "Weather as a Force Multiplier: Owning the Weather in 2025," August 1996. "*2025* is a study designed to comply with a directive from the chief of staff of the Air Force to examine the concepts, capabilities, and technologies the United States will require to remain the dominant air and space force in the future."

13 Tesla: "Man could tap the breast of Mother Sun and release her energy toward Earth as needed, magnetic as well as light."

ELECTROMAGNETIC OPERATIONS

- Ionization of the atmosphere
- Plasma and antimatter "farming"
- Man-made Birkeland currents, Alfvén "whistler" waves, rotating electrical fields (the Hutchison Effect), etc.
- Holography
- Exotic propulsion systems

DIRECTED ENERGY WARFARE OPERATIONS (C4)

- Scalar interferometry (ionospheric heaters, lasers / masers, particle beams, HPMs, etc.)
- Cloaking
- Detection / obscuration of exotic propulsion systems

SURVEILLANCE / NEURAL OPERATIONS

- Artificial intelligence (AI)
- Remote neural monitoring (RNM)
- EM targeting of populations and individuals
- 5G access to DNA

BIOLOGICAL / TRANSHUMANISM OPERATIONS

- "Hive mind" Morgellons delivery
- Nanoparticle delivery: sensors, microprocessors, other electro-optical technology
- Remote genetic engineering of DNA
- Replace Nature with virtual reality

"CLIMATE CONTROL"

The Federal Government has been involved for over 30 years in a number of aspects of weather modification, through activities of both the Congress and the executive branch. Since 1947, weather modification bills pertaining to research support, operations, policy studies, regulations, liabilities, activity reporting, establishment of panels and committees, and international concerns have been introduced in the Congress. There have been hearings on many of these proposed measures, and oversight hearings have also been conducted on pertinent ongoing programs.[14]

Infiltration and co-optation, compartmentalization, nondisclosure agreements, backroom deals, threats, bribes, skewed research, packed peer review committees, embedded international media—one can only marvel at the legerdemain it takes to steer international conferences, committees, publishing houses, news outlets, university and elementary school curricula so as to construct a vast global house of cards built on turning *carbons*, the building blocks of all of life, into the *cause célèbre* that explains away the fact that our atmosphere and weather are being modified by military manipulation of the ionosphere. Carbon taxes and emissions trading are quite the con, given that CO_2 is not far above the *minimum* to sustain plant life[15] and nations should be *increasing* CO_2 instead of being penalized for the CO_2 they do have.[16]

But then, the emperor wears no clothes. The first congressional report on geoengineering did not appear in the U.S. Congress until October 2010, just before the moratorium against geoengineering issued by the 10th Conference of Parties to the Convention on Biological Diversity (COP10) in Nagoya, Japan—a moratorium the U.S. had no intention of ratifying.[17] Were the delegates from 193 nations aware that geoengineering had been going on in the U.S. and other NATO nations for well over a decade?

Four months after the Nagoya moratorium, a geoengineered earthquake struck Japan.

Since then, embedded media have ramped up weather confusion in the public mind, blaming cars and industrial pollutants while assiduously ignoring

---

14 Robert E. Morrison, Specialist in Earth Sciences, Science Policy Research Division, Congressional Research Service, "Chapter 5: Federal Activities in Weather Modification." *Weather Modification: Programs, Problems, Policy, and Potential*. U.S. Senate Committee on Commerce, Science, and Transportation, November 15, 1978.

15 P. Gosselin, "Atmospheric CO_2 Concentrations At 400 PPM Are Still Dangerously Low For Life On Earth." *NoTricksZone*, 17 May 2013.

16 "Deserts 'greening' from rising CO_2." Commonwealth Scientific and Industrial Research Organization (CSIRO), 3 July 2013.

17 Juliet Eilperin, "Geoengineering sparks international ban, first-ever congressional report." *Washington Post*, October 29, 2010.

the greatest polluters and propagandists of all: the over-inflated American military and military-industrial-intelligence complex that runs it. In 2013, the Fifth Assessment Report of the Intergovernmental Panel on Climate Change (IPCC) basically admitted that solar radiation management (SRM) was already underway: "If SRM were terminated for any reason, there is *high confidence* that global surface temperatures would rise very rapidly to values consistent with the greenhouse gas forcing."[18]

Greenhouse gas forcing is the least of our worries. What about a technological military in service to the Deep State now subjecting our atmosphere to full spectrum dominance?

Now and then, scientists like CERN particle physicist Jasper Kirby[19] and NASA Goddard Space Flight Center heliophysicist Douglas E. Rowland ("There's different kinds of chemtrails, as you probably know"[20]) leak a tantalizing tidbit about what is really going on, but it is always tacitly ignored by mainstream media. An Italian senator calls for declassification of chemtrail documents,[21] a Cyprus agriculture and environment minister pledges to look into "chemtrails" a.k.a. aerial spraying[22]—and then, nothing.

Over and over again, agents and agencies near the hub of the "climate change" mafia (NASA, NOAA, EPA, IPCC, etc.) are caught lying, but embedded media roll on. NASA proclaimed July 2012 to be the hottest month on record and NOAA's National Climatic Data Center agreed: the July 2012 temperature average of 77.6°F was 3.3°F above the twentieth-century average and 0.2°F above the previously warmest July of 1936 (during the Dust Bowl years). However, when meteorologist Anthony Watts checked NOAA data, he found July 1936 had been reinstated as the hottest month on record. "You can't get any clearer proof of NOAA adjusting past temperatures," Watts wrote. "This isn't just some issue with gridding, or anomalies, or method, it is about NOAA not being able to present historical climate information of the United States accurately. . .This is not acceptable. It is not being honest with the public. It is not scientific. It violates the Data Quality Act."[23]

David L. Lewis, Ph.D., a former microbiologist for the EPA's Office of Research & Development, wrote in *Science For Sale: How the Government Uses Powerful*

---

18 Rady Ananda, "Solar Radiation Management, Geoengineering and Chemtrails." *Global Research*, November 5, 2013.

19 "Chemtrails Confirmed: Climate Scientist Admits Jets Are 'Dumping Aerosols.'" *Chemtrailsplanet. net*, January 9, 2015.

20 "NASA Scientist Admits 'Chemtrails' Are Real." *Chemtrailsplanet.net*, March 11, 2016.

21 Christina Sarich, "Italian Senator Calls for Declassification of Chemtrail Documents." *Naturalsociety.com*, April 15, 2014.

22 "Minister pledges probe into chemtrails," *Cyprus-mail.com*, February 17, 2016.

23 J.D. Heyes, "NOAA quietly revises website after getting caught in global warming lie, admitting 1936 was hotter than 2012." *Naturalnews.com*, July 1, 2014.

Corporations and Leading Universities to Support Government Policies, Silence Top Scientists, Jeopardize Our Health, and Protect Corporate Profits (Skyhorse Publishing, 2014) that EPA leadership consistently "mishandles science." One bizarre incident among many: in 2003, former Acting Assistant Administrator Henry L. Longest II made mid-level EPA managers read management consultant Margaret Wheatley's *Turning to One Another*, which urged environmentalists "to abandon Western science in favor of 'New Science'. . .the 'space of not knowing' and the 'abyss.' While passing through the abyss, new scientists shed their religious beliefs and sexual inhibitions, then turn to one another."[24] Managerial candidates were then required to fill out a confidential questionnaire about their promiscuity, religion, morality, and willingness to keep secrets.

What exactly was the EPA up to in the Bush II years? What *is* the EPA, really?

Beyond the usual arsenal of propaganda, manipulation of international convocations, sexual confessions to forge bonds of secrecy, and blackmailing nations with geoengineered weather threats, is the possibility of outright murder. U.S. Representative Dennis Kucinich (Ohio) fought hard for HR2977, the 2001 Space Preservation Act "revised" and stalled in committee after committee before its final death. Were the deaths of Kucinich's fifty-two-year-old younger brother (December 19, 2007) and his forty-eight-year-old sister (acute respiratory distress syndrome, November 12, 2008) natural? On August 9, 2010, U.S. Senator Ted Stevens (Alaska) was investigating HAARP at the request of Alaskan bush pilot Theron "Terry" Smith when their aircraft crashed, killing Stevens and Smith but not NASA administrator Sean O'Keefe, who was also on board. Smith's son-in-law had been killed just days before in a C-17 crash at Elmendorf Air Force Base. The National Transportation Safety Board (NTSB) blamed the crash on the pilot's "temporary unresponsiveness for reasons that could not be established."[25]

Strange purges are also underway. Canada has dismissed two thousand scientists and hundreds of programs that monitored smoke stack emissions, food inspections, oil spills, water quality, climate change, etc. while closing seven of the eleven Fisheries and Oceans libraries:

> . . .a document classified as "secret" that was obtained by Postmedia News mentioned "culling of materials" as a main activity in the reduction of libraries. . .reports have emerged of books being strewn across floors and even piled into dumpsters.[26]

Thus it is that decades of subterfuge, manipulation, extreme weather

24 David Lewis, "EPA's disturbing leadership." *The Oconee Enterprise*, May 12, 2016.

25 Alan Levin, "NTSB: Ted Stevens' plane crash remains a mystery." *USA Today*, May 24, 2011.

26 Ari Phillips, "Canadian Government Dismantles Ecological Libraries After Dismissing Thousands of Scientists." *Climate Progress*, January 10, 2014.

experiments, murder and mayhem have preceded the present normalization of geoengineering, and all in full view of nations of citizens who no longer believe in their own perceptions and intuitions. With operational weather weapons and the Space Fence in place, globalists are ready to further subject nations to the will of the United Nations instrument.

THE UN POWER SHIFT

The participation of the U.S. and China is significant, as the two account for more than 40 percent of global greenhouse gas emissions. The agreement goes into force once 55 countries accounting for at least 55 percent of global emissions officially sign. . .Parties to the agreement will still have to go through the process of joining the agreement, which for most will require processes of approval in their home countries . . .[27]

On Earth Day 2016, the United Nations Framework Convention on Climate Change agreement was signed after being hammered out at multiple conferences culminating in the COP21 in Paris (November 30–December 11, 2015). The 2011 UN Climate Change Conference (COP17) in Durban, South Africa had attempted to include an International Tribunal of Climate Justice provision,[28] but by COP21, it was no more.

It was quite a show. Climate mouthpieces had been carefully chosen—the IPCC, geoengineers David Keith and Ken Caldeira, prestigious university Ph.D.s, embedded NGOs, government agencies, the World Bank and IMF, and of course the usual Wall Street-London deep pockets. Scriptwriters worked overtime on the fate of the Earth as cameras panned in on lightning flashes, rolling storms, deluges and droughts, crying babies, hospital emergency rooms filled to overflowing. . .The UN was to be tasked with vast new tax and regulatory powers in the name of keeping global warming below 2°C.

A quieter UN meeting had taken place in New York City two months before COP21: the Sustainable Development Summit concentrating on UN Agenda 2030 (Agenda 21 renamed) that would coordinate with the carbons scam to turn nationhood into a mere environmental address:

To cheers, applause and probably a tinge of relief, the 17 global goals that will provide the blueprint for the world's development over the next 15 years were

---

27 "World Leaders Sign Paris Climate Agreement." *Huffington Post*, April 22, 2016. The article closed with "a group of businesses, including Google, Ikea, Starbucks and General Mills, lent their support to the signing ceremony."

28 Sarah Malm, "UN planning an 'international tribunal of climate justice' which would allow nations to take developed countries to court." *Daily Mail*, 2 November 2015.

ratified by UN member states in New York on Friday. After speeches from Pope
Francis and the Nobel laureate Malala Yousafzai, and songs from Shakira and
Angelique Kidjo, the ambitious agenda — which aims to tackle poverty, climate
change and inequality for all people in all countries — was signed off by 193
countries at the start of a three-day UN summit on sustainable development
... The global goals summit continues until Sunday, after which all eyes will
be on the UN climate talks in November. Asked if the goals will be scuppered
without a strong deal in Paris [COP21], Mogens Lykketoft, the president of
the UN general assembly, was hesitant, saying leaders were making more
commitments than they were in previous COP meetings. "From what we know
and hope for, we will be approaching a better deal."[29]

Weather warfare technology was the teeth Agenda 2030 had been waiting
for (and surely why developing nations had sought to include an International
Tribunal of Climate Justice). Immediately after the two conferences, the Dutch
Defence Joint Meteorological Group (JMG) took the lead "in providing weather
forecasts for every exercise or deployment of [NATO's] Very High Readiness
Joint Task Force (VJTF)."[30] "Weather forecasts" were duly added to the Orwellian
dictionary.

To be fair, some academics have pondered how exactly the naked emperor will
remove CO_2 from the atmosphere "using an infrastructure we don't have and with
technology that won't work on the scale we need, and finally to store it in places
we can't find"[31]? Many see the carbon solution for what it is: a ploy for raking
in disaster capitalist cash—$90 trillion in energy infrastructure investments,
$1 trillion green bond market, multi-trillion-dollar carbon trading market,
$391 billion climate finance industry.[32] The UN Green Climate Fund alone will
clear $100 billion per year, purportedly to support concrete carbons mitigation
in developing countries, but will the money ever make it to those nations after
filtering through multilateral and private banks like World Bank and Deutsche
Bank[33]? After all, the naked emperor is not known for keeping his promises . . .

Traditional bureaucratic foundations like Ford, Rockefeller and Carnegie were
said to be giving way to "philanthrocapitalism," a muscular new approach to

29 Liz Ford, "Global goals received with rapture in New York — now comes the hard part."
The Guardian, 25 September 2015.

30 "The Netherlands takes over meteorology for the NATO Response Force." SHAPE (Supreme
Headquarters Allied Powers Europe) press release, 13 January 2016.

31 Jocelyn Timperley, "Academics call for geoengineering preparation in wake of Paris Agreement's
'deadly flaws.'" *BusinessGreen*, 11 January 2016.

32 James Corbett, "And Now For The 100 Trillion Dollar Bankster Climate Swindle . . ." *The Corbett Report*, February 24, 2016.

33 Tyler Durden, "Deutsche Bank Sued For Running An 'International Criminal Organization' in
Italian Court." *Zerohedge.com*, May 18, 2017.

charity in which the presumed entrepreneurial skills of billionaires would be applied to the world's most pressing challenges . . .[34]

Hopefully, like the slow and steady tortoise that wins the race, the public is awakening to the dismal fact that its institutions, agencies, universities, laboratories, and courts obey the very powers that have milked public assets dry. Worker and food safety, gone. Bill of Rights, gone. Environmental protections, gone. Soon, the Trans-Pacific Partnership (TPP) and Transatlantic Trade and Investment Partnership (TTIP) or facsimiles thereof will lock in corporate feudalism under oligarchic world rule as billionaire Good Club[35] members establish more and more "brain institutes" to support Brain Initiative neuroscience in service to a Transhumanist future.[36]

"SCIENCE IS BROKEN"

For two decades, independent scientists and the science-minded have been attempting to sound the alarm regarding what is going on in our skies and low-earth orbit as university lab scientists buckle to military grants intent on weaponizing everything under the Sun, if not the Sun itself. Rutgers University climatologist Alan Robock relates how CIA-funded consultants asked him two questions: *If we control someone else's climate, would they know about it?* and *Would climate experts be able to determine if another nation was attempting to control the climate?* The CIA—not exactly known for being forthcoming—has funded multiple grants, National Academy of Sciences reports like "Climate Intervention: Carbon Dioxide Removal and Reliable Sequestration" (154 pages) and "Climate Intervention: Reflecting Sunlight to Cool Earth" (234 pages),[37] and defense contractors (Raytheon, Lockheed Martin, L3, SAIC, etc.) to aggressively do whatever it takes to quietly develop geoengineering and the Space Fence while keeping the public ignorant of both—especially when it comes to those pesky chemtrails activists. From CIA Document 1035-960:

> (3)(b) To employ propaganda assets to [negate] and refute the attacks of the critics. Book reviews and feature articles are particularly appropriate for this purpose. The unclassified attachments to this guidance should provide useful background material for passing to assets. Our ploy should point out,

---

34 Jacob Levich, "The Real Agenda of the Gates Foundation." *Aspects of India's Economy*, No. 57, May 2014.

35 Paul Harris, "They're called the Good Club — and they want to save the world." *The Guardian*, 30 May 2009.

36 William J. Broad, "Billionaires With Big Ideas Are Privatizing American Science." *New York Times*, March 15, 2014.

37 Daniel Barker, "Nations are now using weather modification as clandestine warfare, CIA warns." *NaturalNews.com*, December 19, 2015.

as applicable, that the critics are (I) wedded to theories adopted before the evidence was in, (II) politically interested, (III) financially interested, (IV) hasty and inaccurate research, or (V) infatuated with their own theories.[38]

It should therefore not be surprising to learn that the peer review system has been co-opted to banish those theories and scientists who don't "play ball" to the outer darkness of non-publication, stonewalled careers, and worse. Nobel Laureate biologist Sydney Brenner:

> I think peer review is hindering science. In fact, I think it has become a completely corrupt system. It's corrupt in many ways, in that scientists and academics have handed over to the editors of these journals the ability to make judgment on science and scientists. There are universities in America, and I've heard from many committees, that won't consider people's publications in low impact factor journals. . .it puts the judgment in the hands of people who really have no reason to exercise judgment at all. And that's all been done in the aid of commerce, because they are now giant organizations making money out of it.[39]

"Powerful orthodoxy against a marginalized heterodoxy" is how Charles Eisenstein describes the opposition to cutting-edge Electric Universe scientists:

> If you have faith in the soundness of our scientific institutions, you will assume that the dissidents are marginalized for very good reason: their work is substandard. If you believe that the peer review process is fair and open, then the dearth of peer-reviewed citations for [Electric Universe] research is a damning indictment of their theory. And if you believe that the corpus of mainstream physics is fundamentally correct, and that science is progressing closer and closer to truth, you will be highly skeptical of any major departure from standard theories. . .Can we trust scientific consensus? Can we trust the integrity of our scientific institutions? Perhaps not. Over the last few years, a growing chorus of insider critics have been exposing serious flaws in the ways that scientific research is funded and published, leading some to go so far as to say, "Science is broken."[40]

Between 1973 and 2013, six major publishers decided which scientific papers merited publication and which didn't (ACS; Reed Elsevier; Sage; Taylor & Francis; Springer; and Wiley-Blackwell). All were in the back pocket of Big Pharma and the medical industry, which, like the CIA, NASA, EPA, etc., are not what they seem:

---

38 CIA Document #1035-960, "Concerning Criticism of the Warren Report," 1967.

39 Charles Eisenstein, "The Need For Venture Science." *Huntington Post*, August 27, 2015.

40 Ibid.

"As long as publishing in high impact factor journals is a requirement for researchers to obtain positions, research funding, and recognition from peers, the major commercial publishers will maintain their hold on the academic publishing system," added [Professor Vincent Lariviere, lead author of the study from the University of Montreal's School of Library and Information Science].[41]

The danger quotient for scientists working on classified projects is greater than just being stripped of their career and livelihood. In the early days of the "Star Wars" Strategic Defense Initiative (SDI) now culminating in the Space Fence, two dozen scientists and experts working for Marconi and Plessey Defence Systems either disappeared or died under "mysterious circumstances." *Most were microbiologists.* The scientist death toll continued into the 1990s and post-9/11.[42] Now, the targets appear to be naturopathic doctors and health-minded MDs peering behind the curtains of Big Pharma vaccinations, autism, and cancer-for-profit.[43]

NEXT STOP: THE SSS SPACE FENCE

We are a long way from President Kennedy's Space Age dreams and resolution to put an end to chemical polluters and the destruction of soil and biodiversity. Former U.S. Secretary of Defense Ashton B. Carter had a doctorate in physics from Oxford University, and with the stroke of a presidential pen on Thanksgiving 2015, the U.S. Commercial Space Launch Competitiveness Act (HR2262) a.k.a. Space Act of 2015 thumbed its nose at the Outer Space Treaty of 1967 and erased whatever fleeting separation was left between the militarizing of space and corporations bent on making profits from space-based mining of asteroids and the helium-3 isotope on the Moon.

Now that weather can be technically engineered, it is no longer an environmental issue but a political and economic one. Take, for instance, the North American Climate, Clean Energy, and Environment Partnership Action Plan signed in June 2016 by Canada, the U.S., and Mexico. The "sustainable future" for these three nations appears to be as some sort of geographic hemispheric union, as prophesied by Hillary Clinton in 2013.[44] The Partnership Action Plan Leaders Statement describes it like this:

41 Sean Adl-Tabatabai, "Nearly All Scientific Papers Controlled By Same Six Corporations." *YourNewsWire.com*, July 20, 2015.

42 Mark J. Harper, "Dead Scientists and Microbiologists — Master List," February 5, 2005, rense. com/general62/list.htm.

43 Erin Elizabeth, "A Connection with the Holistic Doctor Deaths?" *HealthNutNews.com*, February 1, 2016.

44 Amy Chozick et al., "Leaked Speech Excerpts Show a Hillary Clinton at Ease With Wall Street." *New York Times*, October 7, 2016.

Our actions to align climate and energy policies will protect human health and help level the playing field for our businesses, households, and workers. . .that sets us firmly on the path to a more sustainable future.[45]

In Orwellian, "protect human health" translates to Big Pharma vaccinations for all, and "level the playing field" to the evaporating middle class, while *sustainable future* is something like the "free-range totalitarianism" coined by investment analyst Catherine Austin Fitts—cheaper by far than rounding people up in camps.

And what are we to believe and not believe about the terra incognita of space, now that U.S. Joint Chiefs of Staff Chairman General Henry H. Shelton has confessed that a crucial component of the military doctrine of full spectrum dominance is the use of deception to "defend decision-making processes by neutralizing an adversary's perception management and intelligence collection efforts"[46]? In 2015, the twin LIGOs (Laser Interferometer-Gravitational wave Observatory) were said to have detected a gravitational wave generated by two merging black holes at a distance of 1.3 billion light years.[47] But do "gravitational waves" and "black holes" even exist? And what were the asteroid-type objects near Uranus that the ALMA (Atacama Large Millimeter/submillimeter Array) telescope in Chile detected?[48]

Meanwhile on planet Earth, the price tag for the American military machine in 2016 was $573 billion (not counting the "black budget"), and propaganda continues to frame extreme weather as "acts of God" due to "climate change," while the truth is that weather events are being generated by exceedingly sophisticated electromagnetic technology. Hurricanes, tornadoes, earthquakes, floods, and droughts dominate world news, with FEMA and military security forces descending upon one shattered community after another, real estate agents brokering giveaway property purchases in the wake of disaster after disaster, and insurance companies suing towns for failure to prepare for "climate change."[49]

Geoengineering is a profit-maker for disaster capitalists and a force-multiplier for the military.

45 Patrick Wood, "NAU Reborn As 'North American Climate, Clean Energy and Environment Partnership'." *Technocracy News*, June 30, 2016.

46 Quoted in *Planet Earth: The Latest Weapon of War* by Rosalie Bertell (Black Rose, 2001).

47 Stephen J. Crothers, "A Critical Analysis of Ligo's Recent Detection of Gravitational Waves Caused by Merging Black Holes." *viXra.com*, March 8, 2016.

48 Sarah Kaplan, "Scientists claimed they found elusive 'Planet X.' Doubting astronomers are in an uproar." *Washington Post*, December 11, 2015.

49 John Roach, "Insurer's Message: "Prepare for Climate Change or Get Sued." *NBC News*, June 6, 2014.

The SSS (Space Surveillance System) Space Fence has been constructed by NATO interests. It is a global surveillance machine with many parts above and below the firmament to provide real full spectrum dominance not just of weather and the near-earth environment but of the entire biosphere down to the DNA level—*for generations.*

But despite it all, we must continue to educate ourselves regarding the wireless antenna atmosphere we now breathe and learn how the Star Wars II Space Fence operates for the sake of our own survival and that of the generations still coming. Hopefully, *Under An Ionized Sky: From Chemtrails to Space Fence Lockdown* will give you a leg up as to how the present Space Age has been quietly built around *and in* us for the sake of a Transhumanist Space Age.

▶ CATCHING UP WITH *CHEMTRAILS, HAARP*

Few in the civil sector fully understand that geoengineering is primarily a military science and has nothing to do with either cooling the planet or lowering carbon emissions. While seemingly fantastical, weather has been weaponised. At least four countries—U.S., Russia, China and Israel—possess the technology and organization to regularly alter weather and geologic events for various military and black operations, which are tied to secondary objectives, including demographic, energy and agricultural resource management.

Indeed, warfare now includes the technological ability to induce, enhance or direct cyclonic events, earthquakes, drought and flooding, including the use of polymerized aerosol viral agents and radioactive particulates carried through global weather systems. Various themes in public debate, including global warming, have unfortunately been subsumed into much larger military and commercial objectives that have nothing to do with broad public environmental concerns. These include the gradual warming of polar regions to facilitate naval navigation and resource extraction.

— Matt Andersson, former executive adviser, aerospace & defense,
Booz Allen Hamilton, Chicago

CHAPTER ONE

Project Cloverleaf

▼

Controlling the weather has been a goal of industrialists for 150 years or more. But it was not until the availability of vast numbers of jet aircraft that made it possible to replace huge sections of the atmosphere with chemicals to control the flow of air and moisture. This is not advanced technology but more like farming the sky with a million tractors. *True advanced technology would not require aerosols to produce "climate change." It's brute force climate manipulation for which secrecy cannot be maintained without threatening the potential whistleblower. Now that the public has increasingly realized the climate change cover-up, the media looks more and more foolish in their attempt to protect the masters.*

— Harold Saive, chemtrailplanet.net

As early as 1990, weather force specialists at the U.S. Air Force Academy were hard at work studying how to chemically mix and lay "aerial obscuration" they called "chemtrails." Project Cloverleaf began quietly enough (as all classified projects do) with the 1994 Hughes Aircraft patent for Welsbach Seeding For Reduction of Global Warming. Welsbach seeding called for spreading highly reflective materials in the atmosphere to reflect back into space 1–2 percent of incoming sunlight and thus slow down "global warming." However, it was also about beginning to create a more conductive atmosphere in preparation for Bernard Eastlund's High-frequency Active Auroral Research Project (HAARP) already under construction in Alaska. The reflective material (˜10 microns) to be added to jet auxiliary fuel tanks was the highly conductive compound aluminum oxide (Al_2O_3). The jet's main tanks would be reserved for takeoff and landing, and the auxiliary loaded with Al_2O_3 for cruising altitude. Lawrence Livermore National Labs priced the program at US$1 billion per annum (in 1994 dollars).

By the late 1990s, aerial grids were being laid over chosen regions and cities of the U.S. and other NATO nations. In 1996, the military went public with two documents that obliquely referenced what Cloverleaf was up to: the Pentagon paper "Weather as a Force Multiplier: Owning the Weather in 2025" in *Air Force*

2025,[1] which called the chemical whitening effect "cirrus shielding"; and U.S. Space Command's *Vision for 2020* calling for full spectrum dominance of space, land, sea, and air.[2]

Air traffic controllers (ATC) at major airports were coached to re-route commercial air traffic around military craft engaged in "classified aerial operations" at 37,000–40,000 feet. ATC radar revealed a haze of aluminum and barium—Al_2O_3 for solar radiation management (SRM) geoengineering, barium stearate $Ba(C_{18}H_{35}O_2)_2$ for lubrication, radar imaging, and high-powered RF-microwave beam weapons.

> A wide range of particles could be released into the stratosphere to achieve the SRM objective of scattering sunlight back to space. Sulfates and nanoparticles currently favored for SRM include sulfur dioxide, hydrogen sulfide, carbonyl sulfide, black carbon, and specially engineered discs composed of metallic aluminum, aluminum oxide and barium titanate. In particular, engineered nanoparticles are considered very promising. The particles would utilize photophoretic and electromagnetic forces to self-levitate above the stratosphere. These nanoparticles would remain suspended longer than sulfate particles, would not interfere with stratospheric chemistry, and would not produce acid rain. However, while promising, the self-levitating nanodisc has not been tested to verify efficacy, may increase ocean acidification due to atmospheric CO_2 entrapment, has uncharacterized human health and environmental impacts, and may be prohibitively expensive.[3]

The truth is that the 10 billion SRM discs 10 micrometers across and 50 nanometers thick are engineered with a core of aluminum, a top layer of aluminum oxide, and a bottom layer of barium titanate—aluminum to reflect heat up, heavier barium purportedly to push the discs up (*photophoresis*). Introduce a magnetic component, then spray, and the nanoparticles will follow the Earth's magnetic field into the upper atmosphere.

By 1998, the Ontario Ministry of Environment (Canada) found 7X the safe limit for aluminum in rainwater samples. More and more citizens complained of sudden headaches, joint pains, dizziness, fatigue, acute asthma, gastrointestinal pain, coughs, and feverless flu symptoms.

In 2000, Cloverleaf went public in an offhand way with a comment from an anonymous airline executive to independent scientist Clifford Carnicom, then

---

1 Col. Tamzy J. House et al. "Weather as a Force Multiplier: Owning the Weather in 2025." *Air Force 2025*, "a study designed to comply with a directive from the chief of staff of the Air Force to examine the concepts, capabilities, and technologies the United States will require to remain the dominant air and space force in the future," August 1996.

2 Formed after 9/11 by then-Secretary of Defense Donald Rumsfeld.

3 Utibe Effiong and Richard L. Neitzel, "Assessing the direct occupational and public health impacts of solar radiation management with stratospheric aerosols." *Environmental Health*, January 19, 2016.

the most visible scientist collecting data on the aerosol fallout over northern New Mexico. According to this "Deep Throat," the purpose of Cloverleaf was "to allow commercial airlines to assist in releasing these chemicals into the atmosphere."[4] Military jets simply could not keep up with the "global dimming" now called *solar radiation management (SRM)*, so the entire airline industry was being drafted in the name of national security.

THE IONOSPHERIC HEATERS

The deeper one probes Project Cloverleaf technology, the more obvious it becomes that cosmic processes are being manipulated and employed. For example, Eastlund's last patent, published on October 11, 2007, claims that HF ionospheric heaters can create artificial plasma[5] regions in the atmosphere by means of *cosmic particle ignition*—sort of like "hot spots"—utilizing "ionization trails of cosmic rays and micro-meteors," thus lowering the power requirements for telecommunications, weather control, lightning protection, and defense applications.[6] Dam up one of the five atmospheric rivers of vapor in each hemisphere—rivers 420–480 miles wide, two miles above the Earth—and move 340 pounds of water per second toward the poles. Convert the flow to electricity and steer the jet stream to dump masses of rain on a pre-designated geographic area, such as Storm Frank in England in December 2015.[7]

Even the Sun can be influenced:

> ...the HF modulation transmitted by HAARP in the ionosphere causes an Alfvén (whistler) wave generation, which in turn creates an ELF harmonic in the Earth's atmosphere. This Alfvén wave generation can also induce solar activity due to the quantum vibration (sympathetic resonance) between the two bodies. This is exactly what Tesla was referring to when he stated, man could tap the Breast of Mother Sun and release her energy toward Earth as needed—magnetic as well as light." We're seeing the future.[8]

---

4 "An Airline Manager's Statement," May 22, 2000, carnicominstitute.org/articles/mgr1.htm.

5 Plasma, the fourth state of matter throughout the universe, including our Sun and lightning; a strong electromagnetic field applied by laser or microwave generator.

6 Bernard Eastlund, Patent US20070238252 A1 "Cosmic particle ignition of artificially ionized plasma patterns in the atmosphere," October 11, 2007. (Eastlund died exactly two months later on December 12, 2007.)

7 "UK floods: Storm Frank threatens more misery." *BBC News*, 29 December 2015.

8 Email from Christopher Fontenot, August 1, 2016. See "Solar Perturbation via Ionospheric Manipulation of Schumann Resonance" at his site, amicrowavedplanet.com/solar-perturbation-via-ionospheric-manipulation-of-schumann-resonance/. Fontenot has extensive training in nuclear and electrical engineering. Now, his primary interest is in the Electric Universe Theory and its implications.

But then, there may be blowback chalked up to "climate change":

> It's as if global warming were ringing the Earth's atmosphere like some great, cacophonous alarm bell. The upper level zonal winds are swinging wildly from record high positive anomalies to record low negative anomalies. . .And the Jet Stream now has redefined all boundaries—flowing at times from the East Siberian Sea in the Arctic across the Equator and all the way south to West Antarctica.[9]

Today, we can count more than 260 SDI-inspired installations around the world, all engaged, to one degree or another, in ULF/ELF modulation and manipulation of the ionospheric fluctuation known as the *auroral jet (electrojet)*, where the Van Allen Radiation Belts—the shell of our magnetosphere—intersect our atmosphere. Even before Project Cloverleaf was activated and HAARP built, a network of ionospheric heaters already existed.

The first ionospheric heaters like the Arecibo Radio Telescope ("The Pit") in Puerto Rico (1963), the Platteville Atmospheric Observatory in Platteville, Colorado, and EISCAT (European Incoherent Scatter Scientific Association) in Tromsø, Norway (1981) initially concentrated on interfering high-frequency radio waves for earth-penetrating tomography. The Pit is still the largest dish antenna in the world with three radar transmitters with an ERP of 20 TW[10] at 2380MHz, 2.5 TW (pulse peak) at 430MHz, and 300MW at 47MHz.

Just before HAARP was shut down in summer 2013, the Pit ionospheric heater was upgraded:

> "It is basically the same as HAARP for the science, except that HAARP was in the Auroral Region, where the physics of the ionosphere is quite different with all the energetic particles and magnetic fields," Penn State Electrical Engineering Professor Jim Breakall, WA3FET, told *ARRL*. . .The National Science Foundation and Cornell University, which previously operated Arecibo Observatory, contracted with Penn State's Electrical Engineering Department to construct the "new and enhanced" HF ionospheric instrument. It will be used to study the interaction between HF radio energy and ionospheric plasma.[11]

The Platteville Observatory supposedly shut down in 1984 and is now being used for "wind profiling." EISCAT oversees three incoherent scatter radar

---

9 "Gigantic Gravity Waves to Mix Summer With Winter? Wrecked Jet Stream Now Runs From Pole-to-Pole." *RobertScribbler.com*, June 28, 2016.

10 A terawatt is equal to a trillion watts or 10[12].

11 "HAARP-Like Ionospheric Research Project Underway at Arecibo Observatory." *ARRL*, April 23, 2014. Interestingly, the Arecibo Observatory Amateur Radio Club, KP4AO, is headquartered at the research facility.

systems at 224MHz and 931MHz in Northern Scandinavia, and one at 500MHz at EISCAT Space Centre in Svalbard. Additional receiver stations are located in Sodankylä, Finland and Kiruna, Sweden, transmitting 10MW with an antenna gain of 35 decibels (dB) producing an ERP of 32 billion watts. EISCAT is funded and operated by research institutes and councils of Norway, Sweden, Finland, Japan, China, the UK, France, and Germany,[12] and has recently been recalibrated to serve lower-earth orbit (LEO) Space Fence surveillance.

As for how ionospheric heaters work, picture focused radio waves generated by phased-array radar units striking a targeted area in the ionosphere to ionize (heat) electrons. HAARP's *ionospheric research instrument (IRI)* was constructed with 180 antenna towers in a grid of fifteen columns and twelve rows, with two dipole antennas atop each tower. Beneath the array are transmitter shelters, each housing twelve diesel-powered transmitters capable of generating 10,000 watts of RF power each with an ability to focus *3.6 billion watts* on a single point. Steering this focus and its pulse transmission agility constitutes the secret power of HAARP, according to Eastlund.

Once the electrons in the ionosphere are ionized, they twirl down the magnetic lines of force conjoining with *Birkeland currents*, ready to be steered by various radar instruments for a variety of military-industrial-intelligence *C4 operations (command, control, communications, and cyberwarfare)*, including moving and enhancing weather systems, stimulating seismic plates to generate earthquakes, tomographic penetration of regions of the Earth for mineral or oil exploitation, discovery of underground installations, etc.

The process is not unlike sonar's ability to produce a radar-like echo. Tune the ground mechanism—the phased-array antennas laid out in a hexagonal grid following large copper cables buried underground—to produce a *standing sky wave* that will echo back as a ground wave to be steered and relayed for military operations. HAARP's IRI was thus able to create a "virtual antenna" in the sky called an Ionospheric Alfvén Resonator (IAR). Properly activated, Alfvén waves (whistlers)[13] and magnetosonic (MS) waves in the magnetosphere create a standing wave between the ionosphere and our troposphere that radiates extremely low frequency (ELF) signals around the globe—signals that can be heard in the deepest depths of the oceans.[14] This *resonance* of standing oscillations "behave as strings with the ends fixed in the ionosphere."[15] (Thus, the acronym HAARP.) By constantly priming

12 See *dutchsinse.com/4202015-five-new-haarp-type-arrays-being-built-norway-shutting-down-fm-radio-the-real-reason-why/*, April 20, 2015.

13 Whereas whistler waves are from lightning discharges zapping along the Earth's magnetic lines of force, Alfvén waves are low-frequency and fluid-like—like plasma.

14 Dennis Papadopoulos, "Using Active Experiments to Probe Geo-space." Third International Conference, "The Mechanics of the Magnetospheric System and Effects on the Polar Regions," October 27–November 1, 2013.

15 Ibid.

our atmosphere with conductive metal particulates, a virtual antenna for C4 military operations is thus maintained.

Compare a Space Fence around the Earth with NASA's plan to build a de facto magnetosphere around Mars to protect it from solar wind and debris until a new atmosphere forms.[16] Referencing a 2002 paper,[17] Chris Fontenot says NASA's plan for Mars is entirely possible, thanks to Alfvén waves:

> This technology involves Alfvén wave modulation via HF perturbation by ionospheric heaters located around our planet. The notion of increasing the magnetic field of Mars through a similar process isn't far-fetched in the least. These Alfvén waves are responsible for Earth's magnetic field strength: a naturally occurring process normally fueled by solar activity.

HAARP created the consummate standing sky wave. In HAARP's backyard, Nick Begich and Jeane Manning explained it twenty years ago in *Angels Don't Play This HAARP*:

> This invention provides the ability to put unprecedented amounts of power in the Earth's atmosphere at strategic locations *and to maintain the power injection level*, particularly if random pulsing is employed, in a manner far more precise and better controlled than heretofore accomplished by the prior art. . .[T]he goal is to learn how to manipulate the ionosphere on a more grand scale than the Soviet Union could do with its similar facilities. HAARP [is] the largest ionospheric heater in the world, located at a latitude most conducive to putting Eastlund's invention into practice.[18] [Emphasis added.]

In fact, HAARP's *effective radiative power (ERP)* is so great that when its beam intersects with just one other ionospheric heater beam in the Schumann resonance range of 7–8Hz, the entire ionosphere "rings."

Between HAARP's closure in 2013 and reopening in 2015,[19] its IRI phased array was re-calibrated to synchronize with the vast ground-based network

---

16 "NASA proposes shield around Mars to aid human colonization." *RT.com*, 7 March 2017.

17 E. Kolesnikova et al., "Excitation of Alfvén waves by modulated HF heating of the ionosphere, with application to FAST observations." *Annales Geophysicae*, (2002) 20: 57–67.

18 Begich, Dr. Nick and Jeanne Manning, *Angels Don't Play This HAARP: Advances in Tesla Technology*. Earthpulse Press, 1995, 2002.

19 Emails from Billy Hayes, "The HAARP Man": July 27, 2015: "Just a month ago, I witnessed fellow tower workers onsite at HAARP Gakona making new modifications to the HAARP site: antenna rigging has been replaced, new balum transformers, upgrades of transmitters and trailers (frequency alterations) by P&R Towers subcontracting with DARPA / Raytheon, pursuant to the takeover of the site by University of Alaska Fairbanks (UAF). The site was and has been maintained black [classified] without power from an outside source since August 2013. On June 1, 2015, external power was re-applied (same time the Pecan Project was applied)." August 21, 2015: "They've been doing safety burns around the site this week; can see the smoke plumes by satellite. Don't forget: next turn-on for UAF Gakona is on August 25."

being tuned to resonate with the Space Fence infrastructure, from ionospheric heaters (including many now mobile[20]) to ELF receivers, digital HF ionosondes, magnetometers, photometers, VLF sounders and VHF riometers, HF receivers, HF/VHF radars, optical imagers, rockets with "particle beams and accelerators aboard (e.g., EXCEDE and CHARGE IV), and shuttle- or satellite-borne RF transmitters (e.g., WISP and ACTIVE)," and sounding rockets probing heated regions of the ionosphere and dropping "chemical releases" for "space-based efforts."[21]

The June 6, 1995 "HAARP Research and Applications Executive Summary"[22] put out by the Technical Information Division of the Naval Research Laboratory makes it clear that HAARP was always about the military commitment to controlling the ionosphere so as to generate plasma "lenses"[23] capable of focusing large amounts of HF energy at high altitude, triggering ionospheric processes in the lower atmosphere, etc. *Drawing powering ions from the upper into the lower atmosphere was exactly what the 1980s SDI program lacked.*

> The heart of the program will be the development of a unique ionospheric heating capability to conduct the pioneering experiments required to adequately assess the potential for exploiting ionospheric enhancement technology for DOD (Dept. of Defense) purposes. . .[S]uch a research facility will provide the means for investigating the creation, maintenance, and control of a large number and wide variety of ionospheric processes that, if exploited, could provide significant operational capabilities and advantages over conventional C3 systems.
>
> . . .By exploring the properties of the auroral ionosphere as an active, non-linear medium, the primary energy of the HF transmitter which is confined to a frequency range of 2.8 to 10 MHz can be down-converted to infra-red and visible photons. As a result, the HAARP HF transmitter can generate sources for *remote sensing and communications spanning sixteen decades*[24] *in frequency.*[25] [Emphasis added.]

"Sixteen orders of magnitude" point to HAARP's unique power for manifold operations. As Jerry Smith (1950–2010) commented in *HAARP: The Ultimate Weapon*:

---

20 Ionospheric heaters are even on space-based ionospheric sounders. See R.F. Benson, "Topside sounders as mobile ionospheric heaters." *NASA Technical Reports Server (NTRS)*, January 1, 2006.

21 Executive Summary.

22 www.viewzone.com/harp.exec.html.

23 Peter Koert, "Artificial ionospheric mirror composed of a plasma layer which can be tilted." US 5041834 A, Apti, Inc., May 17, 1990.

24 A *decade* is a factor of 10 difference between two numbers—an order of magnitude measured on a logarithmic scale.

25 Executive Summary.

This shows the deliberate misdirection of the current HAARP literature, which does all it can to downplay the true power of this transmitter in terms of what it can affect. By being able to rebroadcast anywhere in the non-ionizing (non-radioactive) segment of the electromagnetic spectrum, with a potential re-radiated broadcast power equal to, if not greater than, all the TV and radio stations of the world combined, the HAARP array is utterly in a class of its own. Frankly, its potential exceeds the imaginations of many science fiction writers working today.[26]

In summary, with 1.7 gigawatts (billions) ERP and a frequency range of 1–15MHz (millions), HAARP was able to turn the electrojet[27] into a virtual antenna with radio wave "mirrors" (ionospheric lenses) for long-range, over-the-horizon HF/VHF/UHF operations. With the lower atmosphere ionized, all the military had to do was tune the HAARP instrument with one or two other radar beams at other longitudes / latitudes, lay "chaff" loaded with conductive metal nanoparticulates, and whip up a hurricane, earthquake, drought, or flood—or even a directed energy event like 9/11.

In *Chemtrails, HAARP, and the Full Spectrum Dominance of Planet Earth*, I discuss a few such "acts of God." Over the past two decades, hundreds of experiments have been conducted under cover of "global warming" and "climate change." For example, Hurricane Sandy on October 29, 2012, a real Halloween superstorm that inflicted $71 billion in damage on the Ports of New York and New Jersey and killed 159 people. Sandy's size and path were "unusual":

> After moving north from the tropical waters where it spawned, Sandy turned out to sea before hooking back west, growing in size and crashing head-on into the East Coast, gaining strength when it merged with an eastbound mid-latitude storm. . ."These events are somewhat rare in occurrence, but they do exist in nature," [William Lau, research scientist at University of Maryland's Earth System Science Interdisciplinary Center and senior scientist emeritus at NASA's Goddard Space Flight Center] said. "While they're turning about each other, they interact. One just took the energy from the other."[28]

Sandy's "turn out to sea" was to pick up more power from a storm system created in the South Atlantic and pinned offshore for the occasion. Sandy provided the model for dozens of generously funded simulations at NASA Goddard to study what thus far had been learned from Project Stormfury—for

---

26 Jerry E. Smith, *HAARP: The Ultimate Weapon of the Conspiracy*. Adventures Unlimited, 1998.

27 Electrojet: electric current traveling around the layer of the ionosphere 56–93 miles above the Earth.

28 "Simulations of Hurricane Sandy with warmer ocean temperatures resulted in storms more than twice as destructive as Sandy, according to new research." University of Maryland, January 22, 2016.

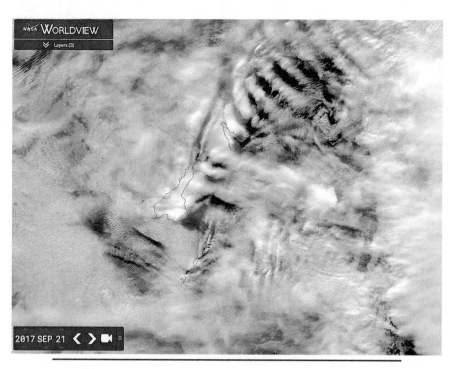

Sometimes, signs of EM pulsing of frequencies can also be observed, as in this September 21, 2017 South Sandwich Islands, Antarctica photograph by V. Susan Ferguson metaphysicalmusing.com/.

example, how warmer ocean-generated storms are 50–160 percent more destructive than Sandy.[29]

Three years to the month after Sandy was built off the Atlantic coast, Hurricane Patricia was built off the Pacific coast; in February 2016 Cyclone Winston devastated Fiji as the most potent cyclone on record in the southwest Pacific; in April 2016 Cyclone Fantala was the strongest known storm in the Indian Ocean.

NBC News hailed Patricia as "the strongest storm ever measured." Mexico was warned of possible thirty-nine-foot waves, flash floods, and mudslides, and 15,000 tourists were evacuated from Puerto Vallarta to Jalisco state before the storm made landfall. On Friday, October 23, Patricia's cyclonic winds were at a frenzied pitch. The sea surged, the rain poured. Four hours after landfall and still moving east, Patricia dropped from a Category 5 to a Category 1, with much of the moisture that would have fed Patricia electromagnetically sent north and east up to Oregon, Washington, and Idaho. Oregonian Ann Fillmore, Ph.D., described it this way while viewing a satellite map:

---

29 Ibid.

. . .first the storm system that appeared on radar to be coming over OR/WA was shown to have more *umph* and wet stuff in it than the hurricane heading into Mexico. It has literally exploded with rain over NE Washington and Idaho, or so radar says. Now watch what should have been an instant pop-up hurricane off the Mexican coast. Yes! It is there, and then it is not there. Even the rain in it is gone. The rain that is now heading into TX is really not an extension of the so-called hurricane. It is simply another pop-up storm generated over the desert of Mexico. There are other small 'hurricane-like' entities over the Pacific, but they seem to be stuck in one spot. I'm surprised the guys who made all of this bizarre stuff didn't sign their names to it. Superbly done, guys! Wow![30]

Texas and Louisiana prepared for Patricia to be amped up by El Niño, the jetstream lobe, and warm moisture from the Gulf of Mexico. But when Patricia moved past Guadalajara, she just—died. Was Patricia about powering a hurricane up in the presence of El Niño, then stopping it dead in its tracks?

THE DELIVERY SYSTEMS

I draw your attention to the little known fact that particulate plumes can also increase the regional surface temperature at great distances from the plume, which is a totally counterintuitive result in that particulates reduce the level of solar radiation reaching the Earth's surface and would be expected to have a cooling effect. The CCS Program however states that by the year 2100 short-lived gases and particles may account for as much as 40 percent of the warming over the summertime continental United States.

— Geophysicist Keith Potts, COP21

Like a vast, random experiment targeting the environment with health doomed to be collateral damage, chemicals have been released into the air, soil, and water since nineteenth-century industrialism. While some may shrug that the aerial release of chemical nanoparticles and nano-sensors, microprocessors, and biologicals under the classified Project Cloverleaf is just more of the same, nanoparticles able to breach the blood-brain barrier make it uniquely diabolical, as does the global conspiracy of power to turn the entire planet into an electromagnetic grid and plug everyone into it. War has gone corporate and all of life reframed as a battlespace of disposable noncombatants (civilians) redefined as potential "terrorists." The military is no longer a protector but partnered with giant transnational corporations and wealthy dynastic cartels like that of Big Pharma and Big Oil.

---

30 Facebook, October 24, 2015.

Fertilizers and pesticides, bombs, "medicine," and jet fuel additives all flow from the self-serving chemical industry of Dow, Monsanto, Siemens, Eli Lilly, Sandoz, Bayer, I.G. Farben, etc. During World War II, I.G. Farben provided nickel, aluminum, and magnesium for Nazi Heinkel and Junkers Stuka bombers, along with fuel, oil, and phosphorous incendiaries.[31] Today, they support chemical warfare against society. Monsanto developed its aluminum-resistant "Terminator" seed in step with the Welsbach patent and Cloverleaf jets furrowing the sky and sowing Al_2O_3 combustion chemicals in soil, oceans, rivers, water reservoirs, gills and lungs. Big Pharma corporations boost cancer, legislate for more vaccinations, and pay off physicians to ply Americans with one drug after another. Like Monsanto seed, fertilizers, and pesticides, "mood stabilizers" and vaccines are designed to work synergistically with the chemicals and nanoparticulates falling from the sky. Profit and population control go hand in hand.

CHEMTRAILS VERSUS CONTRAILS

Our blue skies and golden Sun have turned milky white as the disconnect between what people are observing and what television weather reporters are reading from their cue cards slowly becomes more and more obvious. Occasionally, a scientist admits that what is ridiculed by media is actually true—for example, heliophysicist at NASA Goddard Space Flight Center Doug Rowland admitted over the phone to a citizen: "There's different kinds of chemtrails, as you probably know."[32] The question is not *if* chemicals are in the aerial deliveries but *what* chemicals are being delivered and *why*, and what impact they're having on present-time health and the genetics of future generations.

We stand six miles below what is spewing from the engine and wings of jets and strain our eyes to capture what is coming from the jet to obscure our sunlight and deep blue skies. We read about "climate change," contrails, and carbons, follow "extreme weather" coverage on television, and ponder references to stratospheric aerosol geoengineering (SAG) or solar radiation management (SRM). In 2010, European aerospace engineers presented conclusive evidence of chemical trails over NATO countries in "CASE ORANGE," a 336-page report commissioned by The Belfort Group. American mainstream media buried it to keep pilots who frequent sites like *boldmethod.com* ignorant of more than two kinds of contrails: the aerodynamic and exhaust contrails. But if "contrails are

---

31 James Pressley, "Hitler's Chemists Chased Auschwitz Profits, Financed Mengele." *Bloomberg News*, August 8, 2008; a review of *Hell's Cartel: I.G. Farben and the Making of Hitler's War Machine* by Diarmuid Jeffreys (Metropolitan Books, 2008).

32 Watch "NASA Scientist Answers Questions About Chemtrails," July 18, 2013, www.youtube.com/watch?v=Xz1Gt5moxvw.

'condensation trails,' and they have nothing to do with chemicals,"[33] what about the fact that at least 2 percent of all commercial and military JP-8 fuel is made up of chemical additives?

The truth is that Cloverleaf is so highly compartmentalized and percentages of fuel formulae so highly classified that even pilots and drone operators have no real idea of why and what they are releasing.

The U.S. Air Force distance standard for condensation trails is one-wingspan behind the aircraft, whereas chemical trails begin straight out of the engine compartment. Contrails form at 35,000+ feet at a temperature of -40°C with 67+ percent humidity, whereas chemtrails are generally laid between 28,000 and 34,000 feet, with cargo delivery flights like FedEx above them. Military pilots in pressure suits streak above them all at 39,000+ feet. Given that most chemtrail sightings are of low- and mid-altitude jets, the assumption that kerosene is turning into ice crystals is impossible.

According to the National Center for Atmospheric Research in Boulder, Colorado, the only way to form artificial clouds in warm dry air is to introduce enough particulates into the atmosphere to attract and accrete all available moisture into a visible vapor. Once airborne particles like barium or aluminum are added to create more nuclei for atmospheric moisture to condense around in low-humidity areas, droplets will freeze into persistent plumes at much lower altitudes (and higher temperatures). After acclimation, ice crystals in contrails disappear. Look up and study the clouds and you will see that the cloud cover and bizarrely shaped clouds do not acclimate. This is because they are *not* from contrails but are from chemicalized trails.

JET ENGINES

One burning question of chemical trail students is what is coming directly from the combustion engine and what is being added through modifications to the pylon nozzles above and behind the exhaust output?

In "How Persistent Aerosol Plumes are Being Changed to Short Non-Persistent Plumes to Fool the Public," chemically sensitive activist Russ Tanner (*globalskywatch.com*) suggests that "the contrail deception" depends upon the public remaining ignorant of just how jet engines work:

> . . .multiple professional airline pilots have contacted me and thanked me for my stance against the contrail deception. All of them told me personally that they have *never* seen trails come out of jet engines and that they appreciate my work exposing the disinformation about contrails. Every one of these pilots

---

33 www.boldmethod.com/learn-to-fly/weather/contrails/.

knew that contrails are so rare that most people will never see one in their lifetime, and if they do occur, they are at high altitudes that cannot be seen from the ground. These professional pilots have flown most of their lives and have always had a deep interest in aviation. Some of them fly mainstream commercial jets while others fly large jets for major parcel carriers.[34]

NASA sponsors countless studies of "subsonic aircraft exhaust" and "natural and anthropogenic upper troposphere particles, laboratory studies of ice nucleation and soot particles, and numerical modeling of the formation of cirrus altered by exhaust soot particles."[35] A 2008 patent Abstract posits that control of terrestrial climate "relies mainly on civilian airlines burning. . .sun-shading (sun-blocking / sun-reflective) fuels in the high levels of the atmosphere [troposphere]," the method being "spray-dusting of sun-shading aerial sprays from aircraft, the regulated distribution of sun-shading dust, spray and exhaust fumes in the upper reaches of die [sic] atmosphere mainly by commercial or civilian aircraft."[36] "Sun-shading dust" sounds distinctly like the conductive metal nanoparticles under the rubric of ongoing geoengineering.

Two of the most widely used jet engine models are the *high-bypass turbofan* and *low-bypass turbofan engines*.

The high-bypass flow-cycle turbofan engine[37] operates more like a large-diameter propeller that routes incoming air around rather than through the engine. The propeller delivers 90 percent of the thrust without burning extra fuel, with only 10 percent going through combustion. No extra fuel means less water vapor or CO_2. Given that condensation requires high vacuum (not high pressure), high humidity, and low temperatures, the air-to-exhaust ratio is too high in the high-bypass engine to facilitate condensation. This engine is primarily used by commercial airlines and large supertankers. Lots of air, low fuel, and little moisture in the high-heat combustion chamber. Brief condensation trails may appear during takeoff when water vapor is expelled, often with a faint black trail. Condensation may also form in high humidity behind high-vacuum areas of the wings when the jet is pitching up or using the flaps, lasting only as long as the vacuum zone lasts. Otherwise, the trails we see are of particulate-rich aerosols.

The low-bypass turbofan engine takes air in and forces it through the

---

34 globalskywatch.com/stories/my-chemtrail-story/chemtrail-information/plumes-change.html#. V3fj0lc-gfN

35 Eric J. Jensen and Owen B. Toon, "Cirrus Cloud and Climate Modifications due to Subsonic Aircraft Exhaust." NASA, geo.arc.nasa.gov/sge/jskiles/fliers/all_flier_prose/cirrusclouds_jensen/cirrusclouds_jensen.html.

36 Mark Hucko, "System and Method of Control of the Terrestrial Climate and Its Protection against Warming and Climatic Catastrophes Caused by Warming such as Hurricanes." Patent US20090032214 A1, June 2, 2008.

37 See "High Bypass Turbofan Jet Engine" YouTube, *catagd*, February 12, 2014.

combustion chamber. This engine is less efficient at speeds below Mach 2 but produces great thrust at higher velocities, which makes it perfect for military fighter jets. Under the right conditions (high vacuum, high humidity, low temperature), the increase of water in the exhaust will produce short, non-persistent contrails. Given how high military fighter jets generally fly, sightings of these short non-persistent contrails are rare, and due to the military's desire for stealth, contrail suppression patents make them rarer still.

AIR PHARMACOLOGY I: JET FUEL AND CARBON BLACK DUST (CBD), CARBON BLACK AEROSOL (CBA), AND COAL FLY ASH

Whether one believes that trails being laid from horizon to horizon are due to chemicals added to jet fuel or chemicals being distributed from supplementary tanks and ducts, the end is the same: we are breathing electro-pharmacologically altered air far more destructive to life than the industrial blight of the past century and a half.

Since COP21 in Paris, a flurry of activity has centered on aircraft emissions while carefully sidestepping the very real dangers of jet fuel chemical additives classified under Project Cloverleaf. Meanwhile, environmental watchdogs awakening to the jet fuel factor are carefully guided to view it only in terms of carbons, greenhouse gases, ozone, etc. The global aircraft emissions standards developed by the UN's International Civil Aviation Organization (ICAO) won't go into effect until as late as 2025, and even then will only apply to "large planes like airliners and cargo jets and turboprop aircraft, not to smaller jet aircraft, piston-engine planes, helicopters or military aircraft."[38] The Center for Biological Diversity and Friends of the Earth are suing the EPA for not pressuring airlines to comply with the Clean Air Act.[39] Given that the EPA is a loyal handmaiden to the ICAO, and that the FAA refuses to see itself as responsible for what jets are spewing, it is difficult to believe that legal guidelines will be enforced.

Because domestic airline emissions account for three percent of total greenhouse gas emissions,[40] aircraft makers and fuel experts are working on designing lighter aircraft with more efficient engines, costs which will be passed along to consumers. Fuel and personnel are the two largest operating expenditures of commercial airlines, so the new fuel efficiency standards are forcing domestic airlines to scramble for alternatives. Pratt & Whitney have

---

38 "EPA calls for regulating emissions from US airliners." *Fox News*, June 10, 2015.

39 Michael Biesecker, "Environmental groups sue over pollution from airliners." AP, April 12, 2016. Also see "EPA calls for regulating emissions from US airliners." *Fox News*, June 10, 2015.

40 Bloomberg Business. Activist Harold Saive (chemtrailsplanet.net) begs to differ: "It's worse than 3% since almost all CO_2 emissions are released above 30,000 feet and cannot be readily converted and eliminated by photosynthesis" (June 17, 2015).

developed the high-bypass PurePower Geared Turbofan (GTF) engine, which lowers CO_2 releases "comparable to the amount of CO_2 displaced by 900,000 trees in the same amount of time" and nitrogen oxide by "50 percent below the most rigorous standard set by the United Nations' International Civil Aviation Organization."[41] Ultra-low sulfur diesel fuel is under development, as is biofuel—municipal solid waste converted into jet fuel, diesel, and ethanol, "half a billion gallons of fuel in markets across North America at lower costs compared to conventional petroleum fuel, while reducing carbon emissions by more than 80%," according to Fulcrum BioEnergy Inc. president E. James Macias.[42]

Researcher Jim Lee (*climateviewer.com*) believes the secret to the chemtrail mystery is in the gas tank. Lee has found documents proving that barium fuel additives and aluminum are in jet fuel, and believes that the ICAO, United Nations, U.S. State Department, DOT, EPA, FAA, and NASA seek to chemically alter contrail cirrus to cool the planet. By manipulating *soot* and *sulfuric acid*, heat-trapping pollution can be turned into planet-cooling carbon credit cash. Soot—*carbon black dust (CBD)* and *carbon black aerosol (CBA)*—has been recognized for its weather-altering abilities since the 1970s and is the primary ingredient of artificial cloud production.[43] Both carbon black and soot are fossil fuel waste products, but carbon black is in diesel (diesel oxidation), whereas soot arises only during pyrolysis (jet fuel combustion).

> Soot particles are composed of individual, nearly spherical particles (spherules), which have a number mean radius between 10 and 30 nm.[44]

As Lee puts it, the more soot nanoparticles generated by jet exhaust, the more cloud condensation nuclei (CCN or cloud seeds), the more cirrus cloud cover created.

Lee puts the aviation fuel problem pithily: "The truth is that artificial clouds are destructive to nature, harmful to health, and there is nothing 'normal' about fire-breathing metal tubes spewing nanoparticles at 30,000 feet."

Add to that the two FOIAs in 1994 from the Joint Nonlethal Weapons Directorate (JNLWD) revealing that the U.S. military was still pursuing weather warfare despite the ENMOD ban[45]: the U.S. Naval Research Lab at China Lake,

---

41 "Aircraft Geared Architecture Reduces Fuel Cost and Noise," NASA Technology Transfer Program, *Spinoff* 2015.

42 "Fulcrum BioEnergy proves trash-to-jet fuel method, gets DOD grant." Fulcrum BioEnergy press release, May 28, 2013.

43 Jim Lee, "Chemtrails: The Shady Truth About Contrails." climateviewer.com/chemtrails/

44 "Aviation and the Global Atmosphere: 3.2.3. Soot and Metal Particles," Intergovernmental Panel on Climate Change, www.ipcc.ch/ipccreports/sres/aviation/index.php?idp=35

45 Jim Lee, "1978 ENMOD Convention on the Prohibition of Military and Other Hostile Use of

California was experimenting with weather modification techniques, and the U.S. Air Force Research Lab at Wright Patterson Air Force Base was researching CBD/CBA seeding as a "force multiplier."[46]

In 1996, the oft-quoted "Weather As A Force Multiplier: Owning the Weather in 2025" was released by the USAF Air War College as a roadmap for weather warfare technologies, including a diagram showing the use of CBD by 2005 under "Technologies to be developed by DOD (Department of Defense)."[47] "Owning the Weather" is no "think piece." At the Test Technology Symposium '97, "Session B: Advanced Weapon/Instrumentation Technologies," Dr. Arnold Barnes of the Phillips Lab/GPO at Hanscom Air Force Base reiterated the use of CBD for weather warfare on three slides, including a slide with current capabilities referencing cloud creation.[48]

Harking back to 1973 when atmospheric scientist William M. Gray (Colorado State University) and meteorologist William M. Frank (Pennsylvania State University) proposed aircraft delivery of carbon black soot to heat portions of a hurricane to alter its path,[49] the post-Katrina 2008 Hurricane Modification Workshop sponsored by the U.S. Department of Homeland Security Science and Technology Directorate proposed testing CBD/CBA seeding for hurricane "mitigation." In a presentation called "Collaborative Research: On Hurricane Modification by Carbon Black Dispersion: Methods, Risk Mitigation, and Risk Communication," research scientist Moshe Alamaro (Earth, Atmospheric, and Planetary Sciences, MIT) suggested "the use of carbon black aerosol (CBA) to selectively heat parts of the atmosphere by dispersion of CBA above a hurricane." Among the workshop recommendations was this nugget: "Distribute the CCN (cloud condensation nuclei) aerosols on the periphery of the hurricane using aircraft and ships. When distributing carbon black, release it above the core of the hurricane. Accomplishing the seeding will take further study."

Environmental Modification Techniques," January 22, 2016, climateviewer.com/2016/01/22/1978-enmod-convention-on-the-prohibition-of-military-and-other-hostile-use-of-environmental-modification-techniques/.

46 The U.S. Navy proposal to develop new weather modification weapons, Code C2741 (Warhead Development Branch) NAWCWPNS, China Lake, California, April 1994, www.scribd.com/doc/256210692/Weather-Modification-US-Navy-FOIA-Nonlethal-Warfare-Proposal-1994; and the U.S. Air Force proposal to develop a theater-scale weather modification system using carbon black, Phillips Laboratory (AFMC) Geophysics Directorate, www.scribd.com/doc/184741384/Weather-Modification-Using-Carbon-Black-Phillips-Laboratory-AFMC-Geophysics-Directorate.

47 Tamzy J. House et al. "Weather As A Force Multiplier: Owning the Weather in 2025," Air Force 2025, August 1996; see page 34, Figure 5-2, "A Systems Development Road Map to Weather Modification in 2025." csat.au.af.mil/2025/volume3/vol3ch15.pdf.

48 Jim Lee, "US military discusses future of Weather Warfare despite ENMOD ban," November 16, 2013, climateviewer.com/2013/11/16/us-military-discusses-future-of-weather-warfare-despite-enmod-ban/.

49 William M. Gray, "Feasibility of beneficial hurricane modification by carbon dust seeding," Paper No. 196, NOAA Grant No. N-22-65-73 (G), April 1973; William M. Frank, "Characteristics of carbon black dust as a large-scale tropospheric heat source," Paper No. 195, National Science Foundation, January 1973.

Their plan called for $64 million and 36 months to "conduct large scale test w/evaluation and reports."[50] Jim Lee believes that drones (UAVs) are doing this testing today and offers the Hurricane Aerosol and Microphysics Program (HAMP) and NASA Genesis and Rapid Intensification Processes (GRIP) programs as examples.[51] Former director of the NOAA Atmospheric Modification Program Joe Golden spoke about HAMP to the American Meteorological Society two years before Frankenstorm Sandy.

Lee and four fellow activists spoke at a CSPAN-televised hearing of the EPA in Washington, D.C. on August 11, 2015.[52] The EPA was considering regulating six greenhouse gases produced by jet aircraft. Lee pressed the EPA to include cloud creation and randomized testing of in-flight jet aircraft for metal particulates[53] in upcoming aviation pollution regulations. The EPA is supposedly moving ahead with regulating the six greenhouse gases but still ignores the soot, contrail cirrus clouds, and metal particulates.[54] With the Department of Defense and Department of Homeland Security involved in using carbon black dust for weather modification, it should come as no surprise that the EPA, FAA, DOT, and IPCC *have no soot reduction goals.*

Two studies, one in 2001 and another in 2008, have caused a panic in the aviation industry. The studies prove that the artificial chemical cirrus cloud cover is trapping heat, as much as 5,000X IPCC estimates, and as much as fifty years' worth of aviation-produced CO_2!

> Contrails formed by aircraft can evolve into cirrus clouds indistinguishable from those formed naturally. These 'spreading contrails' may be causing more climate warming today than all the carbon dioxide emitted by aircraft since the start of aviation.[55]

---

50 Jim Lee, "Hurricane Hacking: The Department of Homeland Security enters the weather modification business," November 8, 2013, climateviewer.com/2013/11/08/hurricane-hacking-the-department-of-homeland-security-enters-the-weather-modification-business/

51 "Hurricane Aerosol and Microphysics Program (HAMP)" YouTube of a speech by chairman and cloud physicist William R. Cotton at the 29th Conference on Hurricanes and Tropical Meteorology, May 10, 2010, www.youtube.com/watch?v=z72TJ68zKJE; and Scott A. Braun et al., "NASA's Genesis and Rapid Intensification Processes (GRIP) field experiment." Bulletin of the American Meteorological Society, 94.3 (2013): 345–363.

52 Jim Lee, "World's First EPA Public Hearing on Flight Pollution! Speak Now!" climateviewer.com/2015/07/26/worlds-first-epa-public-hearing-on-flight-pollution-speak-now/.

53 Jim Lee, "My Speech to the EPA about Flight Pollution," climateviewer.com/2015/08/09/my-speech-to-the-epa-about-flight-pollution/.

54 "EPA Finalizes First Steps to Address Greenhouse Gas Emissions from Aircraft Engines," EPA Office of Transportation and Air Quality, July 2016, www3.epa.gov/otaq/documents/aviation/420f16036.pdf.

55 Boucher, O. "Atmospheric science: Seeing through contrails." *Nature Climate Change* 1, 24–25 (2011).

CBD may be a great cloud seed, *but it absorbs heat*. Will the addition of sulfuric acid to these clouds mimic the volcano's cooling effect and increase the aviation industry's carbon credits? Geoengineering SRM advocates think so:

> Options for dispersing gases from planes include the addition of sulfur to the fuel, which would release the aerosol through the exhaust system of the plane, or the attachment of a nozzle to release the sulfur from its own tank within the plane, which would be the better option. Putting sulfur in the fuel would have the problem that if the sulfur concentration were too high in the fuel, it would be corrosive and affect combustion. Also, it would be necessary to have separate fuel tanks for use in the stratosphere and in the troposphere to avoid sulfate aerosol pollution in the troposphere.[56]

Doping commercial flight jet fuel looks like the name of the game, whatever the cost to plants, animals, and human beings breathing it all in. Taking a nod from the geoengineers, atmospheric scientist Ulrich Schumann exposes the real chemtrail conspiracy in his recommendations to the ICAO Colloquium on Aviation and Climate Change in 2010:

- Contrail cirrus contributes a large fraction to the aviation induced climate impact.

- Satellite data analyses suggest observable impact of aviation on cirrus cover and radiation fluxes.

- The climate impact of aviation induced contrail cirrus depends on aircraft properties and routing

- Both aspects [soot and flight routing] offer the potential for aviation to reduce the climate impact of aviation [less soot emissions, LESS WARMING and MORE COOLING CONTRAILS].[57]

The last statement is a bombshell. Schumann is clearly abandoning the idea of ridding the world of contrails; he is, in fact, advocating for creating cooling contrails and using them for geoengineering SRM purposes. Schumann's recommendations combine with the recent mandate for aviation biofuels in the ultimate two-tank geoengineering solution:

> Applying high FSCs [fuel sulfur content] at aviation cruise altitudes combined

---

56 Robock, Alan, et al. "Benefits, risks, and costs of stratospheric geoengineering." *Geophysical Research Letters* 36.19 (2009).

57 Ulrich Schumann, German Aerospace Center, "Recent research results on the climate impact of contrail cirrus and mitigation options." ICAO Colloquium on Aviation and Climate Change, 2010.

with ULSJ fuel [ultra-low-sulfur jet fuel, aviation biofuel] at lower altitudes result in reduced aviation-induced mortality and increased negative RE [radiative effect] compared to the baseline aviation scenario.[58]

In other words, use high-sulfur jet fuel when we get to flight altitude to cool the planet, use no-sulfur jet biofuel on takeoff to kill fewer people around runways. This is exactly what they are doing.

Jim Lee closely follows the alternative jet fuel (biofuels) and sulfur fuel doping now being tested under the FAA's Aviation Climate Change Research Initiative (ACCRI). These tests, dubbed the Alternative-Fuel Effects on Contrails & Cruise EmiSSions or ACCESS-1 and 2, are focused on how different types of fuels make clouds. The three fuels on the drawing board and in the skies are a low-sulfur JP-8 fuel, a 50/50 blend of JP-8, and a camelina oilseed-based HEFA fuel or JP-8 fuel doped with sulfur.[59] The agenda behind sulfur content (FSC) geoengineering is to implement SRM worldwide while circumventing international geoengineering and weather modification laws.

From time to time, sky watchers see "smokers" streak across the sky, leaving behind a dark trail. The oily consistency of carbon black or soot can leave such a trail, as can the sulfur in diesel fuel. Both diesel and carbon black are hard on engines and are sometimes used as short-run solutions for extending old fuel. Carbon black occasionally falls from the sky and coats cars and streets.[60]

Carbon dust is useful to Big Oil and geoengineering in that it absorbs solar energy[61] and can therefore melt the pesky glacial ice slowing up the exploitation of Arctic oil reserves. I already mentioned its use in HAMP to intensify hurricanes, tornadoes/cyclones, etc.,[62] and researcher Harold Saive (*chemtrailsplanet.net*) has posed yet another possible objective of "smokers" laying carbon black:

> Dare we speculate this "smoker" is a desperate attempt by the criminal climate cabal at IPCC to raise the temperature in the upper atmosphere to fool satellite data in order to support the failing evidence for global warming over the past 16 years? As the dark aerosols descend to earth, the ground-based thermometers

---

58 Z.Z. Kapadia, et al. "Impacts of aviation fuel sulfur content on climate and human health." *Atmospheric Chemistry and Physics Discussions* 15.13 (2015): 18921–18961.

59 Bruce Anderson, "Alternative-Fuel Effects on Contrails & Cruise EmiSSions (ACCESS-2) Flight Experiment." NASA LaRC and the ACCESS-II Science and Implementation Teams, January 2015; and Richard H. Moore et al. "In-Situ Measurements of Contrail Properties Measured During the 2013–2014 ACCESS Project," 14th Conference on Cloud Physics, July 2014.

60 Stevie Borrello, "Mysterious Tar-Like Substance Coats Homes in Michigan Neighborhood." *ABC News*, February 17, 2016.

61 W.M. Gray et al. "Weather Modification by Carbon Dust Absorption of Solar Energy." Atmospheric Science Paper No. 225, July 1974.

62 See "Carbon Black Chemtrails Weaponized Hurricane Sandy and Katrina," www.aircrap. org/2015/01/20/carbon-black-chemtrails-weaponized-hurricane-sandy-and-katrina/.

could be expected to read higher temps as solar radiation heats the black carbon near the sensors.[63]

However we look at it, breathing carbon black nanoparticles is bad for human health. The 2011 University of Iowa paper "Induction of Inflammasome Dependent Pyroptosis by Carbon Black Nanoparticles" in the *Journal of Biological Chemistry* sounded the alarm:

> ...the intake of carbon black nanoparticles...caused an initial inflammatory response in lung cells. The surprising results came when the team discovered that these nanoparticles killed macrophages—immune cells in the lungs responsible for cleaning up and attacking infections—in a way that also increases inflammation.[64]

Carbon black doesn't produce cell death by *apoptosis* (the cell shrinking into itself) but by *pyroptosis* (bursting and spreading)—in other words, not by atrophy but by *heated implosion*.

The sulfur coating of soot means sulfuric acid (H_2SO_4), the major component of acid rain, acidification of the oceans, the Greenhouse Effect, etc. The following admission is from 1998, though it is highly doubtful that it was actually the "first" detection:

> Direct detection of total sulfuric acid (SA) has been achieved for the first time in the plume of a jet aircraft in flight. The measurements show the same SA signatures for the case when SA was injected directly into the exhaust jet and the case when sulfur was provided to the engine with the fuel.[65]

Geophysicist J. Marvin Herndon, Ph.D., believes that the soot or carbon black posited by Jim Lee is actually extremely toxic *coal fly ash*, a perfect Welsbach material 10–100 microns thick, the lightest of the coal ashes ("component of flowable fill")[66] in plentiful supply. Captured by electrostatic precipitators, coal fly ash is an anhydrous combustion byproduct containing substantial amounts of silicon dioxide (SiO_2), aluminum oxide (Al_2O_3), and calcium oxide (CaO) as well as arsenic, beryllium, boron, hexavalent chromium, cobalt, lead, manganese, mercury, molybdenum, selenium, strontium, thallium, and vanadium—all of which are readily released upon contact with water.

63 Email, 24 February 2014.

64 "UI study: Carbon black nanoparticles activate immune cells, causing cell death." *nanotechwire.com*, May 19, 2011.

65 J. Curtius et al., "First direct sulfuric acid detection in the exhaust plume of a jet." *Geophysical Research Letters*, Vol. 25, No. 6, March 15, 1998.

66 Cyclone classifiers can be used to reduce aerosol coal fly ash to nano sizes (smaller than 1 micron) for additional loft time.

...Coal fly ash...when subjected to water renders many of its elements partially soluble in water. Some of these are aluminum, barium, strontium, calcium, iron, magnesium, etc. These soluble elements can be measured in rainwater and are in fact found in post-spraying rainwater. Coal fly ash has another advantage to the military — the elements dissolved in atmospheric water make the water much more electrically conducting. This is important for movement by electromagnetic means.[67]

For more than a century, air pollution on the American East Coast has been practically synonymous with the industrial burning of coal (lignite) and 140 million tons of coal combustion waste translating to 70 million tons of fly ash per year, much of which is dumped into 1,100 unlined open storage ponds[68] around the nation contaminating the food chain and local drinking water with toxic heavy metals and radon. Despite the cancer, learning disabilities, neurological disorders, birth defects, and reproductive sterility that arise around these ugly coal ash impoundments, the EPA insists coal ash is a non-hazardous substance[69]:

"Coal ash is basically soil," says Tom Robl, a University of Kentucky geoscientist who serves as a director of the American Coal Association. Not only is coal ash non-toxic, Robl says, it's so safe that you could eat a brimming bowlful without adverse consequences. "Feel free to eat as much coal ash as you want—it's not toxic."[70]

In the same article, Forest Service biologist Dennis Lemly begs to differ: as a highly concentrated byproduct of burning coal, coal ash is hazardous.[71]

In 2015, Herndon published two peer-reviewed papers on how pivotal coal fly ash is to geoengineering and how deleterious it is to human health. The second paper, "Evidence of Coal Fly Ash Toxic Chemical Geoengineering in the Troposphere: Consequencies for Public Health," was published on 11 August 2015 in the *International Journal of Environmental Research and Public Health* and retracted twenty-two days later, followed by a scathing criticism of the first paper in *Current Science*. Herndon's third and most thorough paper, "Human Environmental Dangers Posed by Ongoing Global Tropospheric Aerosolized Particulates for Weather Modification," was published in *Frontiers in Public Health* on June 30, 2016, and retracted two weeks later.

---

67 Email to Harold Saive and cc'd to Elana Freeland, April 29, 2017.

68 Coal ash impoundments were exempted from the 1970s rules monitoring coal slurry storage sites. At slurries, coal is separated from noncombustible ash.

69 "EPA Refuses To Classify Coal Fly Ash As a Hazardous Waste, Primary Toxic Component of Chemical Geoengineering." *State of the Nation*, August 19, 2015.

70 Ben Whitford, "Coal ash: America's multibillion-ton toxic legacy." *Ecologist*, 29 September 2015.

71 Ibid.

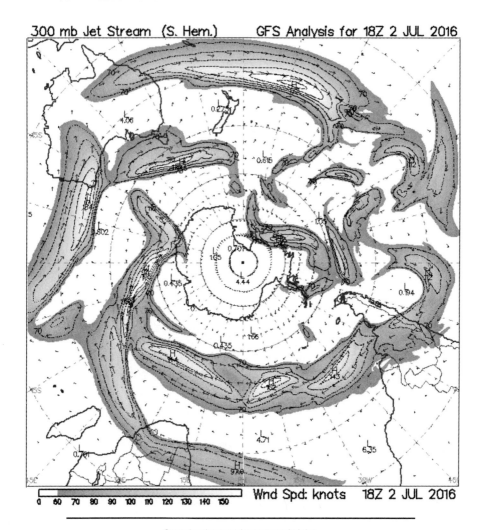

Geoengineering means jet stream manipulation.
Thanks to V. Susan Ferguson metaphysicalmusing.com/.

Unlike the usual procedures employed to fairly address complaints about peer-reviewed and published papers, neither of the two journals involved provided Herndon with verbatim copies of the complaints made by the highly skilled disinformation team so he could refute them. The implications one might draw from this are (1) Herndon is likely correct about coal fly ash being the main aerosolized particulate, and (2) that officials are aware of the highly toxic nature of the material being sprayed into the air we breathe and want to hide that knowledge from the public.

Herndon presents four sets of experiments to back up his thesis that hazardous coal fly ash is "likely the principal aerosolized particulate sprayed in the troposphere by jets for geoengineering, weather modification, and/or climate alteration."[72] Earlier, he'd compared water leach from coal fly ash with rainwater and discovered eight elements in common, especially aluminum/barium and strontium/barium ratios. Rainwater samples in San Diego, California measured at least 30 percent aluminum, and in Chico and Alachua County, Florida, he had high readings of barium, as well. He also compared outdoor dust collected in a HEPA (high-efficiency particulate air) filter with unleached coal fly ash and discovered fourteen elements in common.

Herndon and Florida Public Health medical officer Mark Whiteside, M.D., M.P.H., demonstrated that snowfall can trap and bring down aerosolized particulates, like the co-precipitation technique used in chemical processes.[73] Upon melting, the snow releases the trapped particulates that drain with the melt water into material that might underlie the snow, such as the fibrous, sticky white mesh discovered in melting snow[74] in Wisconsin. Analysis at Northern Lake Services in Crandon, Wisconsin revealed that twenty-three of twenty-six elements corresponded to what is found in unleached coal fly ash. Herndon asks:

> So, how might one account for the strikingly different appearance of the fibrous mesh morphology if indeed it is, as evidence indicates, essentially identical in composition to fly ash?[75]

He then wonders if jet fuel combustion (pyrolysis) is responsible for these fibers, along with environmentally destructive methylmercury (CH_3H_g) and ozone-depleting chlorinated-fluorinated hydrocarbons.

"DUMPING" VIA ENGINE AND WING PYLON DRAIN TUBES

In the Environmental Research Letter 4 (October – December 2009), "Modification of cirrus clouds to reduce global warming," two delivery mechanisms are pointed out, the first being jet fuel and the second being injections into the hot engine exhaust:

> Since commercial airliners routinely fly in the region where cold cirrus clouds exist, it is hoped that the seeding material could either be (1) dissolved or suspended in their jet fuel and later burned with the fuel to create seeding

72 J. Marvin Herndon, "Human Environmental Dangers Posted by Ongoing Global Tropospheric Aerosolized Particulates for Weather Modification." *Frontiers in Public Health*, June 30, 2016.

73 J. Marvin Herndon and Mark Whiteside, "Further Evidence of Coal Fly Ash Utilization in Tropospheric Geoengineering: Implications on Human and Environmental Health." *Journal of Geography, Environment and Earth Science International*, 3 February 2017.

74 One wonders what relationship this "snow" had to chemically nucleated "snow" that burns instead of melts. See YouTube "Shouldn't Snowballs Melt?!?" MrGregory1957, January 30, 2014.

75 Herndon, "Human Environmental Dangers."

aerosol, or (2) injected into the hot engine exhaust, which should vaporize the seeding material, allowing it to condense as aerosol in the jet contrail. The objective would not be to seed specific cloud systems but rather to build up a background concentration of aerosol seeding material so that the air masses that cirrus will form in will contain the appropriate amount of seeding material to produce larger ice crystals.[76]

Among the chemtrail activists studying delivery systems, Harold Saive (*chemtrailplanet.net*) is the most vociferous about a chemical system supplementary to fuel exhaust being necessary to produce "persistent contrails (PC)" and the cirrus cloud cover that follows. Saive makes an excellent point about why the possibility of this delivery system matters so much:

It's important to keep in mind that aerosols intended for synthetic climate change, hurricane engineering (H.A.M.P.) and most likely "biologicals" (Morgellons), are deployed with nozzles that have nothing to do with JP-8 fuel, contrails or engine emissions. This explains why a gap between the engine and visible trail is created (albeit very short) when water vapor is replaced by a vaporized chemical/metal like alumina.[77]

Saive points to two recent Internet videos (and there are many others from years past at his site) uploaded by Jörg Meiring of Maryland: "Jesus Christ help us please – LH501 Chemtrail-Sprayer 'Siegerflieger'" (February 5, 2017),[78] taken by Meiring himself while a passenger on Lufthansa; and "757 chemtrail filmed from cockpit" (February 26, 2017),[79] uploaded by Meiring but filmed by what appears to be an anonymous pilot. In an April 30, 2017 email, Saive describes the cockpit observation thus:

Notice how the solid white line at the bottom of the trail is positioned where most drain tubes are typically mounted. The remainder of the circumferential trail column is a function of turbofan and combustion exhaust turbidity that (in this case) fails to completely capture and homogenize the flow from the drain tubes in order to simulate a water vapor contrail. Notice how the trail becomes less dense as it approaches the top of the column, farthest from point of drain tube injection. A minority of drain tubes mounted above the exhaust would be expected to exhibit differently.

76 iopscience.iop.org/article/10.1088/1748-9326/4/4/045102/fulltext/;jsessionid=8566D916C27507
4AD64F27443B2F1127.c4.iopscience.cld.iop.org

77 Email, April 28, 2017.

78 www.youtube.com/watch?v=EOPo3gHBSHo

79 www.youtube.com/watch?v=xQsdk7iopSU&feature=youtu.be

The cockpit video clearly shows a smaller chemical trail threading beneath the larger, less defined engine exhaust. Critics may say this smaller trail could be a "cargo bay aerosol leak," but what if this "leak" is to supplement fuel emissions so as to super-saturate cloud condensation nuclei (CCN), as was pointed out in an American Meteorological Society video presentation?[80] And what of the bifurcated trails and tw-engine aircraft leaving three trails? *Drain tubes* are often conveniently located (even aftermarket re-engineered) inside the bypass duct so as to create the illusion six miles below that all we are seeing is fuel combustion (pyrolysis) emissions.

Other inconsistencies that question a fuel-only chemical delivery:

- Sudden on-off trails
- Sputtering trails that eventually "go dry" with increasingly wider gaps
- Unique filament, "toothpaste," and "knots on a rope" trails indicating varying particulate ratios and chemical signatures
- Questionable volume and mass of single or intertwining aerosol plumes
- Observations of jets emitting NO trails
- Observations of starboard or port side emitting a different color (or even missing) trail

So far, we've examined chemical delivery systems from the outside. Now, let's look more closely at what might be going on in the engine combustion chamber.

AIR PHARMACOLOGY II:
SPRAY PYROLYSIS AND CHEMI-IONIZATION

In the December 6, 2001 issue of *Columbus Alive*, award-winning Ohio reporter Bob Fitrakis, JD, revealed that according to two Lawrence Livermore National Laboratory scientists, two separate projects in electromagnetic weather modification technology were being conducted at Wright-Patterson Air Force Base, one involving artificial Al_2O_3 cloud cover creation to lessen the effects of global warming, the other with barium stearate for over-the-horizon (OTH)

---

80 "WMO 2015 - Climate Warming Response to Cirrus Cloud Seeding by Trude Storelvmo," speaking at the American Meteorological Society meeting in 2015, youtu.be/cncVgSkjVtM?t=4m37s.

military communications, 3D mapping, and radar.[81]

In his 2013 article "Wireless, Chemtrails, and You," author William Thomas[82] recounts the early work of independent scientist Clifford Carnicom in northern New Mexico and wildlife biologist Francis Mangels, Ph.D., in northern California. Beginning in 1999, Carnicom was documenting high levels of reactive barium salts from aerosol grids over Santa Fe. By 2001, barium levels had nearly doubled in California, despite the fact that industrial barium pollution had all but vanished. With trees and fish dying, Mangels blamed aluminum oxide for soil pH measuring 10X its normal alkalinity. In 2002, an Edmonton, Alberta, Canada landscaper was seeing severe nutrient deficiencies (*chlorosis*) in flowers and trees; electrical conductivity readings for soil should have been no higher than 1 but were showing 4.6 to 7X higher. Snow samples confirmed elevated aluminum and barium. By 2008, a KSLA-sponsored lab test of Louisiana precipitate found barium to be 6X the toxic level set by the EPA; in 2010, Arizona air particulates of barium and aluminum skyrocketed: aluminum was 15.8X the toxic level, barium 5.3X. Carnicom has confirmed that the barium in our atmosphere is now 8X the level deemed safe to breathe.

At the Open Mind Conference in Oslo, Norway on October 27, 2012 and later updated as "The Antennas Within the Body" for the 8th Environmental Conference in Nuremberg, Germany, May 30, 2013, German chemist Harald Kautz-Vella presented his paper "The Chemistry in Contrails: Assessing the Impact of Aerosols From Jet Fuel Impurities, Additives and Classified Military Operations on Nature"[83] in which he discusses the alchemy that occurs in the jet combustion engine as well as a deeper consideration of the biological operations of chemical trails. His background includes university studies in Geology, Physics, Mathematics, and Media Science, with a degree specialization in Geoscience. Since 2001, he's been studying scalar technologies such as zero-point energy, soft weather modification (Wilhelm Reich), electro-fertilization in agriculture, and information-medicine/radionics.

At the outset, Kautz-Vella stresses that the classified Project Cloverleaf is under the jurisdiction of intelligence agencies, and that advanced military and intelligence operations may depend upon some aspects of classical geoengineering but are not limited to it. Transformation of the lower atmosphere into a controllable, artificially engineered plasma is intended to serve weather engineering, yes, but also to serve 3D radar monitoring, range enhancement,

81 Also see Fitrakis' March 28, 2002 *Columbus Alive* article, "Into Thin Air: How Kucinich's 'chemtrails' disappeared ... from right under Congress' nose!" Fitrakis ran for Franklin County Prosecutor in November 2016.

82 Thomas' 2004 book *Chemtrails Confirmed* was one of the first go-to books on chemtrails.

83 This paper can be found at www.aquarius-technologies.de/download/TheChemistryinContrails. pdf minus several pages added to the version in *Dangerous Imagination, Silent Assimilation* by Cara St. Louis and Harald Kautz-Vella, White Lion Press (2014).

advanced DEW systems, and SDI Space Fence objectives of absolute surveillance under artificial intelligence (AI).

Kautz-Vella approaches fuel emissions / spraying from a biological effects perspective concentrating on the electro-optical toxicity of the particulate plasma and how it disturbs cell communication and electrical cell potential in plants (and, by extension, other life forms), leading to biophoton[84] deterioration. He references artificially produced clouds as "piezoelectric particulate plasma" and divides the spraying operation that produces them into three categories:

(1) Jet fuel additives (lead, sulfur, halogens);

(2) Conventional weather manipulation with ions like silver iodide or based on reflective particles like Al_2O_3; and

(3) High-tech aerosols for military and intelligence purposes.

While researching damage to hay harvests in Norway in 2012, Kautz-Vella analyzed soil and rain samples. Heavy metals were within the acceptable range of pollution during the acid rain of the 1970s, so he had to consider what else could have caused grass to stop growing, trees to drop their bark and die, etc.

He pored over the U.S. Air Force Academy "Chemtrails: Chemistry 131 Manual, Fall 1990." Though classified formulae are not mentioned in the manual, enough is present to spell out the chemical processes of *spray pyrolysis*,[85] the thermochemical decomposition of organic material at elevated temperatures (as in aircraft engines) in the absence of oxygen or halogen (the chemically related fluorine, chlorine, bromine, iodine, or astatine); and *chemi-ionization*, the formation of ions through the reaction of a gas-phase atom or molecule. Add metal salts to the combustion process—hydrochloric acid (HCl), ammonium hydroxide (NH_4OH), and sodium hydroxide (NaOH); potassium permanganate (KM_nO_4), aluminum nitrate ($Al(NO_3)_2$), barium nitrate ($Ba(NO_3)_2$), copper II nitrate ($CU(NO_3)_2$), iron nitrate ($Fe(NO_3)_2$), lead nitrate ($Pb(NO_3)_2$), manganese nitrate ($Mn(NO_3)_2$), silver nitrate ($AgNO_3$), sodium carbonate (Na_2CO_3), sodium chloride (NaCl), sodium iodide (NaI), sodium nitrate ($NaNO_3$), etc.—and the jet will create amorphous or monocrystalline non-soluble nanoparticles like, for example, barium-strontium-titanate ($(BaSr_x)TiO_3$).

This is the chemical process of an artificially nucleated condensation trail. In 2005, a paper recounting spectrometer measurements of negative-ion composition and density taken from the exhaust of a J85-GE-5H turbojet at

---

84 Biophotons are DNA send-and-receive single photons responsible for cell communication.

85 See Hendrik K. Kammler et al., "Flame Pyrolysis of Nanoparticles. WILEY-VCH *Chem. Eng. Technol.* 24 (2001) 6; and M. Enhessari et al., "Synthesis and characterization of barium strontium titanate (BST) micro/nanostructures prepared by improved methods." *Int. Journal Nano. Dim.* 2(2):85-103, Autumn 2011.

ground level indicated that "ion nucleation is a probable mechanism for volatile aerosol formation."

> Introduction. Chemiions produced in jet engine combustion are speculated to play a role in ion-induced nucleation of aerosols, possibly followed by condensation, which may result in the formation of contrails, cirrus clouds, and pollutants.[86]

Kautz-Vella realized that recent plant degeneration in Norway was not so much due to the chemistry of non-soluble particles but to their *electrochemical / optical properties*, and that the delivery system seemed to originate from the aerosols in the jet trails overhead. Two compounds were most prevalent: aluminum oxide (Al_2O_3) and barium-strontium-titanate ($(BaSr_x)TiO_3$).

The 1990 Welsbach patent named Al_2O_3 as the primary compound in aerosol spraying:

> A method is disclosed for reducing atmospheric warming due to the greenhouse effect resulting from a greenhouse gases layer. The method comprises the step of seeding the greenhouse gas layer with a quantity of tiny particles of materials characterized by wavelength-dependent emissivity or reflectivity, in that said materials have high emissivities in the visible and far infrared wavelength regions and low emissivity in the near infrared wavelength region. Such materials can include the class of materials known as Welsbach materials. The oxides of metal, e.g., aluminum oxide, are also suitable for the purpose. The greenhouse gases layer typically extends between about seven and thirteen kilometers above the earth's surface. The seeding of the stratosphere occurs within this layer. The particles suspended in the stratosphere as a result of the seeding provide a mechanism for converting the blackbody radiation emitted by the earth at near infrared wavelengths into radiation in the visible and far infrared wavelength so that this heat energy may be reradiated out into space, thereby reducing the global warming due to the greenhouse effect.[87]

Kautz-Vella posited that the Al_2O_3 from military afterburner technology in jets and rockets attaches to the membranes of trees' tiny rootlets and blocks nutrient uptake, causing the roots to die. Aluminum oxide can be created in spray pyrolysis but only in the beginning phase of the engine firing, as monocrystalline structuring needs at least 1700°C (3092°F). Thus the lion's

---

86 Thomas M. Miller et al. "Mass distribution and concentrations of negative chemi-ions in the exhaust of a jet engine; Sulfuric acid concentrations and observation of particle growth." Air Force Research Laboratory, 11 January 2005.

87 "Stratospheric Welsbach seeding for reduction of global warming, US 500316 A," Hughes Aircraft Company, April 23, 1990.

share of Al_2O_3 in the soil must be coming from afterburner technology inbuilt into an endless number of military aircraft.

The second compound of $(BaSr_x)TiO_3$ piezoelectric crystals is formed during the engine combustion process due to the presence of barium, strontium, and titanium salts in the fuel.

In 1996, scientist Jim Phelps of Oak Ridge National Laboratory invented a titanium-based (Ti) aerosol "shield" system that formed less toxic fluor-aluminum compounds, prevented ozone loss, and reduced global warming by seeding reflective particles that blocked sunlight while allowing infrared (IR) through. Unfortunately but not surprisingly, the military promptly abused Phelps' invention in $(BaSr_x)TiO_3$.

Combined with barium (Ba) and strontium (St), Ti piezocrystals photo-ionize to block sunlight during the day while allowing IR at night. Phelps explained:

> The UV sprays use the same techniques to leave a spray of barium and titanium materials in the air. Plain titanium dioxide is an extremely good UV blocker. This brings on the last shield system, on the very surface or Earth and on the most vulnerable UV target organ. The US has arranged for titanium dioxide to be used in things like soap and detergent, which leaves one covered with a film of this material. One can usually see the paint-like scum in showers from the sticky titanium dioxide build up, along with calcium. This same film is all over most persons' skin and helps shield the skin from UV. The body is very tolerant of titanium and is not prone to set off the immune system inflammation response.
>
> This is not so true of aluminum and barium compounds. Even LLNL studies turn down the use of Al in chemspray methods. The combined effects of these compounds with other pollutants can well cause lasting lung damage.[88]

Phelps' titanium-based method and recent reports of methane (CH_4) being used to introduce metal salts to the combustion process[89] point to two new methods of spray pyrolysis: infrasound dispersion of water-soluble salts, and methane-based additives mixed with the jet fuel. But how much and how many of the metal salts raining down on us are bio-available, and how many are insoluble? What *is* obvious is that bioaccumulation of non-soluble crystals with electrochemical / optical effects are now in the food chain and synergistically spell disaster.

Barium occurs naturally in barium-rich soil, but the piezoelectric barium

88 Jim Phelps, DOEWatch 2003 web page; mirrored at www.chemtrailcentral.com/ubb/Forum1/HTML/001839.html. See www.doewatch.com/whistleblowing/ for Phelps biographical information and photos.

89 See YouTube of "New NASA Methane Jet Engine," www.popscreen.com/v/5WeMj/New-Nasa-Methane-Jet-Engine.

nanocrystals aerially distributed over North America are intended to enhance radar signals by refraction for military C4 operations. Sadly, the excited piezoelectric crystals in barium, silver (silver nitrate for cloud seeding), and strontium have been linked to chronic fatigue syndrome (CFS) clusters and other transmissible spongiform encephalopathies (TSEs) like fibromyalgia. But this "collateral damage" does not concern the military, and the medical studies[90] that might issue a wake-up call to the public are marred by ignorance of the role played by the constant 24/7 aerial delivery system.

The synthesis of $(BaSr_x)TiO_3$ and creation of defined nanoparticles (*chemiions*) occurs during spray pyrolysis inside the jet engine chamber:

> Spray pyrolysis involves passing an aerosol of a precursor solution through a graded temperature reactor in which the individual droplets are thermally decomposed to form the oxide particles.[91]

Clouds can then be "switched" on and off by zapping the piezoelectric barium-strontium-titanate compound and altering its crystal geometry. The free current then charges the aerosol particle and creates an ion that attracts moisture and forms a cloud.

Dielectric and piezoelectric barium titanate is excellent for nonlinear optics. High beam-coupling gain means it can operate at visible and near-infrared wavelengths, and its pyroelectric and ferroelectric properties make it useful for thermal camera sensors. In other words, the compound's *optical* capabilities are paramount in the military mind. With photons up-converted and their spin altered, the "sky theater" above can offer holographic screening as well as plasma cloud creation. Kautz-Vella explains:

> With regards to barium-strontium-titanate used as aerosols in the atmosphere, this effect could be used to "switch" clouds "on and off" by applying electromagnetic fields such as those created by HAARP and EISCAT devices or other radar or microwave-radiating antenna systems. These research and military installations have equipment with sufficient strength to alter the crystal geometry of the barium-strontium-titanate while in the atmosphere, which in turn leads to free current that would charge the aerosol particle and create an ion that will attract vapor and instantly lead to the forming of a droplet in the then-forming cloud.[92]

---

90 For example, two studies by Mark Purdey: "Elevated silver, barium and strontium in antlers, vegetation and soils sourced from CWD [chronic waste disease]: Do Ag/Ba/Sr piezoelectric crystals represent the transmissible pathogenic agent in TSEs?" *Medical Hypotheses*, 2004:63(2):211–25; and "Metal microcrystal pollutants: The heat resistant, transmissible nucleating agents that initiate the pathogenesis of TSEs." *Medical Hypotheses*, 2005:65[3]:448–477.

91 Kautz-Vella and Hauksdottir, "The Chemistry of Contrails."

92 Ibid.

Once these non-soluble nanocrystals and their EM properties fall to Earth, however, they are taken up as a whole by plants whose health and growth then suffer. In mammals and humans, these same nanocrystals damage the nervous system and weaken biophotonic activity. Kautz-Vella stresses that because $(BaSr_x)TiO_3$ nanocrystals absorb biophotons, the nanoparticle-plasma mix affects DNA light communication while serving as a matrix for synthetic biology and self-assembling nanomachines (like Morgellons). Electro-optical nanocrystals can produce signals that interfere with DNA, and artificial crystals replace natural ferro apatite crystals that play a major role in central nervous system (CNS) transmission. Kautz-Vella warns of the possibility of mind control via *in vivo* piezocrystals:

> If the natural ferro apatite crystal is replaced by artificially made piezoelectrical crystals, it appears to open the biological system to respond to a greater extent of artificial electromagnetic signals, both low and high frequency.[93]

Many jet propellants and additives are classified while others are open source. Classified JP-8, for example, is suspected of containing chemicals such as 1,2-dibrom ethan (EDB), perfluoroctane sulfon acid (PFOS), perfluoroctane acid (PFOA),[94] and lead tetraethyl.[95] All contain fluorine, sulfur, and brome and are regarded as important sources of the aerosols H_2SO_4, HF, and $HBrO_3$, all of which build highly toxic persistent contrails. The damaging impact of such additives is far more than that of sulfur and aluminum oxide reflective particles.

Media exposure of *aerotoxic syndrome* has drawn attention to the two million military and civilian personnel per year who are occupationally exposed to chemical gases coming from the engine chamber.[96] Unfortunately, the blood tests for aerotoxicity of pilots, cabin crew, and frequent flyers test for organophosphates but not necessarily for heavy metals in the chemical compounds pyrolized in the 60 billion gallons per year of kerosene-based jet propulsion fuels (JP-8, JP-5, and Jet A-1). Add to that no knowledge of the *classified* compounds, pyrolized (heated) jet oil leaks coming straight out of the engine, and nanoparticles leaking in from outside the aircraft (chemtrails, Wigner Effect radiation)—all coming through the bleed air valve.

As for the unclassified chemical gases from the engine chamber and pylon

93 Ibid.

94 See Carsten Lassen, "Survey of PFOS, PFOA and other perfluoroalkyl and polyfluoroalkyl substances." *LOUS Review*, November 27, 2012.

95 Oak Ridge whistleblower scientist Jim Phelps adds SF6, an asphyxiant soluble in ethanol and added to jet fuel because it is pyrophoric. It is particularly hazardous because it replaces oxygen. YouTube "Jim Phelps Scientist Chemtrails," September 1, 2012.

96 "Unfiltered Breathed In: The Truth about Aerotoxic Syndrome," a 2015 documentary by Tim van Beveren, www.unfilteredbreathedin.com.

drain tubes entering the cabin, it is worth quoting at length from a 2003 peer-reviewed paper:

> These exposures may occur repeatedly to raw fuel, vapor phase, aerosol phase, or fuel combustion exhaust by dermal absorption, pulmonary inhalation, or oral ingestion routes. Additionally, the public may be repeatedly exposed to lower levels of jet fuel vapor/aerosol or to fuel combustion products through atmospheric contamination, or to raw fuel constituents by contact with contaminated groundwater or soil. Kerosene-based hydrocarbon fuels are complex mixtures of up to 260+ aliphatic and aromatic hydrocarbon compounds. . .including varying concentrations of potential toxicants such as benzene, n-hexane, toluene, xylenes, trimethylpentane, methoxyethanol, naphthalenes . . .[97]

And the list goes on.

ROCKET PLUME BURNS

Rocket launches release tons of rocket plume burn that then spew over the countryside, ocean, or populated areas as the rocket arcs skyward. With every orbital launch (as opposed to suborbital), 20 to 120 tons of nano-aluminum particulates are dumped with the exhaust plume into the lower and upper atmospheres. Aluminum delivered by chemical trails and rocket plumes is the lightest, most excellent conductive metal we have. (Iron and gold oxides are actually the best but are too heavy and fall too quickly. Aluminum is light as a feather and stays airborne longer.) The electronics necessary for global weather engineering will not work without light aluminum particulates.

The aerosol brew laid by jets above the cloud cover heats regions by fomenting chemical reactions and trapping solar radiation. Previously, it was thought that the critical cloud fraction (CCF) value was the same everywhere, but the truth is that regional CCF values differ according to aerosol chemical signatures.[98] In 2004, NASA's senior research scientist Patrick Minnis admitted that artificial cirrus clouds formed by aircraft exhaust increased Earth surface temperatures between 1975 and 1994 (the Venus Effect) and added to the greenhouse gas effect.[99] In *Chemtrails, HAARP,* I quoted cloud physicist William R. Cotton's insight into how jet-created cloud cover is the opposite of a "global warming" solution:

---

97 G. Ritchie et al. "Biological and health effects of exposure to kerosene-based jet fuels and performance additives." *Journal of Toxicology and Environmental Health,* July-August 2003.

98 "Above-cloud Aerosols Affect Climate, Say Scientists." *The New Indian Express,* 2 April 2015.

99 "Clouds caused by aircraft exhaust may warm the U.S. climate." NASA news release, April 27, 2004.

. . .cirrus clouds contribute to warming of the atmosphere owing to their contribution to downward transfer of LW [long-wave] radiation. In other words they are a greenhouse agent. . .It has even been proposed to seed in clear air in the upper troposphere to produce artificial cirrus which would warm the surface enough to reduce cold-season heating demands (Detwiler and Cho 1982). So the prospects for seeding cirrus to contribute to global surface cooling do not seem to be very good.[100]

Metal oxide nanoparticles are the golden key to the military's C4 objectives. Nano-crystalline metal oxides act as semiconductors[101] to activate acids and bases interacting with light, then bind with chemical and biological agents in strange synergies. With chemical trails feeding radio and microwave frequencies, the military can tune to a frequency and transport it anywhere they want. They can build an antenna in the air and develop a system of computer communication with other systems *through the clouds*. Metal oxides enhance obscurants made from carbon nanotubes to help conceal exotic propulsion craft, jet fighters, and battleships from radar and sharp eyes.

Conductive metal nanoparticles in the atmosphere act as capacitors[102] to store energy and propagate conductive antenna fields. When metals suspended in the air are hit with radio frequency (RF), they resonate the length of the wave, then duplicate it. Fire a laser pulse for one-trillionth of a second and the distance the laser fires will be the distance of the wavelength you can then tune to. If a 10-meter wavelength creates a tunnel or waveguide the length of the pulse in the atmosphere, you can use it to tune a cavity[103] to a specific frequency so tight that it will resonate harmonically and build from that frequency, duplicating itself over and over again, resonating that same wave over and over again, ringing like a bell or guitar string. If there is no resistance, it will go on forever. This is what Tesla meant by the perpetual existence of free energy.[104]

The argument that aluminum is naturally present in the Earth's crust and used in food, water, medicines, vaccines, and cosmetics with no consequences ignores the multiplying electro-chemical effects of aerosol aluminum oxide's disruption

---

100 William R. Cotton, "Weather and Climate Engineering," lecture at the Perturbed Clouds in the Climate System forum, the Frankfurt Institute for Advanced Studies (FAS) in Germany, March 2008.

101 "A semiconductor is a substance, usually a solid chemical element or compound, that can conduct electricity under some conditions but not others, making it a good medium for the control of electrical current. Its conductance varies depending on the current or voltage applied to a control electrode, or on the intensity of irradiation by infrared (IR), visible light, ultraviolet (UV), or X rays." — *WhatIs.com*

102 "A capacitor (capacitance) is a passive electronic component that stores energy in the form of an electrostatic field." — *WhatIs.com*

103 In electronics and radio, microwave cavities consisting of hollow metal boxes are used in microwave transmitters, receivers and test equipment to control frequency, in place of the tuned circuits used at lower frequencies. — Wikipedia, "Resonator"

104 Conversation with Billy Hayes "The HAARP Man."

of "biological self-ordering, energy transduction, and signaling systems, thus increasing biosemiotic entropy."[105] Aluminum forms toxic complexes with fluorine and interacts negatively with mercury, lead, and glyphosate. Central nervous system (CNS) disorders point to aluminum poisoning, in part because the biophysics of water plays such a pivotal role in bio-degeneration. One study used the toxicity of the water flea (*Ceriodaphnia dubia*) to determine the impact of aluminum oxide nanoparticles on fresh water ecosystems.[106]

The average adult human body is 55–65 percent water, with infant bodies at 75–78 percent until one year old, when it drops to 65 percent.

With nano-aluminum, -barium, and -lithium oxides being spewed from jets, rockets, and ships in this nano-chemical warfare era, every community should plan an annual analysis by a trace atmospheric gas analyzer (TAGA) mobile atmospheric lab once only requested for chemical explosions and train derailment or tanker spills. Even eco-friendly Portland, Oregon (population 619,360) has found hexavalent chromium in its air and soil.[107] States like Maryland are passing clean air acts and going after known nano compounds like sulfur dioxide (SO_2) and nitrogen oxide (NO_x). But what of other compounds falling from the sky and synergistically reacting with everything from industry pollutants to ionized radiation?

What is more concerning is how *quiet* atmospheric chemists are about what they're discovering with their advanced aerosol mass spectrometry, particles-into-liquid samplers, and cloud condensation nuclei counters regarding anthropogenic and biogenic releases impacting air quality, human and environmental health, and regional climate. Sulfates spewed by coal-fired power plants and nitrogen oxides emitted by vehicles mixing with oxygen is one thing, but what happens when they interact with ubiquitous non-ionized wireless radiation?[108]

Are chronic sub-lethal synergies transforming life as we've known it?

Meanwhile, space as a business model is taking off. Private corporations profiting from Project Cloverleaf "climate change" ventures are launching private rockets from military and private launch pads. For example, the U.S.-owned Rocket Lab in Auckland, Australia has its eye on an Australian paddock

105 Christopher A. Shaw et al. "Aluminum-Induced Entropy in Biological Systems: Implications for Neurological Disease." *Journal of Toxicology*, Vol. 2014, Article ID491316. Is the rise in cancer connected? "Cancer cases worldwide are predicted to increase by 70 percent over the next two decades, from 14 million in 2012 to 25 million new cases a year, according to the World Health Organization." — *The Guardian*, February 3, 2014.

106 Sunandan Pakrashi et al., "*Ceriodaphnia dubia* as a Potential Bio-Indicator for Assessing Acute Aluminum Oxide Nanoparticle Toxicity in Fresh Water Environment." *PLOS ONE*, September 5, 2013.

107 Rob Manning, "Portland Air Monitors Find More Toxic Chromium, But Source Unknown." *Opb.org*, April 21, 2016.

108 Brett Israel, "Man-made pollutants significantly influence how tree emissions form aerosol particles." *Phys.org*, December 30, 2014.

with an excellent azimuth and launch trajectory and is planning to create orbital rockets for less than US$5 million so that smaller companies, research and environmental groups can launch their own satellites. 3D printers will spit out the engines, and carbon fiber will make the bodies so lightweight that 1-meter rockets will be able to launch low-earth orbit satellites with lifespans of five to seven years instead of 10,000 years.

For a mere $20 million, billionaire Elon Musk's Space X (Space Exploration Technologies Corporation)[109] delivered the first space tourist, multimillionaire Dennis Tito of the nonprofit Inspiration Mars Foundation, to the International Space Station (ISS) in 2001. In January 2018, Tito plans to be on the *Mission for America* to Mars, a modified SpaceX Dragon spacecraft to be launched (given that Earth and Mars are in favorable positions) by a Falcon Heavy rocket. The trip up and back will take 501 days. No mention of the Van Allen Radiation Belts impasse,[110] nor other pitfalls:

> As the joint Russian-European Mars 500 experiment that recently ended proved, we are far from fully understanding how individuals react to long-term isolation. Also, the impact of high-energy particles on the brain during long-duration spaceflight are [sic] only just beginning to come to light. And then there's the degradation of drugs in long-duration spaceflight. And muscle atrophy. And bone wastage (you get the point).[111]

Down below, as we ingest and inhale aluminum oxide and other toxic flotsam and jetsam, rocket deliveries of "dusty plasma" are being laid in near-earth orbit to construct a Space Fence "ring" around the equator. We will take a closer look at this in Chapter 6, "The 'Star Wars' Space Fence Rises Again."

OCEAN DELIVERY SYSTEMS

Beside jet chemical trails and rocket flotsam and jetsam, *cloud reflectivity modification* contributes to the 90 percent cloud cover over water necessary for geoengineering's many military operations. Between the High Seas Alliance and the UN Law of the Sea, two-thirds of the world's international waters lying outside national territories and economic zones—almost half the planet—are being commandeered for geoengineering, as well. Moisture and weather

109 Competition and politics are fierce around Space X. The September 1, 2016 explosion of the Falcon 9 rocket carrying the Amos 6 satellite appears to have been sabotage, a possibility Western media have buried. Asian press believes it was an X-37B "UFO." See the YouTube "Space X UFO Explosion – Slow Motion!" by Graphics King, September 1, 2016.

110 See "Orion: Trial By Fire," www.nasa.gov/press/2014/october/nasa-premieres-trial-by-fire-video-on-orion-s-flight-test.

111 Ian O'Neill, "Private Mars Mission in 2018?" *NewsDiscovery.com*, February 21, 2013.

systems over oceans are grist for the mills of three ocean delivery systems: iron particulates, marine brightening, and ship tracks.

In 2012, Russ George, former CEO of Planktos, Inc., dumped iron sulfate particulates off the British Columbia coast, claiming that quickened phytoplankton growth would replenish Haida salmon. However, quickening the alpha-penene in plankton to increase a biogenic aerosol reaction to magnify cloud mass may have been closer to the truth. The resulting 10,000 square kilometers of phytoplankton blooms could be seen from space,[112] along with significant cloud mass.

Trees are the basis of terrestrial life while oxygen-producing phytoplankton is the basis of ocean life. Phytoplankton is also the chief driver of organic matter and bacterial enzyme activity in submicron sea-spray aerosols that contribute to cloud formation.[113] Ironically, much of the decline of phytoplankton is due to the aluminum toxicity raining down on biotopic communities. (Nano-crystals remain as long as eighteen months in the atmosphere before falling on the ocean surface.) Less phytoplankton means CO_2 above the current 404 ppm. The more alumina, the less phytoplankton; the more carbon dioxide, the less oxygen.[114]

It is possible that scientists who are making connections about what's really going on are remaining silent for self-preservation: Research scientist Tiffany Moisan, 48, an expert on phytoplankton and climate change, was employed at NASA's Wallops Flight Facility until Sunday, June 5, 2016, when she was found murdered behind a store in Princess Anne, Maryland.[115]

Regarding the other two ocean delivery systems, *marine cloud brightening* is basically spraying seawater toward the upper atmosphere to make clouds reflect sunlight back into space, while *ship tracks* are cloud releases from ship exhaust (and wet surface air cooler technology). Marine cloud brightening is used to feed and direct "extreme weather events" like hurricanes (cyclones). Kerry Emanuel, who teaches meteorology at MIT, offers a clue as to how *sea spray aerosols* can be useful:

> Hurricanes get their energy from evaporated seawater. We all know this from experience: you climb out of the swimming pool and shiver even on a warm day, especially when it's windy. Because the evaporation of water on the skin draws warmth away from the body. This is exactly what happens with a hurricane: the

---

112 "Iron fertilization project stirs West Coast controversy," *CBC News*, October 16, 2012.

113 Simon Yu, "Scientists Find Link Between Ocean Bacteria and Atmosphere." *UCSD Guardian*, May 31, 2015.

114 See physicist Paul Beckwith's video "Oxygen Level Decrease in Air from Climate Change," February 26, 2016.

115 Her research papers can be found at www.researchgate.net/profile/Tiffany_Moisan/publications. She was a friend of Billy Hayes, "The HAARP Man." Authorities insisted there was no foul play.

wind causes the water to evaporate and draws warmth out of the ocean. The water vapour condenses to a cloud in the wall of the eye of the hurricane. The stronger the wind, the more the water evaporates, and the stronger the storm becomes. We think it is possible to intervene at precisely this point so as to ensure that the water evaporates more slowly.[116]

Ship tracks make and steer weather, keep the ship hidden, and ease and block communications. Since the 1967 Clean Air Act, ship track plumes from ship exhaust have been loaded with sulfur particulates (in marine diesel) and CO_2, which in the atmosphere provide nuclei for cloud droplets to condense around to form bright clouds that are more reflective, carry more water, and withhold precipitation.

> In general, the air above the oceans suffers from less turbulence and convection than the air above land. The lower atmosphere is especially calm over the eastern Pacific in the summertime due to a layer of hot air that settles in 500 to 700 meters above that region of the ocean, [James Coakley, atmospheric scientist at Oregon State University] explained. This effect creates a temperature inversion, placing a cap on the cooler air below, trapping pollutants and water vapor. While the inversion is responsible for the smog that reduces air quality in Los Angeles, it also allows for the formation of long lasting ship tracks. The particles bellowing from the ships' smokestacks enter the air above the eastern Pacific and create long, thin clouds that remain there for days.[117]

Some ship tracks arise from commercial ships, others from U.S. Navy (and other military) ships. In June 1994, the Monterey Area Ship Track (MAST) experiment was conducted off the California coast, its purported objective being to see how ship tracks modify cloud albedo, "the effects of ships on the microphysics and radiative properties of marine stratocumulus clouds . . ."[118] But the MAST experiment was really about concocting sea spray aerosols to ice nucleate particles and produce artificial clouds.[119]

In 2003, the Arabian Sea Monsoon Experiment (ARMEX-2003) was conducted aboard the *ORV Sagarkanya* off the west coast of India. Once again, the engine's combustion chamber revealed itself to be key to the production of chemical clouds via chemiions:

---

116 The source seems to have been removed from the Web. However, articles on Emanuel's studies can be easily found, such as "Cooking up a Storm — A Recipe For Disaster?" *PRI's Environmental News Magazine*, May 28, 2010.

117 John Weier, "Every cloud has a filthy lining." NASA, September 30, 1999.

118 Philip A. Durkee et al. "The Monterey Area Ship Track Experiment." Naval Postgraduate School, August 15, 2008.

119 See Paul J. DeMott et al., "Sea spray aerosol as a unique source of ice nucleating particles." *Proceedings of the National Academy of Sciences*, December 17, 2015.

In addition to sulphate and water vapour emissions, the carbonaceous emissions are known to add to the nucleation/growth of the chemiions to intermediate ions. . .Formation of intermediate ions in the engine's exhaust has been associated with the generation of chemiions within the engine combustors which provide centers for the rapid growth of molecular clusters and the formation of electrically charged sulfuric acid/water aerosols. . .the engines of aircraft, motor vehicle, and ships all emit chemiions and these chemiions may play an important role in particle formation.[120]

The base nano-metal in ship tracks is lithium oxide (Li_2O). Lithium is highly water-reactive,[121] and by combining it with superheated saltwater it becomes not just battery-ready but prolific.

Lithium is usually mined from ore, but brine evaporation could become its lowest-cost source. . .Salt water could become the best source for lithium extraction.[122]

In 2015, Oregon activist Ann Fillmore, Ph.D., wrote an article[123] about how lithium was hitting coastal towns along the Oregon Coast and the resultant health complaints of residents: lethargy, thirst, stomach distress, sudden weight gain, muscle and joint pain, twitching, loss of appetite, slurred speech, blurred vision, confusion/hallucinations, ersatz goiter, impotence, endocrine disruptions causing severe menses, kidney pain, skin rashes, hair loss, etc. Her Facebook comments drew two letters from an anonymous whistleblower who chose the moniker "Locke" (as in John Locke), "an employee of a weather data collection company and, by proxy, a subcontractor for the National Weather Service office in [central Oregon]. . .I collected data, e.g. soil samples, that were used to direct spraying operations for the last three years."

He described how the "psychoactive chemical" lithium was being dispersed "to manufacture air stagnation in the Rogue and Umpqua Valleys as well as much of the Oregon Coast south of Florence," and how the Sociological Research Division "has operatives throughout the region gathering a massive amount of data regarding the test population's behavioral traits like consumer habits, political engagement levels, *and awareness of geoengineering programs.*" (Emphasis added.) Ann writes:

120 V. Gopalakrishnan et al. "Intermediate ion formation in the ship's exhaust." *Geophysical Research Letters*, Vol. 32, May 2005.

121 See the YouTube "Lithium Reacts with Water," Sully Science, October 4, 2009.

122 Jon LeSage, "Lithium from salt water could make batteries, and EVs, less expensive." *AutoBlog*, September 30, 2012.

123 Ann Fillmore, Ph.D. "Aerosol Experiments Using Lithium and Psychoactive Drugs Over Oregon." *PositiveHealthOnline*, Issue 228, February 2016.

Up until now [September 2015], they have had to manufacture a weather inversion or stagnant air inversion to apply the lithium. . .I noticed the new method of holding the lithium haze cloud in place first over the area in northern California where the massive fires have hit . . .

"Holding the lithium haze cloud in place" sounds distinctly like ship tracks. The most active part of ship tracks lies *above*, where a plasma-induced iCloud "computer" can be pinned and wedged between the frequencies of higher altitude chemclouds and low-altitude chemclouds whose ice particle count has been increased by laser- and radio frequency-induced plasma. Introduce anhydrous ammonia (NH_3) at specific points during the lithium/water reaction and the mid-atmosphere can be stimulated. Supercharge the two layers of chemclouds and run a laser beam between them to create a super-antenna that gathers and stores messages. Ergo, set up networks, communicate with satellites and other processors, etc. Strike the first beam with another beam (scalar interferometry) and you can store hard data on a virtual "CD."

Utilizing clouds as optical systems is exactly what will be occurring in space as rockets deliver their payloads of "dusty plasma."

WSAC STEAM RELEASE

Direct steam condensation for generating power—steam releases from nuclear reactors, power plants, etc.—often depends upon evaporative *wet surface air cooler (WSAC)* technology that can also be used to fuel engineered weather systems in the South Pacific or on land.

There are seven thousand power plants in the U.S., most of which are east of the Mississippi River and all of which have large cooling towers and/or WSACs that can produce thousands of gallons of water vapor per minute. Researcher WeatherWar101 makes an important observation about *water vapor generation fueling*: "Clearly these immense facilities and all of this open-loop water are not there solely to cool closed-loop fluid. . .at a coal power plant."[124] As with cell towers built to produce far more power than cell receptivity requires, WSACs are built for "dual use."

In an email, WeatherWar101 explained how the WSAC closed-loop / open-loop technology works:

WSACs are both closed loop and open loop. That is to say they are used to cool the closed-loop working fluid driving the turbines to create electricity, whereas *the water sprayed on closed loop tubes is an open loop system*. Hence the billowing

---

124 "Manufactured Louisiana Flood 8-14-16" YouTube, 15:07. Please read his e-book *No Natural Weather: Introduction to Geoengineering 101*.

steam seen from satellite is produced from the open loop portion. In most instances, cooling tower and WSAC water is as close to "pure" as they can get it, primarily to ensure that the equipment stays clean. This applies only to cooling tower stacks, of course, and not to actual power plant stacks whose contents mix with the cooling tower water vapor on ascending to mix with the descending chemtrails / nanoparticles that facilitate NexRad frequency manipulation.[125]

Thus we see how power plant and nuclear stacks add particulates to the chemical stew overhead, synergizing with the descending conductive metals and Mylar nanoparticles. As steam particles ascend, the moisture feeds what is descending and multiplies the chemical effect.

Western Canada GOES Satellite feeds captured how billowing water vapor from industrial cooling and power plant stacks and WSACs in East Texas, Louisiana, Mississippi, and Alabama fed the storm front that became the Louisiana flood of August 14, 2016.[126] That 100,000 people in Louisiana and southwest Mississippi lost everything was simply the collateral damage that accompanies successful weather warfare operations.

Just as the U.S. Navy and recruited merchant ships produce ship tracks in the South Pacific that can be harnessed to feed and direct weather systems, so industrial plants are recruited to do the same on land, proving once again that the military and industry work hand in hand. The floods in Louisiana, West Virginia (two weeks before), and Macedonia in the Balkans (one week before) were not due to any 1,000-year cycle or "inland sheared tropical depression." They were engineered with in-place water vapor generation. Spray aerosols into the atmosphere to capture moisture, then release billions of tons of rainwater with electromagnetic pulsing.

CLOUD SEEDING AND IONIZATION (PLUVICULTURE)

Next to rain dances in the Southwest United States, cloud seeding may be the oldest rainmaker technology. Whereas high-tech geoengineering creates original chemical clouds on a *global* scale, cloud seeding since 1946 has depended upon silver iodide (AgI) flares to produce rain from already-present clouds on a local scale. Nanosilver is in anti-bacterial agents and food packaging, and its nanoparticles end up penetrating the intestinal cells and wreaking havoc.[127]

Despite the fact that nanosilver is damaging to humans and animals, state governments have passed codes allowing aerial spraying and given their ecology

125 Email, August 17, 2016. Please view WeatherWar101's NexRad YouTubes, as well.

126 The YouTube "Louisiana (August 2016) – Geo-engineered Flooding" by, August 15, 2016 exposes minute-by-minute power plant vapor releases feeding the Louisiana flood system.

127 Joe Whitworth, "Nanosilver concerns raised by researchers." *Food Quality News*, 10 March 2014.

agents carte blanche to issue licenses and permits for "weather modification operations," the (wrong) assumption being that weather modification refers only to cloud seeding for rain and not to geoengineering.[128]

Might on-the-books cloud seeding/ionization laws just as well provide legal ballast for prosecuting businesses and state agencies that sign off on broader geoengineering agendas? Nevada and Texas come to mind. Nevada has invested in a DAx8, an eight-rotor UAS (unmanned aircraft system) whose attached cloud-seeding cargo is remotely controlled by advanced software and GPS guidance.[129] Then there is SOAR (Seeding Operations and Atmospheric Research), that during the week of April 6–10, 2015 fired sixteen silver iodide flares into passing clouds from an aircraft and produced a tornado that killed two people and wounded a dozen others:

> The tornadoes created by this group annually have led to the deaths of thousands, the displacement and inconvenience of millions, and unspeakable property destruction through "tornado alley". . .As there is no such thing as a natural cloud, fog dissemination a/k/a geoengineering or chemical terrorism, and the processes of cloud seeding work together as part of a worldwide carbon credit scheme, cashing in on people's lives and property.[130]

In *Chemtrails, HAARP, and the Full Spectrum Dominance of Planet Earth*, I discussed how rainmakers like SciBlue and Aquuiess are about far more than delivering rain:

> To be blunt, electromagnetic weather engineering that can create and steer extreme weather promises untold profit and power. At the close of the YouTube "The Story of Artificial Clouds," NASA climate scientist Jim Hansen is even more blunt: "Human-induced climate change is a great moral issue on par with slavery."

The cunning misapprehension that cloud seeding is a synonym for geoengineered weather modification is hoodwinking every state and nation. For example, what was going on in Tasmania when Hydro Tasmania conducted cloud seeding over the Derwent River catchment as huge storms approached and flooding, death, and destruction followed?[131] And was China's gesture to share

---

128 Clint Richardson, "Geoengineering and Cloud Seeding Statutes." *RealityBlog*, November 25, 2011.

129 "Autonomous cloud seeding aircraft successfully tested in Nevada." *Globe Newswire*, February 24, 2016.

130 "As tornadoes cause deaths, cloud seeders claim success." *Stop Chemical Terrorism*, April 2015.

131 Will Ockenden, "Hydro Tasmania asked to explain cloud seeding in catchment day before flooding." *ABCNews.au*, 10 June 2016.

its cloud seeding technology with drought-torn Maharashtra, India,[132] a BRICS partner, a generous offer or a political move? The Technical Management Unit of Artificial Rain in Indonesia is spending US$45 million to purchase seven Casa CN-235s, C-212s, and N-219s to plump up its weather mod task force.[133] Is this more "dual use"? India, Indonesia, and Iraq have used the SU-30K as a dual use, all-weather aircraft with extended range and mid-air refueling capability—well suited for high-altitude/high-speed aerosol delivery in service to weather warfare.

Cloud seeding is now a handmaiden to global geoengineering as "climate change" carbon credit scams and political muscling march on with companies like SOAR, Weather Modification, Inc., Pacific Northwest National Laboratory, etc. Disasters spell money to be made, human society be damned.

RADIATION AND THE WIGNER EFFECT

The complexity of nuclear experimentation is beyond the pale of postmodern human comprehension. It also reveals, although we would like to believe otherwise, our inability or unwillingness to consider the unseen. Whether it is invisible because of ethereal origins or because it is nano-sized poison does not matter; collectively we tend to obfuscate the unseen. Nuclear experimentation also reveals our collective inability to conceptualize time, and to understand just how long nuclear radiation lasts in our environment . . .

— Ethan Indigo Smith and Andy Whiteley,
"Geoengineering and the Nuclear Connection"

Meanwhile, the weaponization of space continues to violate the 1967 UN Treaty on Principles Governing the Activities of States in the Exploration and Use of Outer Space. In 1965, SNAP-10A was launched, the first space nuclear reactor with a controlled fission reaction inside and enough uranium to produce 600 watts of power for a year. Forty-three days after launch, the reactor miraculously shut down, but SNAP-10A is still orbiting as it falls apart, spreading radioactivity with its debris.[134]

After a twenty-eight-year hiatus, NASA has revived production of powder

132 "China offers its cloud seeding technology for parched Maharashtra," *India TV News*, May 29, 2016.

133 "More airplane needed for weather modification in Indonesia: BPPT." *Antara News*, 26 March 2014.

134 Sarah Zhang, "For 50 Years Now, the U.S. Has Had a Nuclear Reactor Orbiting in Space." *Gizmodo*, April 4, 2015.

plutonium-238 in tandem with the Kilopower Project's uranium-fueled (U-235) Stirling engines for deep-space missions.[135] The half-life of plutonium-238 is 88 years while the half-life of uranium-235 is *700 million years*. "Radioisotope power systems" for "deep-space exploration,"[136] or for space wars?

Prowling nuclear submarines, aging silos, radioactive waste dump fires,[137] and Big Oil's reckless releases of radioactive radon via fracking (hydraulic fracture and shale gas extraction processes) are what most people think of when the topic of radiation hazards rears its ugly head. *Ionized radiation technology is big business*, protected by a sprawling international industry in turn protected by agencies like the U.S. Department of Energy and Nuclear Regulatory Commission (the Atomic Energy Commission until 1974). For more than 70 years, nuclear energy has marched on, despite public opinion that it spells death for the planet and all living beings.

On any day in America, radioactive I-131 particles with a half-life of eight days are settling into thyroid glands as strontium-90, cesium-137, zirconium and other radioactive isotopes nestle into other tissues. Cows eat fallout grass, Americans drink fallout milk and eat fallout cheese and meat as the Sun's cosmic rays penetrate the atmosphere and intensify the ionized radiation. This witches' brew is then absorbed by soil, rock, water—and this is before considering the new synergies since chemtrail fallout.

Because galactic and solar radiation filter through our magnetosphere shield, commercial aircrews are classified as radiation workers, though the levels of radiation to which they are subjected are mysteriously undocumented. The National Council on Radiation Protection and Measurements reported in 2009 that aircrews have the highest annual dose of radiation of all radiation-exposed workers in the U.S.[138] While it is well known that a solar storm or frequent high-latitude flight increases radiation exposure, the extent of what it means for human health remains unquantified, despite the spontaneous abortions, tissue damage, breaks in DNA strands, chemically active radicals altering cell function, etc. Nowcast of Atmospheric Ionizing Radiation for Aviation Safety (NAIRAS) and Automated Radiation Measurements for Aviation Safety (ARMAS) claim they will rectify this oversight.

> Quantifying the levels of atmospheric ionizing radiation is of particular interest to the aviation industry since it is the primary source of human exposure to

135 "Kilopower for space exploration," *World Nuclear News*, December 2014.

136 Jeanna Bryner, "Space Fuel: Plutonium-238 Created After 30-Year Wait." *Live Science*, December 30, 2015

137 Henry Brean and Keith Rogers, "Investigators taking close look at radioactive waste dump fire." *Las Vegas Review-Journal*, October 19, 2015. "In 2010, the EPA fined US Ecology nearly $500,000 for 18 hazardous waste violations."

138 Julia Calderone, "Here's why airline crewmembers are classified as radiation workers." *Tech Insider*, November 19, 2015.

high-linear energy transfer (LET) radiation. High-LET radiation is effective at directly breaking DNA strands in biological tissue, or producing chemically active radicals in tissue that alter the cell function, both of which can lead to cancer or other adverse health effects.139

But it is not just solar storms and galactic radiation pouring through our HAARP-damaged magnetosphere that are making pilots pass out. Three years after the horror of the initial 2011 Fukushima Daiichi release,140 a team of independent researchers—geoscientist Leuren Moret (*www.leurenmoret.info/*), former U.S. Navy engineer Laurens Battis, and former NIH Clinical Study Coordinator and researcher Christina Consolo (*climateviewer.com/radchick/*)—began collecting and documenting aircraft mishaps that seemed to point to the *Wigner Effect* or *discomposition effect,* the displacement of atoms in a solid caused by neutron radiation. Their twelve YouTube videos document their findings, beginning with "MAYDAY: The Wigner Effect" (September 16, 2014) and ending with "The Wigner Effect: Full Interview" (March 25, 2016).

Fukushima is an ongoing tragedy not just for Japan but for the entire planetary biosphere. In Japan, radiation levels exceed 40,000 Becquerels141 per square meter (Bq/m2) over 54,054 square miles. *Ten million* Japanese citizens are still living in the "radiation-controlled" area. Nuclear engineer Hiroaki Koide believes that Fukushima should be declared uninhabitable and citizens evacuated; that this remains undone "is a serious crime committed by Japan's ruling elite." Former Prime Minister Naoto Kan says half of Japan's entire population of 127 million has been exposed, including Tokyo (50 million), and believes that if all were evacuated, "it would have virtually meant the end of Japan." Over 70 percent of Japan's agricultural land is contaminated.142

I repeat: Fukushima is an ongoing global event—and synergy experiment.

Even now, the spent fuel pools that broke through the pressure vessel and entered the ground and ocean—initially 210 quadrillion Bq of cesium-137, which reacts rapidly with water to become hygroscopic cesium hydroxide, plus 2,000 other nuclides (isotopes) and breeder reactor "Wigner black dust" (nanoparticles)—are daily adding 30 billion Bq of cesium-137 and 30 billion Bq

139 Christopher J. Mertens, NASA Langley Research Center, et al. "Atmospheric Ionizing Radiation from Galactic and Solar Cosmic Rays," Chapter 31 in *Current Topics in Ionizing Radiation Research*, ed. Mitsuru Nenoi, March 9, 2012.

140 The HAARP creation of the March 11, 2011 Fukushima event was exactly sixty-six years after Hiroshima and Nagasaki. View YouTube "Leuren Moret: Fukushima radiation is intentional extermination that HAARP Tesla technology can reverse," Alfred Lambremont Webre, June 9, 2014.

141 A becquerel (Bq) is equal to one disintegration or nuclear transformation per second.

142 "Top Official: Over 60 million Japanese irradiated by Fukushima – Nuclear Expert: 50,000 sq. miles of Japan highly contaminated . . . Many millions need to be evacuated . . . Gov't has decided to sacrifice them, it's a serious crime – TV: More than 70% of the country contaminated by radiation (VIDEOS)." *ENENews*, April 12, 2016.

of strontium to the ocean. Add the neutron-pulsing plume and Japanese debris heading ever east around the globe via ocean and atmospheric fluid dynamics, and it becomes evident how imperative it is to analyze the impact of the synergy between the radiation and our chemicalized plasma atmosphere.

Consolo points out that a Fox News investigation discovered that the U.S. military saw a 48 percent increase in non-combat aviation crashes in 2014 and 2015 and that in part it is blamed on maintenance issues.[143] Almost daily, reports of runway delays and crashing aircraft due to structural anomalies are occurring. Windshields are cracking, helicopter cross tubes breaking, tail rotors failing, dropping oil pressure, melting batteries, lithium batteries exploding on the ground, faulty hydraulics, panels falling off, wings partially detaching, electrical generator and fluid power system failures, nose gear and landing gear issues, engines catching fire or failing, strange explosions, loud bangs, smoke, smells, cabin depressurization, de-icing systems kicking on, metal particles found in failed engine filters, malfunctioning computers, deployment of emergency slides in midflight, water pipes breaking. Is repetitive flying through radioactive air the cause?

While these failures are clear indicators of entropy (breakdown of compounds), it is the *speed* at which they are occurring—even with budget-cutting maintenance taken into consideration—that points to radiation's cumulative crystallizing effect on metal,[144] similar to how nuclear reactors eventually submit to thermal embrittlement. Look a little further and we see young people dropping dead from cardiac arrest and children as young as two having in-flight heart attacks.[145] Fissures may be behind hydrovolcanic explosive events underground, but "metal fatigue" is failing hydraulics in cars, rollercoasters, and jetways, water main breaks, nuclear plant transformer fires, even "freak accidents" like that at the Brooklyn Bridge.[146]

Are these signs of entropy connected to the synergy going on between ionized chemtrail nanoparticles, "Wigner black dust" radioactive nanoparticles, static electricity, and other ionizing and non-ionizing sources? The interaction between the nano-chemical ionizing of the atmosphere and the Fukushima catastrophe—both under a silent shroud of "national security"—may be escalating a planetary decay and disintegration of collapsing bridges, failing buildings, and unnaturally deteriorating bodies.

Moret emphasizes that only 15 percent of the radioisotopes from the 1,200 bomb tests at Nevada Test Site for the past half-century has been precipitated, which means

---

143 Lucas Tomlinson, "Military jet crashes on rise as some cite training and fleet issues." *FoxNews. com*, September 2, 2016.

144 Swordmakers temper steel with heat and pressure to align the steel's crystalline structure and add carbon.

145 "More Young People Dying of Cardiac Arrest." *ABCNews*, March 1, 2016.

146 Matt McNulty, "Collapse at Brooklyn Bridge injures 5 people." *New York Post*, July 2, 2014.

that *85 percent is still suspended in the atmosphere*—and this is not counting ionized chemtrail nanoparticulates! High overhead, moisture is gradually condensing around these highly charged particulates and forcing them earthward. Not only are they all desiccants, but wherever they fall as rain or snowflakes, radiation remains (the Van der Waal force), and there is no way to decontaminate it.

Fighter jets fly higher and faster than commercial jets and are therefore subject to higher radiation levels. The static electricity they (and helicopters) generate leads to a massive voltage buildup that contributes mightily to multiplying nanoparticles trapped in electrical fields subject to ionized radiation. What happens to pilots traveling at Mach 1.5+? Stealth F-22 Raptor[147] pilot Josh Wilson with the Virginia Air National Guard announced on CBS News *60 Minutes* that the U.S. Air Force was threatening pilots with career reprisals if they talked about oxygen system problems and hypoxia.[148] Captain Jeff Haney died in November 2010 when his oxygen was cut off and he crashed at faster than the speed of sound in the Alaska wilderness. The Air Force blamed Haney, but his sister disagreed.[149]

Above, I mentioned the possibility of Fukushima being a synergy experiment. Is the creation of an entropic environment intentional? Not only do chemical trails loaded with conductive metal oxides provide an excellent *multiplier* effect on nuclear nanoparticles (Wigner black dust) created by the fission/fusion of plutonium and uranium, but ionospheric heaters like HAARP can heat radioactive plumes and eddies to create "fences" around populations or conduits for direct targeting of populations.

Waveform energy can be used to exert force at a distance. Pulse high energy into matter and it will change.

RADIATION AND FRACKING

Jets and rockets lay trails of metal nanoparticles that eventually filter into our lungs and brain, food and water, then orbit above the Earth in a Saturn-like "ring." Big Oil's hydraulic fracturing (fracking) is the practice of pumping water, sand and chemical lubricants to keep the sand flowing underground to split rock containing profitable oil and gas. But it plays its part in the Space Fence as well once compounds are injected deep in the Earth and "thumped" (pulsed) with a corresponding resonant frequency.

---

147 The Air Force had paid Lockheed Martin $79 billion for 187 Raptors, each jet costing $420 million.

148 Bill Bartel, "Pilot's career stalls after criticizing oxygen system." *The Virginian-Pilot*, April 20, 2014.

149 Lee Ferran and Megan Chuchmach, "Fighter Pilots Claim Intimidation Over F-22 Raptor Jets." *ABC News*, May 7, 2012.

Al + Fe_2O_3 AlO_2 + Fe_3 is used to perforate oil well casings prior to fracking with a C-4 (explosive)[150] detonation-forced ignition. Compare this compound with the $Ba(NO_3)_2$ + AlO_2 + Fe_2O_3 + O in the atmosphere hours after chemical spraying[151] and you'll see that the chemical "trade secrets" of aviation fuel and hydraulic fracturing fluids are in sync. Environmentalists have tried for decades to discover the formulae for the fracking lubricants polluting land and water much as jet fuel and aerial chemical spraying are polluting the air and entire biosphere.[152] In 2014, North Carolina went so far as to pass the Energy Modernization Act (SB786) making disclosure of fracking trade secrets a Class I felony.[153]

Fifty-five percent of the 40,000 fracking wells drilled in the U.S. since 2011 are in areas of prolonged drought, thanks to the billions of gallons of water fracking requires—and not counting the billions of gallons of oilfield wastewater fluids loaded with chemicals that have to be disposed of. Is it any wonder that fracking has been implicated in the increase of earthquakes, sinkholes, and salt dome collapses? In Oklahoma, where there are 35,000 active wastewater disposal wells, earthquakes measuring 3.0 or greater reached 890 in 2015 alone. In 2008—the year before oil companies began using fracking—only two earthquakes occurred.

Then there are methane releases:

> A Cornell University scientist's claims that oil and gas development is so harmful to the climate that methane emissions and oil and gas production in general need to be cut back immediately to avoid a "global catastrophe". . .[Environmental biology professor Robert Howarth]. . .concluded that the climate impact of natural gas produced from shale—most of which involves hydraulic fracturing, or fracking—may be worse than that of coal and crude oil. That's because methane leaks from natural gas production have a greater effect on the climate than carbon dioxide emissions, Howarth said.[154]

Fracked gas contains radioactive radon, and venting radon means releasing methane. It takes ten half-lives (8.3 days = 1 half-life) for radon to decay enough to be safe, and yet it's being pumped to market immediately for people's homes, while radon "well brine" is spread over roads before snow and ice storms.[155] Tons of hot radioactive waste in the "orphan" waste stream (brine, sludge, rock, soiled equipment, etc.) ends up in landfills in trash bags. Why?

---

150 C-4 is a plastic explosive with a modeling clay texture.

151 Thanks to Billy Hayes "The HAARP Man" for this chemical insight.

152 "Fracking chemicals to stay 'trade secrets.'" *RT*, 28 March 2013.

153 "North Carolina law would make discussing fracking chemicals illegal." *RT*, 20 May 2014.

154 Bobby Magill, "'Catastrophe' Claim Adds Fuel to Methane Debate." *Climate Central*, May 15, 2014.

155 Thanks to June 21, 2016 comment by Paul Robert Roden at "Hot mess: states struggle to deal with radioactive fracking waste" by Jie Jenny Zou, The Center for Public Integrity, June 20, 2016.

Because Big Oil has always had the privilege of self-reporting, self-regulating, and writing off taxes.

But the worst news yet may be that burying radioactive waste in the hydrocarbon-poor shale of already-fractured deep underground rock is under consideration. The 2014 proposal by U.S. Geological Survey hydrologist Chris Neuzi[156] is made to sound like it's "in the future," but the truth is it may already be underway. And it's been done before. In "Oilmen Help Dump Radioactive Waste" (*San Antonio Express/News* of May 3, 1964), we learn that Halliburton Company and Union Carbide Corporation

> . . .combined the oil well cementing technique with the hydraulic fracturing production stimulation technique [i.e. fracking] to entomb radioactive wastes in an impermeable shale formation a thousand feet underground. . .Union Carbide Corporation which operates facilities at Oak Ridge [National Laboratory] for the U.S. Atomic Energy Commission, and Halliburton, which provides specialized oilfield services such as cementing and fracturing worldwide, have collaborated on the project since 1960.[157]

Radiation in hidden places synergizing with what falls from the sky to be breathed, grown as food and imbibed as water. Evil in the guise of profit-hungry corporations stalks the land . . .

---

156 Eric Roston, "Not Into Fracking? How About Some Nuclear Waste?" *Bloomberg.com*, March 17, 2014. Also see the 2012 patent US 20140221722 A1, "Abyssal sequestration of nuclear waste and other types of hazardous waste" by Leonid Germanovich et al.

157 Phoenix, "Radioactive Waste 122 — Halliburton and the U.S. Defense Department Were Fracking Nuclear Waste in the 1960s." *Nucleotidings*, March 17, 2015.

CHAPTER TWO

Æther, Plasma, and Scalar Waves

We shall say "Our space has the physical property of transmitting waves" and so omit the use of the word [æther] we have decided to avoid.
— Albert Einstein, *Evolution of Physics*, 1938

One cannot escape the feeling that [James Clerk Maxwell's] mathematical formulas have an independent existence and an intelligence of their own, that they are wiser than we are, wiser even than their discoverers in that we get more out of them than was originally put into them.
— Heinrich Hertz (1857–1893)

There are five elements, four of which are held in the matrix of the fifth which is unaffected by them, of ETHER.
— Michael Gabriel, *The Holy Valley and the Holy Mountain*, 1994

Imagine the Sun's magnetic field divided from pole to pole like sections of an orange, each section oriented to its opposite section, so that when the Sun rotates, the Earth's fields change with the Sun's changing fields. Now imagine that everything is connected and speaking from a variety of levels to whoever is listening at that particular level—an intelligent and conversational universe waiting for us to listen and familiarize ourselves with its cipher language.

The universe may indeed be Hilbert's Space of infinite dimensions. Quantum mechanics implies an infinity of different shapes and topologies that fit together as waves, each interfering with another. Intercept the charges around the terminals of generators or batteries and you may discover endless, hidden-in-plain-sight free energy.

Æther was the primordial god (protogenos) of light and the bright, blue ether of the heavens. His mists filled the space between the solid dome of the sky (ouranos) and the transparent mists of the earth-bound air (khaos, aer). In the evening his mother Nyx drew her dark veil across the sky, obscuring the ether and bringing night. In the morn his sister and wife Hemera dispersed night's mist to reveal the shining blue ether of day. In the ancient cosmogonies, night and day were regarded as elements separate from the Sun.
– www.theoi.com/

Æther was the upper air; the middle air was *aer* or *khaos*, a colorless mist enveloping the mortal world; the lower air, *erebos*, enveloped the dark places in the realm of the dead under the Earth. *Æther's* female counterpart was Æthra, Titaness of the clear blue sky and mother of the Sun and Moon.

Here in the Pergamon Altar (second century B.C.) in Berlin, Æther battles a lion-headed Giant.
(Ahriman a.k.a. geoengineering?)

The term we have needed for the untapped "free energy" potentials of pure *æther* energy is *scalar*, not electromagnetic. Æther and scalar energy have shared many epithets: *prana, chi* or *qi*, zero-point energy, tachyon energy, biophotonic energy, Rife energy, orgone, *kundalini*, life force, etc. Both æther (from the Greek αἰθήρ, to light up or kindle) and *scalar waves* (*scalar:* a quantity with magnitude but not direction) have been suppressed for a century. Why? Basically, it is the same old story: the elite bend and suppress inventors and scientific genius so as to control revelations of scientific truths and inventions according to their own timetable.

Nikola Tesla's experiments revolved around scalar waves, not EM waves, as he sought to transform electrons into vortex electrons. Faster-than-light—yes, that is right: faster than light—scalar waves are actually *spiraling* waves (as in wind rotation) continually branching out into recurring fractal patterns. Scalar waves "coagulate" as particles of matter (plasma) when they slow down to slower-than-light (subluminal) speeds. Once they converge with EM waves, alter the structure of EM energy. Unlike the usual side-to-side oscillating *transverse* waves (like on the ocean) we see in RF-zapped plasma cloud cover, scalar waves are *longitudinal* and more like compression waves.

Ionospheric heaters like HAARP are scalar transmitters. Cross or "interfere" two scalar beams in the stratosphere over a target area and the resultant superheating will produce a plasma "ring" over the interference zone. Pump (pulse) this "ring" with radar and storm intensification will follow, as Michael Janitch (Dutchsinse) explains.[1] One scalar beam strips the electrons to form superheated plasma, the other beam pumps up and sustains the plasma.

Konstantin Meyl, Ph.D.,[2] professor of power electronics and alternative energy technology at the University of Applied Sciences in Furtwangen, Germany, clarifies the existence of two kinds of scalar waves and the usual electromagnetic wave:

Electric scalar wave [longitudinal / standing] propagates at 1.5X the speed of light in the direction of the electric pointer (90°)[3];

Magnetic scalar wave [longitudinal / standing] propagates at 1.5X the speed of light in the direction of the magnetic pointer (90°); and

Electromagnetic wave [transverse] propagates in the direction of both (zero).

Western physics posits four forces of the universe that address matter: gravitational, electromagnetic, weak and strong nuclear. What has been left out since the beginning of the twentieth century is æther, the sub-quantum field that appears as empty space but is actually the ground state of the universe.[4] Like "junk" DNA—now proven to not be "junk" at all[5]—the "quantum vacuum" isn't a void but a highly charged plenum whose subtle order measurably influences the spacetime motion of the entire physical universe. Like their predecessors, nineteenth-century scientists recognized that all-pervasive space was filled to overflowing with soniferous æther, the soul of the cosmos by which divine thought manifests in matter and therefore the common origin of all matter, as Michael Faraday (1791–1867) believed. Less than a century before Faraday, the

---

1 Michael Janitch, "3/08/2016 — US Military confirms HAARP 'ring' formed by Radio Waves hitting the Atmosphere / Ionosphere."

2 Listen to "Konstantin Meyl Scalar Wave Interview with William Alek," AdomHussein69, January 12, 2011. Meyl's paper "Scalar Waves: Advanced Concepts for Wireless Energy Transfer" can be found at www.k-meyl.de/go/Primaerliteratur/Wireless-Energy-Transfer.pdf.

3 Sound waves propagate similarly because they are a form of scalar wave, as is a gravitational wave or plasma wave. For more on the speed of light no longer being considered a constant as in E=mc2, see Umer Abrar, "Speed of light not so constant after all — New experiment proves." *Physics-Astronomy. com*, January 2015.

4 Robert B. Laughlin, *A Different Universe, Reinventing Physics from the Bottom Down*. New York: Basic Books, 2005.

5 Alice Park, "Junk DNA — Not So Useless After All." *Time*, September 6, 2012.

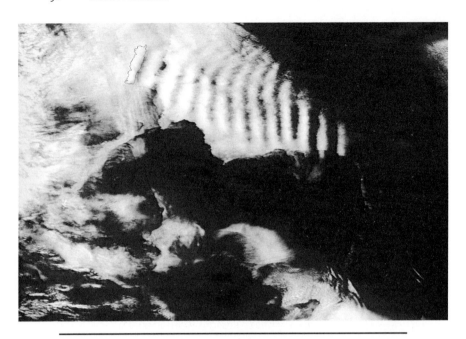

go.nasa.gov/2lWxO4h
Over Auckland Island, NZ on February 24, 2017. Photo by V. Susan Ferguson metaphysicalmusing.com/.
What scalar can look like in plasma cloud cover.

polymath Gottfried Wilhelm Leibniz (1646–1716) claimed, "There is no vacuum. . .space without matter is something imaginary."

Æther was also viewed as extending to the *subtle bodies of light* that all biological creatures have (including planet Earth)—bodies vulnerable to ionized and non-ionized electromagnetic radiation. Austrian scientist Rudolf Steiner delineated *four ethers* (note the difference in spelling for subtle bodies) binding the physical to the psychic (ψυχή Greek for "soul" or "spirit"):

The *warmth ether*, the most primordial, manifests centrifugally as heat and appears as SPHERICAL.

The *light ether*, centrifugal, manifests as gas. Its primary quality is LUMINOSITY.

The *chemical/sound ether*, centripetal, manifests as fluid and is DISC-FORMING.

The *life ether*, centripetal, immediately precedes matter and is INDIVIDUALIZING.[6]

If we think of these four types of ether in terms of *plasma*, the fourth state of matter making up most of the matter in the universe, it seems evident that plasma and æther are one and the same but in different phases or forms of spherical, luminous, disc-forming, and individual. The fourth state of matter means just that: heat a solid sufficiently and it becomes liquid; heat a liquid and it becomes a gas; heat a gas (or subject it to a strong electromagnetic field) and

6 Each ether stage can be made to *devolve* into the previous stage or *evolve* into the next stage.

it *ionizes* to become something no longer like ordinary gas in that it is strongly influenced by electric and magnetic fields. This is plasma, and the physics of plasma is about electrons and ions, conduction, high-voltage discharges, the mirroring ionosphere, the "dancing plasma" of the aurora borealis, our Sun, other stars . . .[7]

Now, plasma is being "farmed" in lower ionized atmospheres.

On Earth, a healthy biological / emotional / mental life requires a balance of all four ethers supported by vital food, air, water, and consciousness, all of which are now under assault by a blizzard of radio frequency and microwaves, ionized metals, polymers, upper-atmosphere fungi, and genetically engineered biologicals.

From the sixteenth to the early twentieth century, gravitational phenomena were modeled on the æther medium. Einstein retained the term until 1920:

> We may say that according to the general theory of relativity space is endowed with physical qualities; in this sense, therefore, there exists an æther. According to the general theory of relativity, space without æther is unthinkable; for in such space there not only would be no propagation of light, but also no possibility of existence for standards of space and time (measuring-rods and clocks), nor therefore any space-time intervals in the physical sense. But this æther may not be thought of as endowed with the quality characteristic of ponderable media, as consisting of parts which may be tracked through time. The idea of motion may not be applied to it.[8]

Somewhere between Darwinian evolutionary dogma, good-old-boy peer reviews, and powerful *sub rosa* occult societies, a decision was made to eliminate æther from mainstream science and replace it with space as a vacuum while scalar waves and the rest of Tesla's work and that of Maxwell and E.T. Whittaker (1873–1956)[9] were suppressed.

What happened was this: In 1873, James Clerk Maxwell (1831–1879) linked electricity to magnetism and discovered three components in electromagnetic waves he called *EM mass-entities* (nonphysical hyperspace particles), now called *evoked potentials*—very fine scalar waveforms existing at right-angle rotations to electromagnetic fields. Hyperspace flux energy potentials flow unharnessed like waves in a cosmic sea of intense power *in dimensions adjacent to ours*, which is why they are not subject to our spacetime unless disturbed by our planetary

---

7 Clifford Carnicom recommends *The Fourth State of Matter: An Introduction to Plasma Science* by S. Eliezer and Y. Eliezer (Institute of Physics Publishing, 1989).

8 Einstein, Albert. "Ether and the Theory of Relativity." Address delivered May 5, 1920 at University of Leyden.

9 In his 1910 *History of the Theories of Æther and Electricity*, Whittaker credits Henri Poincaré (1854–1912) and Hendrik Lorentz (1853–1928) with developing special relativity and E=mc2, not Einstein.

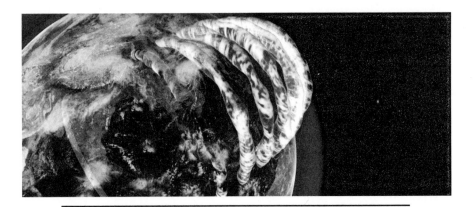

Twenty-three-year-old undergraduate student Cleo Loi of the ARC Centre of Excellence for All-sky Astrophysics (CAASTRO) at the University of Sydney found a way to use a radio telescope to take 3D images of these giant plasma tubes existing inside our magnetosphere and following the magnetic lines of force that Bernard Eastlund addresses in his 1987 HAARP patent. Ions from the ionosphere flow through these plasma "waveguides" and into our atmosphere and magnetic field.

magnetic lines of force—the same magnetic lines of force so pivotal (no pun intended) to leaching ions from the ionosphere. Thus it is that *electromagnetic effects can be initiated from outside our electromagnetic spectrum.* The fact is that a variety of fields and waveforms exist at right-angle (*orthogonal*) rotations in hyperspace outside our electromagnetic fields. Generate and manipulate these orthogonal components and they can invisibly generate electromagnetic effects to influence the Earth, human biology and consciousness.

Maxwell died at forty-eight of stomach cancer, after which two assaults were initiated: his field equations were edited to misrepresent scalar waves as zero (like EM waves), and æther was "disproven" by the dubious 1881 Michelson-Morley experiments. Mathematician Oliver Heaviside (1850–1925), chemist Josiah Willard Gibbs (1839–1903), and physicist Heinrich Hertz (1857–1894)[10] then buried æther, along with Maxwell's hyperspatial Quaternions (tensors) and scalar equations—the very foundations of electromagnetics and unified field theory— and replaced them with vector analysis and the insistence that gravitation and electromagnetism are mutually exclusive instead of interdependent. Instead of five dimensions—X, Y, Z, time, and gravity—mainstream twentieth-century physics was subsequently confined to four dimensions, with hyperspace superluminal signals viewed as "imaginary components." In the 1920s, English theoretical physicist Paul Dirac (1902–1984), one of the founders of quantum mechanics and quantum electrodynamics, insisted that science would have to return to the ether theory, but of course no one listened, scientism being entrenched.

---

10 Are these the Three Assassins of Freemasonry, which views itself as the gatekeeper of what the profane should and shouldn't know? *Scientia est potential.*

About the time relativity was being pushed as a truism, studies of radioactivity began showing that the empty vacuum of space had a spectroscopic structure similar to that of ordinary quantum solids and fluids. Subsequent studies with large particle accelerators clarified that space is more like a piece of window glass than Newtonian emptiness, filled with "stuff" that is normally transparent. Once it became obvious that Leibniz had been right about space, the term *dark matter* was proposed to explain away why galaxies contain so much more mass than can be accounted for by visible matter.[11] Physics Nobel Laureate Robert B. Laughlin (1950-) thought that the æther cover-up for the sake of bolstering special relativity was "unfortunate" enough, but then to import a term like "dark matter" for æther?

> It is ironic that Einstein's most creative work, the general theory of relativity, should boil down to conceptualizing space as a medium when his original premise [in special relativity] was that no such medium existed. . .The word 'ether' has extremely negative connotations in theoretical physics because of its past association with opposition to relativity. This is unfortunate because, stripped of these connotations, it rather nicely captures the way most physicists actually think about the vacuum. . .Relativity actually says nothing about the existence or nonexistence of matter pervading the universe, only that any such matter must have relativistic symmetry . . .[12]

Only now that the atmosphere is ionized and the Space Fence infrastructure complete are Maxwell's equations being returned to their original form. The makings of life itself—æther, plasma, and scalar waves—are being weaponized. Maxwell's evoked potentials are being artificially manufactured and weaponized in three modes: pulse, energy extraction, and explosion. To proponents of classical electrodynamics, conductivity happens via metal wires, whereas in Maxwell physics, conductivity occurs in dielectric æther-filled hyperspace when conductors serve as waveguides for crossed scalar beams (interferometry) that suck the energy out of airspace and cause a *cold explosion*, or just as easily direct an invisible hot beam to instantaneously drop a target. Cold explosion preserves machines and buildings but not life forms; hot energy forces buildings, machines and bodies to explode / implode and "melt" as the nucleus of each atom *disintegrates*. (Shades of 9/11.)

> These weapons involve beams. Two beams overlapped will couple into a particle-ion beam that will bounce off of a remote target and send a holographic image back to the satellite for remote spying operations. When you cross two strong beams, you can supposedly* create scalar energies. These energies can

11 Zeeya, Morali, "Ether returns to oust dark matter." *New Scientist*, 27 August 2012.

12 Laughlin, *A Different Universe*.

be used as untraceable weapons for nuclear size explosions or for defense. These crossed-energies can be used to cause a person's physical electrical system to fail or with a lower frequency, administer a kind of remote electro-shock. Visualize touching a positive and negative electric cable to each other on top of your head. Scalar energies can be utilized in hand-held military guns and on tanks. They can dud-out electronics or cause large, electrical blackouts. Scalar energies are practically impossible to shield against. You need lead, ceramics, and a deep underground facility to not be affected by these weapons. Or, you need to be up and above the field of battle.[13]

When interfering scalar transmitters cross beams over coordinates on the other side of the Earth, they utilize artificial evoked potentials and earth-penetrating tomography to create earthquakes at the points of interference. Natural seismic waves don't travel through the Earth; only man-made scalar waves do. Two timed pulses are sent along two or three beams meeting over the target area, heated energy is extracted from the airspace, a cold explosion occurs, and the beams swing back to their originating transmitters with none the wiser.

Transmitters, receivers, and transceivers under the scalar HAARP system are now tuned to the scalar harmonic and transmitting at 90° angles to each other, following waveguides around and through the planet. Because scalar weapons produce gravitational waves, a worldwide network of gravitational (longitudinal) wave detectors is constantly and quietly tracking and measuring weapon capabilities—like GEO600, a ground-based interferometric gravitational wave detector near Hannover, Germany, Virgo in Italy, KAGRA in Japan, and the twin Laser Interferometer Gravitational-wave Observatory (LIGO) detectors in Livingston, Louisiana and Hanford, Washington (operated by CalTech and MIT).

To counter public knowledge of these scalar-produced gravitational waves, cosmic cover stories are being conjured out of the same whole cloth as "meteors," the Big Bang and the Higgs boson "God Particle." For example, the 2016 story about the LIGO interferometer detecting "ripples in the fabric of space time" (Einstein) produced by the union of two black holes 1.3 billion years ago,[14] "shocking confirmation that the waves emanated from a cataclysmic collision 29 years ago in Pontiac, Michigan, an irresistible force meeting an immovable object" inside a "dome-like structure." Astrophysicist Kip Thorne looks into the distant past and sees gravitational waves produced by a colossal mass slammed to the ground by a powerful star, "93,173 bodies, by my calculations—before rippling outward into the universe."[15]

---

13 Carolyn Williams Palit, "What Chemtrails Really Are," November 9, 2007.

14 "China to Study Gravitational Waves in Domestic Research Project." *Sputnik News*, February 14, 2016.

15 "Physicists pinpoint source of gravitational waves to Pontiac Michigan 1987." *Kayfabe News*, February 12, 2016.

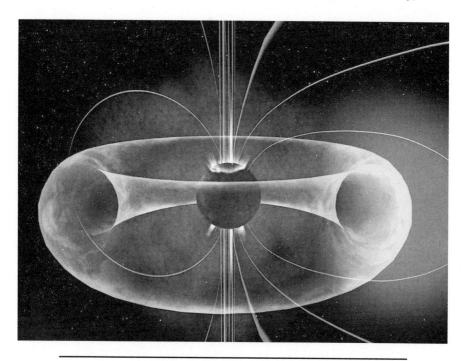

Here, we see the Birkeland current spewing through and forth from the Earth at the North and South magnetic poles. From Christopher Fontenot:

Known to exist at Earth's poles and equatorial plane, Birkeland currents were first proposed by their namesake Kristian Birkeland to be electromagnetic conduits transferring energy between celestial objects over great distances in space, such as between the Earth and Sun. At their core, Birkeland currents are longitudinally compressed waves of electric force, their core being the center point of concentric counter-rotational magnetic sheaths of transverse *Alfvén* wave propagation. These transverse waves are responsible for the "plasma tubes" of magnetic sheaths observed from pole to pole on Earth.

Alfvén waves are named for Hannes *Alfvén* and are filaments of oppositely charged ions pulsing in opposite directions. The force that binds these oppositely charged ionic *Alfvén* waves is called the Lorentz force. Three rings or electrojets consisting of two twisted *Alfvén* waves encircle the Earth at near the speed of light.

While weaponizing evoked potentials is plowing full steam ahead, physics doctoral programs are still not offering electrogravitics courses; electromagnetic waves are all still categorized as Hertzian, despite the fact that ELF, ULF, and biosystems waves aren't Hertzian; and classical Newtonian physics still pooh-poohs evoked potentials as imaginary. Nuclear engineer Thomas Bearden calls modern electrodynamics "a piece of tripe."

With an understanding of electrodynamics, Tesla weapons like the HAARP Howitzer would be seen for what they are: hyperspace weapons. People would understand how the quantum potential in nature can be built and steered by "using a modified Whittaker EM to implement David Bohm's hidden variable theory of quantum mechanics."[16] They would learn how HAARP passes scalar waves through matter and *dislodges the target from spacetime itself*—like the 1943 Philadelphia Experiment and the 2001 World Trade Towers. They would learn that *scalar waves hearken from the same realm as the antimatter that CERN is attempting to access.* HAARP Howitzers disintegrate soft tissue and destroy every living cell until the body falls like a limp rag but does not decay, not even for forty-five days. The right frequencies pulsed on scalar waves can be made to influence thoughts as well as physical functions like vision, glands, and musculature.

Scalar interferometry spells the arrival of global C4 domination, given that scalar beams can be delivered by aircraft, satellite, or vectored craft. Artificial evoked potentials shaped into plasma orbs / Tesla globes can be directed like cannonballs against a target by two or more transmitters and can fail the electronics in airplanes, jets, helicopters, and missiles. Normal and encrypted communications can be tapped while evoked potentials can also be used to transmit impenetrable communications inside ordinary carrier waves, making them perfect for two-way communication with nuclear submarines, ships, aircraft, and satellites. By shaping a spherical interference shell (plasma shield / Tesla dome) in the searching radar bandwidth, radar invisibility can be created.

The Laser Developed Atmospheric Lens (LDAL) turns "the Earth's atmosphere into a lens-like structure to magnify or change the path of electromagnetic waves such as light and radio signals"[17] by means of high-powered lasers. The lens can also be used as a plasma shield against enemy laser. We are living in the Star Wars matrix.

Space Fence S-band radar (2.2–2.3 GHz) in the now-conductive lower atmosphere can generate plasma. (See Chapter 8 for much more on S-band radar.) It is not difficult to envision how Tesla's protective 3D plasma shield or dome might be zapped into being in a flash by linking wide scalar beams from two or more transmitters in the airspace over even a vast geographic area like the Arctic Circle. Multiple layers of plasma would create a "nesting" effect in which each layer's sensors would probe for various levels of identification. Illicit attempts to breach this force field could result in the craft or missile becoming inoperative or blowing up while human nervous systems are in meltdown. Raytheon is already constructing large luminous plasma shields over oceans and

---

16 Thomas Bearden, "Scalar Electromagnetics and Weather Control." Esoteric Physics homepage, May 26, 1998. Bearden's background: Lieutenant Colonel U.S. Army (Retired); MS Nuclear Engineering, Georgia Institute of Technology; BS Mathematics, Northeast Louisiana University; graduate of Command & General Staff College, U.S. Army; graduate of Guided Missile Staff Officers' Course, U.S. Army (equivalent of MS in Aerospace Engineering); numerous electronic warfare and counter-countermeasures courses.

17 "A space laser could turn the Earth's atmosphere into a giant magnifying glass and be used to spy on enemies in future wars." *BT.com*, 18 January 2017. See "BAE Systems – Laser Developed Atmospheric Lens Force Shield Combat Situation" youtube.

installations in the North and South Poles (including over Gakona HAARP).[18]

In X-ray mode, scalar waves become remote viewing radar, synthetic telepathy or voice-to-skull (V2K) mind control. On a scalar carrier, an advanced quantum potential weapon can mimic the signature or frequency of a disease and recreate it. Collective kills can be remotely achieved in a fifty-mile radius, individual kills on high-intensity pulse mode. In mood mode, a phase-locked 10Hz modulation pulsed by scalar transmitters can subject a population in a seventy-five-mile radius to a quiescent, hypnotic state[19] *because the body functions not on electromagnetic waves but on scalar magnetic waves.*

As invisible scalar weapons make missiles and nuclear bombs obsolete, near-invisible nanotechnology tethers our bodies and brains to the trillions of sensors, microprocessors and genetically engineered nanobots now being loosed in our compromised biosphere.

---

18 Conversation with a passenger in a C-130 who saw a plasma shield go up over Gakona HAARP in 2015.

19 See Joanna Lillis, "The village that fell asleep: mystery illness perplexes Kazakh scientists." *The Guardian*, 18 March 2015.

CHAPTER THREE

The Nano Assault

▼

Metamaterial (Greek, meta, beyond) - engineered to have a property not found in nature

Micron (μm) - micrometer; one millionth of a meter (0.000039 inch)

Nanometer (nm) - one billionth of a meter (0.000000001 m)

Oh, and as for the smart dust concept, the whole notion was created by the Defense Department, and UC Berkeley scientists are currently working on projects that include such applications as "battlefield surveillance, treaty monitoring, transportation monitoring, scud hunting." CNN headlined a story on the technology this way: 'Smart dust' aims to monitor everything.

— "Cornell sounds nanoparticle health warning,"
eats shoots 'n leaves, February 17, 2012

Our Earth is surrounded and suffused by an electric circuit. More than a century ago, Norwegian scientist Kristian Birkeland postulated huge electric currents powered by solar wind being drawn through the ionosphere by Earth's magnetic field. These Birkeland currents connect the Earth to the Sun. This is a natural fact, as lightning, volcanic eruptions of sulfur, and Alfvén waves ("whistlers")[1] once were.

Multiply what was once natural by radio waves and microwaves now 200 million times those of a century ago: FM radio, television, cell phones, computers, radar, etc. Next, add the stealthy "Star Wars" Strategic Defense Initiative (SDI) silently resurrected and morphed into the wireless AI-run Space Fence infrastructure dependent upon electro-optical chemicals loaded with nanoparticles dosing our skies—all without public debate or knowledge.

---

1 Low-frequency oscillations of ions in plasma and magnetic field perturbations.

For over a half-century, "national security" technology has been quietly overturning democracy and the public transparency it relies on.

Weather engineering is key to all other electromagnetic military operations. Not only is high-frequency "sky beam" technology daily adjusting the temperature-dependent conductivity of the lower atmosphere by heating portions of the ionosphere,[2] but it has been weaponized for the military doctrine of full spectrum dominance via C4 (command, control, communications, and cyberwarfare) dependent upon (1) a conductive lower atmosphere, and (2) space- and ground-based systems that assure 24/7 conductivity.[3] Along with conductive metal nanoparticles, trillions of nano-sensors and microprocessors have been and continue to be released into the stratosphere and troposphere via jets, drones, ships, and rockets.

In short, our atmosphere is no longer conducting natural charge, current, and voltage. Instead, we are now breathing from a ramped-up amplifier / condenser / antenna built of conductive nano-metals and ionized electrons. Tiny and almost weightless for maximum long-lasting loft, metal nanoparticles offer multiplied surface area and attract moisture for generating storms. Besides attaching to and ionizing molecules of oxygen, those in the upper atmosphere bond with and draw unknown organisms down into our atmosphere, creating chemical synergies our immune systems know nothing about. Yet we are forced to breathe this chemical witches' brew while the nanoparticles (including "smart dust" and "dusty plasma") breach our blood-brain barrier.

Thus it is no longer only the unseen world of wireless radio waves and microwaves pulsing through our bodies and brains that we must take into account but nano-synergies that our Earth and immune systems are so far ill-equipped to handle.[4]

The military-industrial-intelligence complex gave the green light to nanotechnology more than twenty years ago, about the time that jets began spewing nano-metals. Nano-scale materials, tools and devices now exceed profits of $1 trillion per year.

The nano-revolution—like the non-ionized radiation revolution—occurred very quietly so as to avoid public attention and government safety trials and standards. Owning the weather was the overarching military-industrial-intelligence objective, not health. Nor was it the first time that the biosphere had to serve as collateral damage for open field "research" under Section 1520a Chapter 32 of U.S. Code Title 50, "Restrictions on use of human subjects for testing of chemical or biological agents":

2 M.B. Cohen et al. "Geometric Modulation: A more effective method of steerable ELF/VLF wave generation with continuous HF heating of the lower atmosphere." *Geophysical Research Letters*, Vol. 35, 19 March 2008.

3 ELF/VLF radio waves (300Hz–30kHz) are so long (10–1,000 kilometers) that they are difficult to generate (much less sustain) with ground antennas alone, which is what makes a "sky antenna" essential.

4 Thanks to William Thomas, "Nano Chemtrails," July 17, 2014, chemtrailsmuststop.com/2014/07/an-explanation-of-nano-chemtrails/.

Title 50 defines the role of war and national defense, and Chapter 32 sets limits on chemical and biological warfare programs. While the Secretary of Defense may not conduct any chemical or biological experiments on civilian populations, the loophole lies in allowing for medical, therapeutic, pharmaceutical, agricultural, and industrial research and tests, including research for protection against weapons and for law enforcement purposes like riot control.[5]

It is not just the size and chemical signatures of nanoparticles that makes them a planetary hazard, it is their *unnatural* relationship with Nature. Nanotechnology is about taking apart, reconstructing, and condensing Nature at a molecular level measuring 1/80,000th the diameter of a human hair. Polystyrene carboxylated nanoparticles are in food additives and vitamins (all FDA-approved, of course).[6] "Nano-enabled products" span medicines, textiles, laundry detergent and fuel additives, dental fillings, food packaging, and skin care. A campfire naturally produces carbon nanotubes, but titanium dioxide (TiO_2) added to processed foods, toothpaste, gum, paint, paper, plastic, etc., is not "natural" but engineered, which places corporations like Kraft, Nestle, Hershey, Campbell, and Unilever squarely in the nano-food business. It is *not* just Monsanto GMOs that we need to research.

But the Food and Drug Administration (FDA) views nanos as fundamentally no different from the silver or titanium they come from and so allows corporations to simply label nanoparticles as "additives." The truth is that *Nature at the nanoscale has far different properties*. Copper becomes transparent, aluminum explosive, solids liquid. At the molecular level, the laws of physics and chemistry governing color, solubility, strength, reactivity, toxicity, etc. work differently and may represent a danger to the biosphere. More research is needed.

Both the Center for Food Safety and the Energy & Environment Legal Institute are suing the EPA for not protecting citizens or the environment from what boils down to military-industrial-intelligence complex agendas dependent upon the unregulated use of nanotechnology.

> "Nanotechnology is a novel technology that poses unique risks unlike anything we've seen before," said Jaydee Hanson, policy director at the International Center for Technology Assessment. "Scientists agree that nanomaterials create novel risks that require new forms of toxicity testing. EPA's use of a conditional registration could not be more inappropriate in this context."[7]

5 Elana Freeland, *Chemtrails, HAARP, and the Full Spectrum Dominance of Planet Earth*. Feral House, 2014.

6 Anne Ju, "Nanoparticles in food, vitamins could harm human health." *Cornell Chronicle*, February 16, 2012.

7 "Groups Sue EPA over Faulty Approval of Nanotechnology Pesticide." Center for Food Safety press release, July 27, 2015.

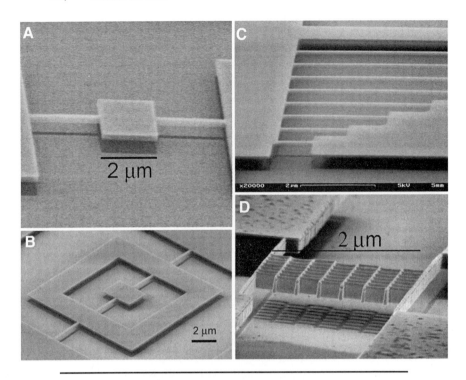

MEMS – Microelectromechanical Sensors or Systems
GEMS – Global Environmental MEMS Sensors or Systems
NEMS – Nanoelectromechanical Sensors or Systems

Electron micrograph of NEMS objects fabricated in single-crystal silicon by using electron beam lithography
and surface micromachining. A torsional oscillator, a compound torsional oscillator, a series of silicon
nanowires, and an oscillating silicon mesh mirror. – H.G. Craighead, "Nanoelectromechanical Systems,"
Science, Vol. 290, Issue 5496, 24 November 2000.

Neither lawsuit takes into account the electromagnetically charged nanoparticles that have been aerially spread throughout the atmosphere for the past two decades. Whatever the "reasonable" rationale, the truth is that we are being heavily dosed inside and out with ionized nanoparticles we have yet to learn the effects of.

Then there's the PM2.5 atmospheric particulate matter safety measure set in 1997 for the National Ambient Air Quality Standard. The science behind this "safety measure," like the science behind other national standards, appears to have been set by the military: the EPA bribed researchers working for the Clean Air Scientific Air Advisory Committee with $190+ million in grants.

> EPA began regulating PM2.5 in the early 1990s, and today says there's no safe level of exposure to the air pollutant. . .These small particles can get into people's respiratory system and can harm human health and even lead to

death after just short-term exposure, according to EPA. In 2011, former EPA Administrator Lisa Jackson told Congress that PM2.5 "causes premature death. It doesn't make you sick. It's directly causal to dying sooner than you should," she said.[8]

Look up and you may see camouflaged DC-10 tankers or ghostlike low-flying drones compressing and releasing classified chemical compounds in which swarms of smart nano-machines are self-adjusting their size, temperature, and polarity to enhance their dispersal rate and refine their buoyancy as they seed storm fronts and hurricanes, communicate weather patterns to supercomputers, and increase or decrease the storm's size and intensity as radio frequency steers them to a specified target area. The U.S. Air Force document "Weather As A Force Multiplier: Owning the Weather in 2025" credits nanotechnology with the total weather modification process becoming "a real-time loop of continuous, appropriate, measured interventions, and feedback capable of producing desired weather behavior."[9]

Meanwhile, we breathe in these nano-machines, along with everything else.

Not only is nanotechnology's stance toward Nature unnatural, but the nano sciences operate outside the laws of classical physics in the uncharted realm of quantum mechanics. As theoretical physicist Richard Feynman (1918–1988) said, "If you aren't deeply disturbed by quantum mechanics, you clearly haven't understood it." Like antimatter, nanobots may be from parallel dimensions that obey different laws. (See Chapter 7.) For example, tiny nano-scale molecular rotors don't interact with air and so cannot be influenced by friction, as normal rotors in an aircraft engine are. A subtler example is how the Sun's photons convert semiconducting heavy metals into hot energy charge carriers from *inside* plasmonic[10] metal nanoparticles:

> Classical models of photo-induced charge excitation and transfer in metals suggest that the majority of the energetic charge carriers rapidly decay within the metal nanostructure before they are transferred into the neighboring molecule or semiconductor, and therefore, the efficiency of charge transfer is low. Herein, we present experimental evidence that calls into question this conventional picture. We demonstrate a system where the presence of a molecule, absorbed on the surface of a plasmonic nanoparticle, significantly changes the flow of charge within the excited plasmonic system. The nanoparticle-absorbate

8 Michael Bastasch, "EPA's 'Independent' Science Advisers Got $190 Million In Agency Grants." *The Daily Caller*, May 17, 2016.

9 Col. Tamzy J. House et al., "Weather As A Force Multiplier: Owning the Weather in 2025." *Air Force 2025*, Chapter 3, "System Description," August 1996.

10 Wikipedia: In physics, a plasmon is a quantum of plasma [ionized gas] oscillation. As light consists of photons, plasma oscillation consists of plasmons.

system experiences high rates of direct, resonant flow of charge from the nanoparticle to the molecule, bypassing the conventional charge excitation and thermalization process taking place in the nanoparticle . . .[11] (Emphasis added.)

We really don't know the end result of nanotechnology marching to a quantum drummer, but continuing to ignore its complex invasive and invisible aspects everywhere inside and outside biological bodies and brains is perilous.

Lawmakers and nano-materials physicists have been irresponsible. Feeble warnings followed the 1996 introduction of the term *smart dust* by UC Berkeley professor Kris Pister in reference to nanoparticles fitted with computing power, sensing equipment, tiny wireless radios, and self-sustaining batteries as the planetary nerve endings that would monitor everything and everyone, then report back to corporate headquarters of Hewlett-Packard's Central Nervous System for the Earth. The 1999 National Science and Technology Council's *Nanotechnology Research Directions: Vision for Nanotechnology in the Next Decade* admitted: "The risks to human health for particles on this length scale have not been assessed. In some cases, as for silica and asbestos fibers, the hazard potential is clear; in others, it appears that the hazard potential may be lower. *Nanoscale aerosol particles are constantly involved in complex chemical processes in the atmosphere.*"[12] (Emphasis added.)

Was there a congressional investigation? No.

A major hurdle for the neuroscientists, geneticists, and Pentagon visionaries committed to achieving an "enhanced" Transhumanist humanity has been how to mount mass-scale brain-machine interfaces (BMIs) for a "hive mind." The answer is to disseminate *neural dust* composed of complementary metal oxide semiconductor circuitry (CMOS) and sensors that lodge in the brain. But while entering the brain from the bloodstream, nanoparticles and nanobots overstimulate brain cells, form blood clots, and punch holes in cell membranes.

Are chemical trails (and possibly GMOs and vaccinations) delivering neural dust?

> Each particle of neural dust. . .is coupled to a piezoelectric material that converts ultra-high-frequency sound waves into electrical signals and vice versa. The neural dust is interrogated by another component. . .powered from outside the body. This generates the ultrasound that powers the neural dust and sensors that listen for their response, rather like an RFID system.[13]

---

11 Calvin Boerigter et al., "Mechanism of Charge Transfer from Plasmonic Nanostructures to Chemically Attached Materials." American Chemical Society *ACS Nano*, June 7, 2016. Also see "Plasmonic Nanoparticles and Nanostructures (Ivan Smalyukh)," NanoBio Node, October 28, 2015 (Smalyukh's talk at the BioNanotechnology Summer Institute '15, July 29, 2015, 1 hour 17 min).

12 *Nanotechnology Research Directions: Vision for Nanotechnology in the Next Decade*, National Science and Technology Council, 1999; Chapter 10, "Nanoscale Processes and the Environment," pp. 143–144.

13 "How Smart Dust Could Spy On Your Brain." *Technology Review*, July 16, 2013. Also see Dongjin

After September 11, 2001, the new Department of Homeland Security (DHS) activated Oak Ridge National Laboratory's SensorNet program to begin integrating nano- and microsensors into real-time detection and surveillance.

> It is [in transportation and commerce] that the full scope of surveillance integration can be seen as a management strategy that merges legislation, federal inspection systems, international standards, security threat assessments, and the latest in nanotechnology.[14]

At the close of 2003, Public Law 108-153, the "21st Century Nanotechnology Research and Development Act," quietly made its way through Congress. "The President shall implement a National Nanotechnology Program. . .The activities of the Program shall include (1) developing a fundamental understanding of matter that enables control and manipulation at the nanoscale . . ." Congress was assured that nanotechnology was the "science of the future."

In 2005, the Woodrow Wilson International Center for Scholars announced the moneymaking "Internet of Things" (IoT) angle in its Project on Emerging Nanotechnologies:

> To document the marketing and distribution of nano-enabled products into the commercial marketplace, the Woodrow Wilson International Center for Scholars and the Project on Emerging Nanotechnologies created the Nanotechnology Consumer Products Inventory (CPI) in 2005. . .The revised inventory was released in October 2013. It currently lists 1,814 consumer products from 622 companies in 32 countries. The Health and Fitness category contains the most products (762, or 42% of the total). Silver is the most frequently used nanomaterial (435 products, or 24%); however, 49% of the products (889) included in the CPI do not provide the composition of the nanomaterial used in them. About 29% of the CPI (528 products) contain nanomaterials suspended in a variety of liquid media and dermal contact is the most likely exposure scenario from their use. The majority (1,288 products, or 71%) of the products do not present enough supporting information to corroborate the claim that nanomaterials are used . . .[15]

Seo et al., "Neural Dust: An Ultrasonic, Low Power Solution For Chronic Brain-Machine Interfaces," University of California, 8 July 2013. Ultrasound works far better than direct electromagnetic waves in that there is less buildup of body heat, less signal-to-noise ratio, and it can transmit at least 10 million times more power than EM waves at the same scale.

14 Michael Edwards, "How Close Arc We to a Nano-based Surveillance State?" *Activist Post*, February 21, 2011.

15 Marina E. Vance et al., "Nanotechnology in the real world: Redeveloping the nanomaterial consumer products inventory." *Beilstein Journal of Nanotechnology*, August 21, 2015. All the research for this paper was done at Virginia Tech.

Nano research boomed at Lawrence Berkeley National Laboratory (LBNL), the Department of Energy research facility run by University of California Berkeley. Once the National Nanotechnology Program was in place, all caution was consigned to the outer darkness of "national security." In 2007, when the City of Berkeley requested that the LBNL (and UC Berkeley) comply with a city ordinance requiring corporations working with engineered nanoparticles to submit a toxicology report and "how the facility will safely handle, monitor, contain, dispose, track inventory, prevent releases and mitigate such materials," neither institution complied.[16] A month later, UC Regents approved major expansion of LBNL, virtually ignoring city and community outcries about toxic compounds in the soil and groundwater, including polynuclear aromatic hydrocarbons, hazardous metals, tritium, etc.[17]

For the most part, critical warnings and second thoughts about nanotechnology withered on the vine of scientific journals, though the *Journal of Nanoparticle Research* did manage to publish "Nanotechnology and the need for risk governance" about the "governance gap" between nano- and micro-technologies. (1 nanometer *nm* = 0.001 micron μm, a quantum world of difference.) From the Abstract:

> . . .The novel attributes of nanotechnology demand different routes for risk-benefit assessment and risk management, and at present, nanotechnology innovation proceeds ahead of the policy and regulatory environment. In the shorter term, the governance gap is significant for those passive nanostructures that are currently in production and have high exposure rates; and is especially significant for the several 'active' nanoscale structures and nanosystems that we can expect to be on the market in the near future. Active nanoscale structures and nanosystems have the potential to affect not only human health and the environment but also aspects of social lifestyle, human identity and cultural values . . .[18] [Emphasis added.]

But as is usual among sciences shanghaied by the military-industrial-intelligence complex, the warning went nowhere. In fact, the sixty-page 2011 National Nanotechnology Initiative Strategic Plan mandated that each government agency do its part to converge society with a nano-based "integrated technology":

- Department of Defense – persistent surveillance
- Intelligence Community – unmanned aircraft
- Department of Energy – energy and climate change

---

16 Judith Scherr, "UC, Lab Opt Out of Nanoparticle Report." *The Berkeley Daily Planet,* June 5, 2007.

17 Richard Brenneman, "UC Regents Expected to Approve Lab's Expansion." *The Berkeley Daily Planet,* July 13, 2007.

18 O. Renn and M.C. Roco, "Nanotechnology and the need for risk governance." *Journal of Nanoparticle Research,* April 2006.

- Department of Homeland Security – sensor platforms

- Department of Justice – criminal justice

- Department of Transportation – modify travel behavior

- Environmental Protection Agency – environmental sensing

- Food & Drug Administration – biological systems

- National Institute of Food and Agriculture – global food

- National Institutes of Health – precise control for predictable outcomes

- Department of Treasury – economic sanctions

- National Science Foundation – education

Nanotechnology is seamlessly merging with the biosphere to take the place of what was once called Nature.

> Nanoparticles, because of their ultramicroscopic size, readily penetrate the skin, can invade underlying blood vessels, get into the general bloodstream, and produce distant toxic effects.[19]

On the seemingly benign medical side, nano-devices are being engineered to detect molecules, enzymes, proteins, and genetics. "Microparticulate delivery systems" are in vaccines[20] and on the tips of vaccination needles. Nanobots made of graphene are programmed to swim in the bloodstream to release drugs. Tiny Janus particle motors made of gold and platinum can repair microcircuits when propelled by a chemical reaction;[21] engines the size of atoms under cones of electromagnetic energy heated by lasers can be pulsed in a heat-cool pattern to behave like pistons; and even biological nano-rockets are being engineered by attaching strands of DNA as "catalytic engines" to a gold and chromium polystyrene bead:

> When placed in a solution of hydrogen peroxide, the engine molecules caused a chemical reaction that produced oxygen bubbles, forcing the rocket to move in the opposite direction. Shining a beam of ultra-violet light on one side of the rocket causes the DNA to break apart, detaching the engines and changing the rocket's direction of travel.[22]

---

19 "A Serious Warning about Nano-Technology in Cosmeceuticals." *Mercola.com*, April 24, 2010.

20 Eric Farris et al., "Micro- and nanoparticulates for DNA vaccine delivery," *Experimental Biology and Medicine*, 2016.

21 Todd Jaquith, "These Nanobots Can Repair Circuits All by Themselves." *Futurism*, March 30, 2016.

22 Shutterstock, "Meet the nanomachines that could drive a medical revolution." *The Conversation*, April 21, 2016.

The *genotoxicity* (damage to DNA) of engineered nanoparticles (ENPs) may be the most concerning, particularly for future generations. A 2013 paper in *Environmental Science & Technology* points to metal oxide nanoparticles, fullerenes (molecules of carbon in the form of hollow spheres, ellipsoids, tubes, etc.) and carbon nanotubes—all of which are implicated in chemical aerosols. Fullerenes assemble and replicate, delivering and embedding their programmed metamaterials and pharma agents directly into the host's DNA.

> . . .it is currently postulated that ENPs cause nonspecific oxidative damage and that the resulting stress may be the predominant cause of DNA damage and subsequent genotoxicity.[23]

This self-assembly process is called *Teslaphoresis* and sounds disturbingly like descriptions by Morgellons sufferers of the self-assembling fiber networks building grids throughout their bodies and at times erupting from their scalps, gums, and skin. Teslaphoresis depends upon a Tesla coil force field for remote directives to carbon nanotubes to self-assemble and extend in long wires that can be magnetically reeled in and out:

> "Electric fields have been used to move small objects, but only over short distances," [Rice University chemist Paul] Cherukuri said. "With Teslaphoresis, we have the ability to massively scale up force fields to move matter remotely. . .There are so many applications where one could utilize strong force fields to control the behavior of matter in both biological and artificial systems."[24]

Telephoresis nanowires "grow and act like nerves," simultaneously assembling and creating circuits powered by force field energy. Cherukuri mentions how "patterned surfaces and multiple Tesla coil systems could create more complex self-assembling circuits from nanoscale-sized particles"—like what is being delivered by chemical trails, GMO foods, and vaccinations.

Disseminating nanoparticles makes sense if your objective is an "integrated technology" matrix for a socio-biological "battlespace" AI-programmed for surveillance, detection, and two-way communication. Think gigaflop microprocessors the size of molecules, MEMS (microelectromechanical systems) and GEMS (global environmental MEMS) "smart" sensors and microprocessors (computers), and magnetoelectric nanoparticles (MENs) gravitating to neurons in the brain.

We are now in the realm of *optogenetics* and deep brain stimulation (DBS),

23 Huanhua Wang et al., "Engineered Nanoparticles May Induce Genotoxicity." *Environmental Science & Technology*, November 14, 2013.

24 Mike Williams, "Nanotubes assemble! Rice introduces 'Teslaphoresis.'" Rice University, April 14, 2016. See www.youtube.com/watch?v=wid0Lg6wuvc.

the manipulation of brain cell activity by switching brain cells on and off with remote light or magnetic fields. All that is needed is optical nanofibers programmed to deliver light to the brain, plus gold nanoparticles and nano-rods that absorb light and convert it to heat. Iron oxide nanoparticulates (and other heavy metal oxides) give off heat when exposed to an alternating magnetic field, "causing the neurons to fire long trains of nervous impulses."[25]

Thus we can begin to piece together why military intelligence has bent over backwards to deceive the public about nanotechnology *and* the chemical trails above. Wires, RFIDs and implants are no longer necessary; remote access can be obtained and maintained by creating an external magnetic field in which magnetic nanoparticles stimulate and respond to other magnetic fields outside (*in vitro*) and inside (*in vivo*) the body to create signaling networks that control ion channels, neurons, and behavior.[26]

The campaign to convince the public that controlling machines by thought alone is "progress" flies in the face of the shouting inverse: that *all of it*—the chemical trails loaded with conducting metals and nanobots, the nano GMO food additives, the cell phones and towers, the Internet of Things—is blasting us toward a Transhumanist future in which tiny machines and their AI gods remote-control our brains.

The deployment of trillions of nanobots above and in our atmosphere with no proof of safety is a planetary crime whose magnitude is difficult to measure—a crime for which neither national nor international judiciary systems are prepared, a crime that makes twentieth-century Nazi crimes of extermination and experimentation seem juvenile and clumsy.

We are being besieged by "metamaterial assemblies" that debilitate the planetary biosphere built and maintained for untold eons by an extraordinary spiritual power we call Nature. The unstudied synergies forming among chemicals, radiation, nanos, and living beings are leaving chronic immune-deficiency illnesses and dying species in their secret wake. Nano-scale chemicals combine and share ions with carbons, sunlight, radio waves, bacteria, mold, fungi, and algae, then mutate and integrate with DNA, mutating it in the process. Mutations are inevitable when nanoscale thorium, strontium, aluminum, barium, lithium, silver, styrene, polymers, liposomes, hydrogels, etc. overload cell communication.

"Smart dust" detects earthquakes and tracks weather systems, predicts traffic flow, monitors energy use, and measures vibration, sound, temperature, and chemical signatures, and wireless sensor networks *(netscapes* or *mesh)* monitor

---

25 See Carvalho-de-Souza, J.L. et al., "Photosensitivity of Neurons Enabled by Cell-Targeted Gold Nanoparticles" in *Neuron*, Vol. 86, Issue 1, 8 April 2015; Chen, R. et al. "Wireless magnetothermal deep brain stimulation." *Science*, March 27, 2015.

26 Ellen Goldbaum, "With Magnetic Nanoparticles, Scientists Remotely Control Neurons and Animal Behavior." University of Buffalo at SUNY, July 6, 2010.

farms, factories, data centers, airports, and the atmosphere. Then there are the tiny magnetometers, cameras, LiDAR and radar for tracking and surveillance. Then there is the "smart dust" already in us, being daily delivered. All of it together is the "hive mind" communicating with itself by wireless frequencies and with non-human AIs at 3-space bases, servants of the World Wide Web[1] in service to Space Fence lockdown.

To paraphrase George Washington regarding dangerous servants and fearful masters, are our tiny servants our future masters?

---

1 John D. Sutter, "'Smart dust' aims to monitor everything." *CNN*, May 3, 2010.

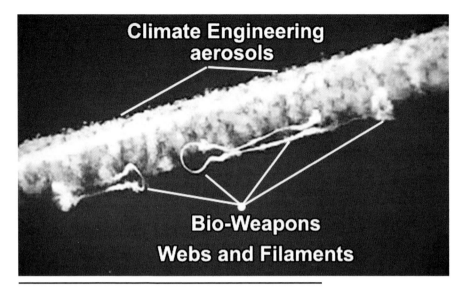

Plasma clouds and cloud cover simply do not look like the moisture-particulate clouds of yesteryear. Chemical signatures indicate that clouds are no longer simply moisture and random volcanic or micro-dust particles. Nanoparticles have changed all of that.

Figure 5. Scalar potential interferometry (between the two sets of bidirectional longitudinal EM wavepair functions) produces all EM force fields and waves.

"Others [terrorists] are engaging even in an eco-type of terrorism whereby they can alter the climate, set off earthquakes, volcanoes remotely through the use of electromagnetic waves... So there are plenty of ingenious minds out there that are at work finding ways in which they can wreak terror upon other nations...It's real, and that's the reason why we have to intensify our [counterterrorism] efforts." — Secretary of Defense William S. Cohen (1997-2001), Q&A at the Conference on Terrorism, Weapons of Mass Destruction, and U.S. Strategy, University of Georgia, Athens, April 28, 1997

From the Tom Bearden Website, www.cheniere.org/images/weapons/.

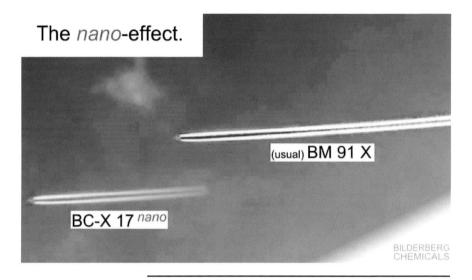

The NanoEffect.

Conductive military "chaff" has been modified into nanoparticles that easily make their way into our lungs, bloodstream, and past the blood-brain barrier.

The once pristine Olympic Peninsula of Washington State is now a US Navy electronic warfare (EW) zone. I am constantly aware of operations being conducted overhead. In December 2015, a friend suggested I hold a flashlight up to the night sky to see the particles the Navy was spraying to enhance their radar and EW weapons. The beam of light picked up a thick stream of tiny particles flowing downward. Later, when I turned the flashlight back on again, I was stunned and horrified to see that the particles were still adhering to the lens. This is the photo I took. – V. Susan Ferguson metaphysicalmusing.com/

Sun simulator (U.S. Patent No. 3,239,660)? Or plasma mirroring in the lower atmosphere may lie behind the widely witnessed illusion of two Suns.

Man could tap the Breast of Mother Sun and release her energy toward Earth as needed, magnetic as well as light. – Nikola Tesla

...of all his inventions, what may become the most beneficial are those dealing with creating electricity from magnetic energy waves created by the Sun. In very simple laymen's terms Tesla concluded that the Sun itself is a white-hot ball of pure energy with a positive electrical charge. The Earth, spinning on its axis once every 24 hours, displays a negative electrical charge. This electrical "relationship" between the Sun and the Earth greatly interested Tesla and resulted in his declaring that he could create electricity by harnessing radio magnetic waves in the atmosphere.

Tesla's personal diary contains explanations of his experiments concerning the ionosphere, and the ground's telluric currents via transverse waves [electromagnetic] and longitudinal [scalar]l waves. At his lab in Colorado, Tesla proved that the earth was a conductor, and he even produced artificial lightning, creating discharges consisting of millions of volts, and up to 135 feet (41 5 meters) long. Tesla also investigated atmospheric electricity observing lightning signals via his receivers . . . There must have been something disturbing about Tesla's work to American energy tycoons, as ones like JP Morgan locked up Tesla's most advanced power generator and threw away the key. Morgan then had Tesla's Power Generator for broadcasting electricity through the air dismantled.
— Maurice Picow, "Were Tesla's Solar Innovations 'Buried' by Big Oil?"
Energy, Green Tech and Gadgets, March 10, 2011

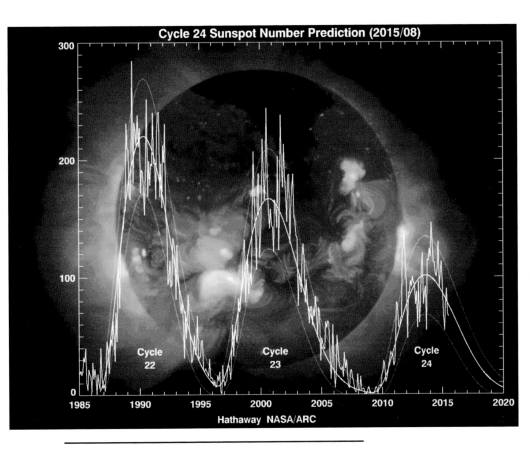

Cycle 24 Sunspot solar minimum.

Tesla: ... One summer evening in Budapest, I watched the sunset with my friend Sigetijem.
Thousands of fires were turning around in thousands of flaming colors. I remembered Faust
and recited his verses, and then, as in a fog, I saw a spinning magnetic field
and an induction motor. I saw them in the sun!
— "Interview with Nikola Tesla from 1899,"
www.stankovuniversallaw.com/2015/05/nikola-tesla-everything-is-the-light/

"The Chemical Corps is the branch of the United States Army tasked with defending against chemical, biological, radiological, and nuclear (CBRN) weapons." — *Wikipedia*

Elementis Regamus Proelium is Latin for "Let us rule the battle by means of the elements."

"The United States Army CBRN School, located at Fort Leonard Wood, Missouri, is the primary American training school specializing in military Chemical, Biological, Radiological, and Nuclear (CBRN) defense. Until 2008, it was known as the United States Army Chemical School." — *Wikipedia*

"More important by far to the geoengineering matter at hand are two intelligence agencies: the National Reconnaissance Office (NRO) and the National Geospatial-Intelligence Agency (NGA) . . . The National Geospatial-Intelligence Agency (NGA) (2003) develops imagery and map-based intelligence (IMINT) and is horizontally integrated with the NSA, one being the eyes, the other being the ears. The NGA employs 15,400 people, many working in its four football fields-long headquarters south of Washington, D.C. In conjunction with the U.S. Air Force, it cuts deals with telecom corporations for dual-use tech. See James Bamford's article "The Multibillion-Dollar US Spy Agency You Haven't Heard of" (*Foreign Policy*, March 20, 2017) and Mohana Ravindranath's "Mission Possible: A Spy Agency Builds A Tech Scout Network" (*Nextgov*, October 19, 2016)."

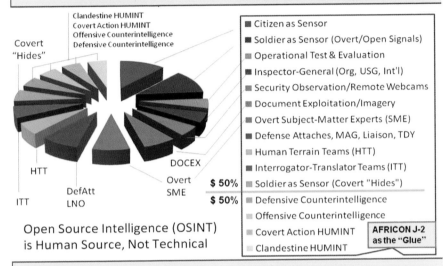

"Full Spectrum" Human Intelligence

Covert "Hides"
Clandestine HUMINT
Covert Action HUMINT
Offensive Counterintelligence
Defensive Counterintelligence

HTT
DefAtt
ITT   LNO

DOCEX
Overt $ 50%
SME $ 50%

- Citizen as Sensor
- Soldier as Sensor (Overt/Open Signals)
- Operational Test & Evaluation
- Inspector-General (Org, USG, Int'l)
- Security Observation/Remote Webcams
- Document Exploitation/Imagery
- Overt Subject-Matter Experts (SME)
- Defense Attaches, MAG, Liaison, TDY
- Human Terrain Teams (HTT)
- Interrogator-Translator Teams (ITT)
- Soldier as Sensor (Covert "Hides")
- Defensive Counterintelligence
- Offensive Counterintelligence
- Covert Action HUMINT
- Clandestine HUMINT

AFRICON J-2 as the "Glue"

Open Source Intelligence (OSINT) is Human Source, Not Technical

Education, Lessons Learned, Research, & Training are Foundation for Intelligence

HUMINT - intelligence gathered by humans
SIGINT - signals intelligence (SIGINT)
IMINT - imagery intelligence
MASINT - measurement and signature intelligence
GEOINT – geospatial intelligence

Human Intelligence (HUMINT) J-2 Central

"This is a variation of the HUMINT graphic created in Sweden, one that shows the J-2 for a region, in this case AFRICOM, being central to the provision of leadership across all of the aspects of HUMINT represented within DoD and the rest of government, all of them largely 'out of control' right now."
— Earth Intelligence Network, 8/15/2008,
phibetaiota.net/2008/08/graphic-human-intelligence-humint-j-2-central/

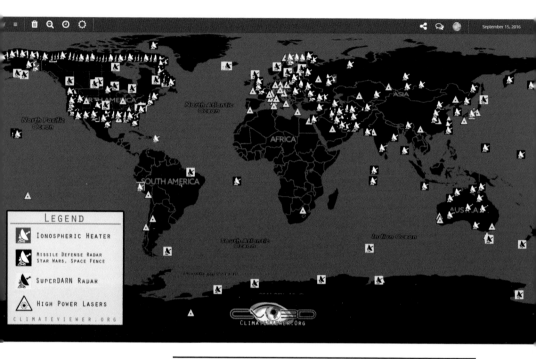

"Space Fence Map" by Jim Lee, available at ClimateViewer.org. This map reflects the major elements that comprise the Space Fence:

RED SQUARE / WHITE RADAR = Ionospheric Heaters. Facilities like the High- Frequency Active Auroral Research Project (HAARP) in Gakona, Alaska are used by scientists and military (including defense contractors) to control radio frequency and radar operations through the combination of satellites and sounding rockets deploying chemical payloads (ion clouds, dust ice, metal-oxide clouds, barium, strontium, trimethylaluminum [TMA], sulfur hexafluoride [SF6], and lithium) and illuminating these plumes with extremely focused, high-powered microwaves.

BLACK SQUARE / WHITE RADAR = Missile Defense Radars similar in nature and power to ionospheric heater facilities like HAARP. These radars are always on unless urgent maintenance is needed.

YELLOW SQUARE / BLACK RADAR = Super Dual Auroral Radar Network (SuperDARN) installations used to monitor conditions in the ionosphere and verify modifications made by ionospheric heaters.

YELLOW TRIANGLE / BLACK STAR = Lasers. Some of these installations are military-operated while others are typically operated by universities conducting "dual use" research. GEODSS, USAF Satellite Control Network, and the International Laser Ranging Service (ILRS) LiDARs (Light Detection and Ranging) are used to track satellites and near-earth objects.

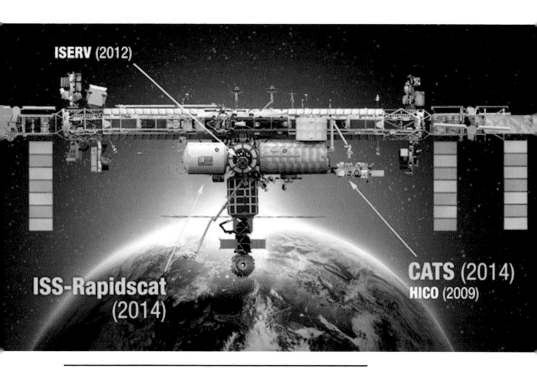

ISERV (2012)

ISS-Rapidscat
(2014)

CATS (2014)
HICO (2009)

Are industry and research instruments "dual use" for planetary surveillance? Certainly, the
International Space Station (ISS) is used to "test" various instruments. Here, the Cloud-Aerosol
Transport System (CATS) employs LiDAR to measure location, composition and distribution
of aerosols, pollution, "dusty plasma," smoke, and other particulates, while the ISS-RapidScat
radar scatterometer gauges solar winds.

Laser systems like LaWS on the sea and Starfire on land are key to *space situational awareness*: "To maintain space situational awareness, the Air Force conducts research in laser guided star adaptive optics, beam control, and space object identification." ("Starfire Optical Range at Kirtland AFB, New Mexico," Kirtland AFB Fact Sheet, March 9, 2012)

Ennylabegan

Ronald Reagan Ballistic Missile Defense Test Site

8°47'51.82"N 167°37'10.88"E

© 2016 Google

Image © 2016 DigitalGlobe

Google earth

2005   Imagery Date: 10/13/2014   8°47'52.68" N 167°37'15.24" E elev   10 ft   eye alt 1151 ft

"Nice touch, that tip of the hat to Ronald Reagan whose administration initiated the 'Star Wars'
program now culminating in the latest addition to the ground-based system upon which the
Space Fence depends going up on the Kwajalein Atoll at the old
Ronald Reagan Ballistic Missile Test Site."
— *Under An Ionized Sky: From Chemtrails to Space Fence Lockdown*

NexRads and the look-alike SBX radomes offshore operate at similar frequencies, as both are
super high-frequency (HF) Doppler radar systems. NexRads by land, SBXs by sea.
— *Under An Ionized Sky: From Chemtrails to Space Fence Lockdown*

The Aztec / Mexica calendar and the Large Hadron Collider (LHC) at CERN,
the European Organization for Nuclear Research

"On May 25, 2007, Bernard Eastlund, Tom Bearden, Fred Bell, a filmmaker and I had a
conversation in Texas about the possible development of Saturn rings around the Earth like a
celestial space collider. It seemed like every time the LHC was activated, there were cyclotronic
reverberations like shockwaves of earthquakes, volcanic and massive spontaneous gravity
(scalar) wave activity. Bernard Eastlund died 12/12/2007."
— Billy Hayes, "The HAARP Man"

CERN's Globe of Science and Innovation. This wooden "bee hive" is about the size of Saint
Peter's Basilica dome in Rome, 27 meters high (9 X 3), 40 meters in diameter (10 X 4).

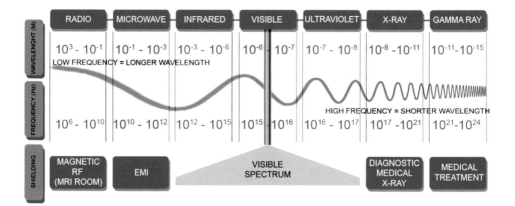

RADIO	MICROWAVE	INFRARED	VISIBLE	ULTRAVIOLET	X-RAY	GAMMA RAY

WAVELENGHT (M)

$10^3 - 10^{-1}$ $10^{-1} - 10^{-3}$ $10^{-3} - 10^{-6}$ $10^{-6} - 10^{-7}$ $10^{-7} - 10^{-8}$ $10^{-8} - 10^{-11}$ $10^{-11} - 10^{-15}$

LOW FREQUENCY = LONGER WAVELENGTH

FREQUENCY (Hz)

HIGH FREQUENCY = SHORTER WAVELENGTH

$10^6 - 10^{10}$ $10^{10} - 10^{12}$ $10^{12} - 10^{15}$ $10^{15} - 10^{16}$ $10^{16} - 10^{17}$ $10^{17} - 10^{21}$ $10^{21} - 10^{24}$

SHIELDING

MAGNETIC RF (MRI ROOM)	EMI	VISIBLE SPECTRUM	DIAGNOSTIC MEDICAL X-RAY	MEDICAL TREATMENT

Start studying invisible wavelengths, frequencies, and pulses. Invest in detectors. Memorize how narrow our visual spectrum is. *Licensed under Creative Commons Attribution-Share Alike 3.0 Unported at Wikimedia Commons.*

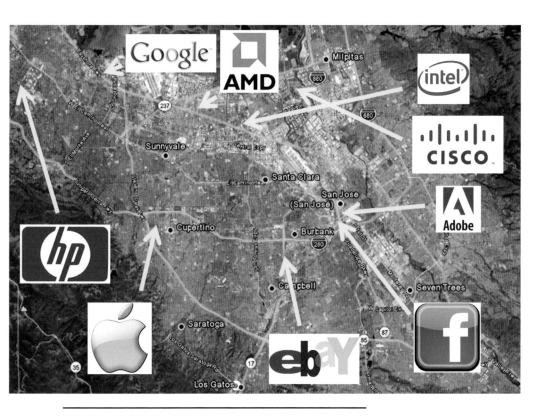

Capital city of the Technocracy: Silicon Valley.

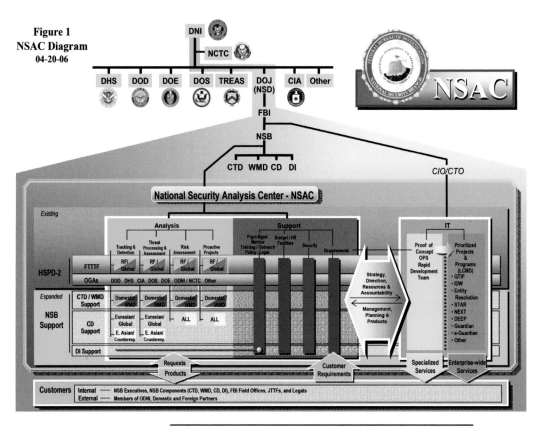

Yet another quiet agency with devastating power over our lives,
the National Security Analysis Center (NSAC).

"Today, through a series of high-level classified authorities and commercial relationships,
the Center has access to over 130 databases and datasets of information comprising some
two billion records, over half of which are unique and not contained in any other government
information warehouse. The Center is, in fact, according to interviews with government officials,
the sole organization in the U.S. government with the authority to delve deeply into the activities
and associations of foreigners and Americans alike."
— phasezero.gawker.com/this-shadow-government-agency-is-
scarier-than-the-nsa-1707179377

CHAPTER FOUR

Sky Anomalies

▼

A system for facilitating cloud formation and cloud precipitation includes a controller and a beam emitter that is responsive to the controller. The beam emitter is configured to emit a beam to form charged particles within an atmospheric zone containing water vapor. The charged particles enhance the formation of cloud condensation nuclei such that water vapor condenses on the cloud condensation nuclei forming cloud droplets. The system further includes a sensor configured to detect a cloud status and output a signal corresponding to the cloud status to the controller.

— "System for facilitating cloud formation and cloud precipitation," US 9526216 B2, Kenneth G. Caldeira, December 27, 2016

At last count, there were sixteen different mixtures of chemicals for aerosol distribution—twelve electrically different from each other and four being time tags logging how the mixtures are working at all layers in the vertical wall of connectivity between them. Ultimately, the military plans twenty-four layers extending from Earth into the higher frequencies of space. Thus the electromagnetic Space Fence, with its cloak of invisibility dependent upon *terahertz (THz)* wavelengths so small they can manipulate individual atoms, is to be contiguous with the witches' brew in our atmosphere.

Not surprisingly, the number of "new clouds" being generated by geoengineering is the same as the number of chemical mixtures—a baker's dozen decided by the Cloud Appreciation Society for the World Meteorological Organisation to include in its revised International Cloud Atlas first published in 1896. Joining Nature's clouds—cirrus (L., feathers), cumulus (L., heaped up), stratus (L., layered and smooth), and nimbus (L., rain-bearing cloud)—are asperitas (L., roughness), cauda (L., tail), fluctus (Kelvin-Helmholz), murus (L., wall), flumen ("beaver's tail," associated with a supercell severe convective storm), and the species volutus (L., rolled).

The inclusion of rainbows, halos (sun dogs), snow devils and hailstones is odd—"All types of optical effects can be defined as clouds," says BBC meteorologist John Hammond, who is looking forward to "many new

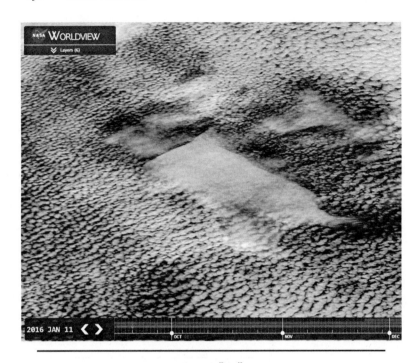

go.nasa.gov/2yomill
South of Greenland in the Labrador Sea, January 11, 2016. Photo by V. Susan Ferguson
metaphysicalmusing.com/.

entries in the future"[1]—but no odder than lumping anthropogenic plasma cloud production via chemicals and radio frequency and microwaves in with Nature's clouds. As the World Meteorological Organization puts it, "19th century tradition, 21st century technology":

> The International Cloud Atlas also proposes five new "special clouds": cataractagenitus, flammagenitus, homogenitus, silvagenitus and homomutatus. The suffix *genitus* indicates localized factors that led to cloud formation or growth, while *mutates* is added when these caused the cloud to change from a different form. These special clouds are influenced by large waterfalls, localized heat from wildfires, saturation of air above forests and humans. Thus, a common example of homogenitus is contrails sometimes seen after aircraft.[2]

As I indicated in Chapter 1, even with a telephoto lens, it is difficult to discern just *where* on the jet the chemical spray is issuing from. As for contrails,

1 Matt McGrath, "'New' wave-like cloud finally wins recognition." *BBC,* 23 March 2017.

2 "New International Cloud Atlas: 19th century tradition, 21st century technology." World Meteorological Organization, 22 March 2017 [World Meteorological Day].

water-injection engines are extremely rare, so if you see a short trail behind a large jet, you may be seeing the new short non-persistent chemical trails. One or two streams out of each wing making as many as four trails; one or two trails directly behind the engine; sometimes no trail at all. Excess fuel is often released from pylon drain tubes before landing, but what if it occurs at 26,000 feet? Does that make it an intentional chemical dump? At what point are we seeing supplementary chemicals not "cooked" (pyrolyzed) in the engine chambers, such as the Airbus A320 passenger jets with pipes or ducts in the pylon just above the engine? And when we see the "on-off" staccato trail, is one chemical configuration being switched for another? And the photographs of onboard modular aerial spray systems in the aircraft fuselage—is the equipment "dual use"?

Massive experimentation is going on in our skies. The old chemical trails of the past two decades are being changed. The barium in the old trails was invisible from one direction and foggy white to grey from another. Some observers now report *blue* chemtrails, not the classic milky white or grey—blue meaning transparent or translucent. Even nano-sized Mylar polymers are now translucent in terahertz. Could this be due to Aerochem Corporation's new manganese ion fuel additive?

> The absence of the typical whitish coloring is due to a kaleidoscopic refraction of depolarized barium particles exposed to a significant temperature swing (from 400–500°F, which is -55° to stratosphere output). Reversing the hygroscopic effect (a substance's ability to attract and hold water molecules) does away with visible condensation.[3]

"INVISIBLE"

What if chemical trails go completely invisible and future generations believe that perpetually milky baby-blue skies in the day and no stars at night is natural?

The military has thought long and hard about not just mitigating the heat-producing "contrails" their C4 agendas require but about making aircraft invisible to sight and radar. (See General Electric's 1964 patent US3127608 A "Object camouflage method and apparatus.") In the 1980s, stealth programs concentrated on electromagnetics for building radar-evading bombers. For example, the McDonnell Douglas F-4 Phantom fighter bombers were designed so that pilots could activate cryogenic superconducting magnets to create an EM "bubble" around the aircraft. Light striking the EM field would divide and pass around the aircraft and reunite on the other side—thus, invisibility. But

---

3 Email, Mark Porrey, June 2, 2015.

the pilot's inability to see beyond the invisibility field ("a cloud of ionization"[4]) forced the military to keep looking for more perfect visual camouflage and radar-masking. Is it possible to cloak without layer masking? Apparently so. At certain frequencies, light waves scattered by resonant and non-resonant mechanisms go into opposite phases and cancel each other out, thus producing invisibility.[5]

Nanoparticles are the name of the invisibility game, whether it's transparent optical displays on cockpit windows or "obscurants" hiding warfighters from plain sight. Both applications have to do with engineering the size, shape, and composition of nanoparticles in accordance with the attenuation properties of light.

> Recent design upgrades can now hide a warfighter from infrared and other sophisticated types of viewing, thanks to a range of metallic nanoparticles [with optical properties arising from localized surface plasmon resonance (LSPR)[6]] including gold and silver that enhance the attenuation of light in a given region of the electromagnetic spectrum. . .[P]articles are used to absorb or scatter light in order to block a warfighter's visibility over several bands of light.[7]

The latest is the dielectric *metasurface* cloak, a super-thin (3 mm) Teflon substrate with tiny embedded ceramic cylinders that manipulate EM waves.[8] This ultra-thin metamaterial[9] can work at 1/10 of missile guidance and marine radar wavelengths, and can be used for EM waves as small as those of visible light (400–700nm). Radio waves can't detect an aircraft if radio waves don't bounce back to a receiver, and seeing needs light to bounce off the object. Thus, if you can manipulate the waves, you can obtain invisibility. However, a 6° angle of sunlight can compromise the cloak, and the cloak can't cover for visual *and* radar at the same time (due to the narrow range of wavelengths).

So you may see a chemical trail laying itself, the jet or drone disappearing before your eyes, even if you're looking through IR night vision goggles. The effort to make contrails invisible "to a size below a humanly visible range" has been going on for decades, the problem being the chemicals needed to

---

4 Arnold L. Eldredge, "Object camouflage method and apparatus." General Electric Patent US3127608 A, 1964.

5 Mikhail V. Rybin et al., "Switching from Visibility to Invisibility via Fano Resonances: Theory and Experiment." *Scientific Reports*, 2015:8774.

6 Plasmonic nanoparticles are used in tagging, tracking, and smart bar code applications.

7 Donald Kennedy, "Tiny Nanoparticles—A Big Battlefield Impact?" *Defense AT&L*, September-October 2014.

8 Kyle Jahner, "Breakthrough in cloaking technology grabs military's attention." *Military Times*, September 21, 2015.

9 According to Harald Kautz-Vella, the ancient Greeks distinguished between *form forces* and matter forces. These form forces are now called metamaterials and dominate nanoparticles.

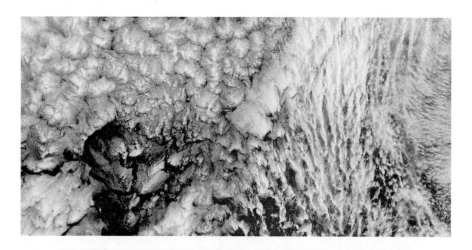

go.nasa.gov/2yuB7mx
Guadalupe Island & Baja CA / Sept. 21, 2017. Photo by V. Susan Ferguson metaphysicalmusing.com/ :

The more I see these particular clouds, the more they resemble a hive form. Creating a database of images
from NASA Worldview has only reinforced my conclusion that our planet has now been successfully invaded.
A 'foothold' Invasion! Beginning in December 2016, I have now collected over 3000 images. The formations of
the electrically charged clouds – by 88 and more transmitters around the planet – have changed their shape
over time. The technicians seem to be evolving their techniques. Recently there are more of what I term
"HIVE" shapes, cloud formations that resemble hives.

depress the freezing point of water. Toxic additives mixed with the chemical trails to make them dissipate quickly so they are invisible are even worse for those who must breathe them in. Even Rolls-Royce's ultrasonic wave method depends upon chemicals ("hydroscopic materials") like chlorosulfonic acid and sulfur trioxide:

> US 20100043443 A1, "Method and apparatus for suppressing aeroengine contrails," February 25, 2010, Rolls-Royce.
>
> Claim: An aircraft comprising a gas turbine engine that exhausts a plume of gases in use, the aircraft is characterized by comprising an ultrasound generator having an ultrasonic actuator and a waveguide to direct ultrasonic waves at the exhaust plume to significantly reduce the formation of contrails.

Other patents over the years:

- US3517505 A, "Method and apparatus for suppressing contrails," June 30, 1970, U.S. Air Force.

- US4766725 A, "Method of suppressing formation of contrails and solution therefor," August 30, 1988, Scipar, Inc., Williamsburg, New York.

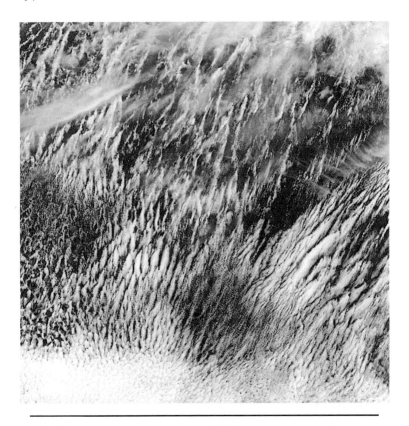

go.nasa.gov/2yiiMsZ
The eastern Pacific off Colombia & Ecuador / Sept. 17, 2017. Photo by V. Susan Ferguson
metaphysicalmusing.com/ : "This is an even better example of the 'ringed coiled tufts.' You can easily see the
rings or spiral around the 'tuft' clouds."

• US5005355 A, "Method of suppressing formation of contrails and solution there for," April 9, 1991, Scipar, Inc., Williamsburg, New York.

CHEM DUMPS AND "BOMBS"

In *Clouds of Secrecy: The Army's Germ Warfare Tests Over Populated Areas* (Rowland & Littlefield, 1988), Leonard A. Cole discusses the first Operation LAC (Large Area Coverage) in 1957–1958 when zinc cadmium sulfide nanoparticles were sprayed from the Rocky Mountains to the Atlantic Ocean and Canada to the Gulf of Mexico, cutting a swath through the San Francisco Bay area to South Dakota, Minneapolis, Corpus Christi, and back.

Harold Saive (*chemtrailsplanet.net*) has created his own acronym for what he's seeing over southeast Florida:

A TERRA satellite view from the early afternoon of January 28th, 2017 about 1 PM reveals an unmistakable north-south aerosol operation over South Florida that drifted with the cirrus level flow out into the Atlantic as the day progressed. These are not chemtrails. They are large volume aerosol plumes or LVAP. This technique is increasingly used to dump aerosols in much larger volumes than chemtrails alone could provide. A ground observer in Southeast Florida posted a video that captured the event early in the morning. This is evidence the LVAP dumps were deployed under cover of darkness before sunrise . . .[10]

Like endless war outside of Congressional jurisdiction, we are now subject to an endless Operation LAC or LVAP. After the COP21 Paris climate conference at the end of 2015 and since the "green light" for geoengineering at the beginning of 2017,[11] nanos producing endless chemical cloud cover have been barraging the skies—and this is after *two decades* of commercial and military jet trails diffusing into chemical *cirrus contrailus*. Fleets of Boeing 747 supertankers carpet the skies not in long discrete chemical trails but from massive dumps in one pass. From satellite, the dump pattern looks like hundreds of fountain or waterfall cascades. Less and less are we seeing the crosses, tic-tac-toes, and linear fade-outs.[12] In a morning or night, large geographical areas are inundated with nary a trail—like the continental-scale carbonaceous aerosol plumes over Southeast Asia described by geophysicist Keith Potts in his 2008 paper "Aerosol Plumes: The Cause of Droughts and El Nino Events By Regional Dimming."[13]

Two Evergreen Air patents make "superdumps" not only feasible but probable: Evergreen International Aviation Patent US7413145 B2 (2008) "Aerial Delivery System," and Evergreen International Aviation Patent US20100314496 A1 (2010) "Enhanced Aerial Delivery System." The patents reference "powders" for cloud seeding and forest fire retardants (U.S. Navy patent US3899144 A "Powder contrail generation," 1975), but it is not difficult to imagine supertankers like the new Boeing 787 and Airbus A350 in service to the "dual use" credo.[14]

---

10 Harold Saive, "Florida Climate Engineering Captured by Satellite and Ground Observer," February 3, 2017, chemtrailsplanet.net/2017/02/03/florida-climate-engineering-captured-by-satellite-and-ground-observer/.

11 Jamie Condliffe, "Geoengineering Gets the Green Light from Federal Scientists." *Technology Review*, January 11, 2017.

12 See Harold Saive's YouTubes at "Chemtrails Not Enough: Climate Engineers Dumping Large Volume Aerosol," April 13, 2016, chemtrailsplanet.net/2016/04/13/chemtrails-not-enough-climate-engineers-now-dumping-large-volume-aerosol-plumes/.

13 See Harold Saive's chemtrailsplanet.net/2016/01/03/scientist-el-nino-created-and-sustained-by-aerosols/.

14 Evergreen International Airlines is a CIA carrier. It was originally based outside McMinnville, Oregon near Portland before declaring bankruptcy and shutting down its specially outfitted fleet of 747s in January 2014. By December 2014, Evergreen supertankers were flying in and out of Arizona's Pinal Airpark of Marana Army Airfield. For the questionable history of Evergreen Air and its relationship to

Besides Evergreen supertankers, new C-130Js will be replacing the 910th Airlift Wing's current fleet of C-130H aircraft at Ohio's Youngstown Air Reserve Station (YARS) "to ensure that the 910th's DoD specialized aerial spray mission continues safely and without interruption."[15] Are those C-130J's delivering chemical dumps?

Then there are reports of "chembombs." Saive wonders if they are adding to, or gradually replacing, the chemical trails, which would fit with mitigating "contrails" as "unintended geoengineering"; stop spraying through the nozzles and let the less visible fuel pollution continue the aerosol flooding. Some chembombs seem to be producing individual plasma clouds all on their own, while others may be plasma experiments in now ion-rich skies. Bruce Douglas of Maui Skywatch inserted the term into the chemtrail movement when he spoke at the unforgettable Consciousness Beyond Chemtrails conference in Los Angeles in 2012.[16] Chembombs are yet another way to insert chemicals into our geoengineered skies.

The recent U.S. Air Force announcement of four-inch CubeSats that "bomb the sky" by carrying "massive amounts of ionized gas to the ionosphere to create radio-reflecting plasma" explains how by heating vaporizing metal beyond its boiling point (shades of 9/11), metals react with atmospheric oxygen to produce radio-reflecting plasma:

> Another project from a team at Enig Associates and the University of Maryland plans to heat metal [nanoparticles?] by detonating a small [chemical?] bomb and converting the blast into electrical energy. And the shapes of the plasma clouds could be fine-tuned by altering the form of the initial explosion, *New Scientist* explains.
>
> In the past, researchers with the High-frequency Active Auroral Research Program in Alaska have attempted to create plasma using radiation from ground-based antennas to stimulate the ionosphere. The new plan from the USAF aims to find a more efficient way.[17]

Is radio frequency being used to vaporize metal nanoparticles in the lower atmosphere to produce better communications for C4 objectives?

tanker distribution, see "How a Few 747 Supertankers Can Dump Tons of Aerosols as 'Climate Change'," October 11, 2014, chemtrailsplanet.net/2014/10/11/evergreen-air-fleet-of-747-super-tankers-involved-in-chemtrails-and-climate-modification/.

15 "Senate Approves Portman, Brown Amendment Urging Air Force To Prioritize C-130 Upgrades For the Youngstown Air Reserve Station," Rob Portman press release, June 7, 2016.

16 "From Chemtrails to Chembombs, Aerial Aerosol Explosions are the New Dispersion Method," September 8, 2012, www.youtube.com/watch?v=dMSRPem4n7s

17 Cheyenne MacDonald, "The US Air Force reveals radical plan to 'bomb the sky' to improve radio reception." *Daily Mail*, 22 August 2016.

The ENIG space plasma generator for low Earth orbit insertion has flux compression generators that convert explosive chemical energy into electromagnetic energy:

> "We're going to take mega-amps of energy and then joule-heat a light metal load through multi-phase transitions to generate artificial man-made plasma cloud in the ionosphere," said ENIG President, Eric N. Enig. . ."[T]he space plasma generator can be used to smooth out ionosphere disturbance to assure reliable communications and navigation in-theater, or to provide novel capabilities for RF systems."[18]

Given that our lower atmosphere is now basically a mini-ionosphere, we will be seeing not just plasma clouds masquerading as "natural" but other plasma events, as well.

SHADOWS

Shadows that either follow or precede jets laying chemical trails are not necessarily carbon black trails. They may be optical effects taking place in an ionized "cloud chamber" of the upper troposphere loaded with conductive and reflective metal nanoparticles. Also, take the angle of the Sun into account as sunlight albedo can reflect skyward and cast a trail shadow on the chem-haze above it. If the Sun is *behind* the observer, then the shadow may appear *ahead* of the aircraft. Thus when sunlight reflects from the Earth (*albedo*), it can project a reverse shadow on the chem-haze "screen" above, whether of a cloud or a trail. (Clouds also cast shadows into the sky when sunlight bounces from the Earth skyward.) Not only barium turns the sky into a projector screen; the albedo effect does, too.

SUNDOGS

Once upon a time, a sundog was a bright, rainbow-colored patch of light on either side of the Sun when it was low on the horizon. Now, a "sundog" seen through the haze of chemical trails may actually be a supercontinuum white light produced by laser pulses busy propagating plasma channels (filamentation) and remote sensing fluorescent chemical and biological signatures of dissolved metallic ions.[19] Patent US 20080180655 A1, "Remote Laser Assisted Biological

18 "U.S. Air Force Awards Contract to ENIG for Space Plasma Generator (...) for Artificial Modification of Ionosphere." ENIG press release, July 8, 2015.

19 Huailiang Xu, "Femtosecond Laser Filamentation for Atmospheric Sensing." *Sensors* 11(1):32-53,

Aerosol Standoff Detection in Atmosphere," describes how a femtosecond laser or LiDAR system (light detection and ranging, combination of "light" and "radar") can directly produce a white light supercontinuum from the ultraviolet (UV) to the visible (VIS), near infra-red (NIR) and middle infra-red (MIR) in a particle cloud for a full spectrum rainbow effect.

> The heart of the new device is a sheet only nanometers thick made of a semiconducting alloy of zinc, cadmium, sulfur, and selenium. The sheet is divided into different segments. When excited with a pulse of light, the segments rich in cadmium and selenium gave off red light; those rich in cadmium and sulfur emitted green light; and those rich in zinc and sulfur glowed blue.

HOLE-PUNCH CLOUDS (FALLSTREAK HOLE)

I did not find the once-rare hole-punch cloud (fallstreak hole) among the new clouds in the revised International Cloud Atlas. Hole-punch clouds point to an atmospheric release of electrical charge potential, subtle perturbations following from how the technology serving geoengineering multiplies and releases ions into the lower atmosphere to increase power density for military operations. Control over the electromagnetic potential of the Earth's atmospheric layers is key to fulfilling the doctrine of full spectrum dominance. Phased-array Doppler radar facilities (ground-, air-, and satellite-based) can focus the microwave energy needed for this cascade of atmospheric ions.

A DOUBLE SUN

In 2010, two Suns appeared in the sky over New York City, and in 2011 ABC News commented on two Suns videoed over China. Interviewed astronomers who said it was just lens reflection, an evening curvature of light through pollution, an ice particle mirage, a rare optical refraction due to a "blob" or "thick patch" of atmosphere wandering in front of the Sun.[20] In the same time period, two bodies of light were also seen over Bangkok, Dubai, and Russia,

---

December 2010. Also view "Lightning in the lab: Femtosecond laser generating plasma in air." YouTube, Dave Sheludko, March 29, 2009: "A 100GW 35fs 1kHz laser is focused down to a narrow waist, which creates an electric field large enough to ionise the air. You can see the resulting plasma glowing (the little white floating blob). The process is essentially the same as lightning, which of course is accompanied by "thunder": the 1kHz buzz you can hear is the thunder, repeated 1,000 times per second (the laser pulses at 1kHz). The coloured pattern on the wall is the expanded beam, the moving colours generated by the nonlinear effects in the plasma."

20 "China's 'Two Suns' Video Unexplained By Science." *Space.com*, March 7, 2011.

one body of light always smaller than the Sun. A double sunrise in 2013, more dynamic duos in 2014, 2015, and 2016. What's going on, other than Internet declamations of Niburu and Planet X? Is the second Sun China's artificial star created in 2016?[21]

Or has a mirroring body been quietly ensconced in space? In 1993, the Russians claimed to be testing *Znamya* or Banner, a sixty-five-foot-diameter mirror coated with aluminum plastic film to reflect sunlight down to Earth from their Mir space station.[22] In 2007, *The Guardian* brought up giant space mirrors:

> The US government wants the world's scientists to develop technology to block sunlight as a last-ditch way to halt global warming, the Guardian has learned. It says research into techniques such as giant mirrors in space or reflective dust pumped into the atmosphere would be "important insurance" against rising emissions . . .[23]

We know about "reflective dust pumped into the atmosphere," thanks to chemical trails, but what about mirrors? Is the James Webb Space Telescope—twenty-one hexagonal mirror segments made of beryllium, each mirror coated with 0.12 ounces of gold in layers 120 nm thick—a giant mirror become "the world's most powerful space observatory"[24]—or DARPA's Space Surveillance Telescope (SST) at White Sands Missile Range sweeping up its terabyte of data per night?

The Sun is not a nuclear furnace filled with gamma rays but a positively charged plasma in a negatively charged environment. Plasma mirroring in the lower atmosphere appears to lie behind the illusion of two Suns.

PLASMA ORBS

In 2004, independent scientist Clifford Carnicom observed and called attention to lighted spheres or orbs in the New Mexico sky:

> The physics of motion of the ["ball of light"] defy common explanation. There is no obvious propulsion system visible, and the movement of the object is generally non-linear. . .In the original video or higher resolution formats of the

21 Aditya Iyer, "China creates artificial star 8,600 times hotter than sun's surface." *Hindustan Times,* February 11, 2016.

22 Warren E. Leary, "Russians To Test Space Mirror As Giant Night Light for Earth." *New York Times,* January 12, 1993.

23 David Adam, "US answer to global warming: smoke and giant space mirrors." *The Guardian,* 26 January 2007.

24 Clara Moskowitz, "NASA Completes Giant Mirrors for Hubble Successor Telescope." *Space.com,* September 16, 2011.

video, an interaction of the object with the surrounding atmospheric medium can be seen. This interaction occurs in periodic pulses, always on the same side of the object. . .It would appear that this interaction is of a plasma nature. There also appears to be a pulsation within the light source itself . . .[25]

From time to time, observers see plasma orbs guiding or accompanying jets laying chemtrails,[26] "shooting"[27] or ejecting plasma to form horseshoe-shaped clouds.[28] It is possible that lasers are used to create those orbs, given that they can be used to create "artificial stars" and

...beacons to guide the process of atmospheric compensation. When astronomers use the method, they aim a small laser at a point in the sky close to a target star or galaxy, and the concentrated light excites molecules of air (or, at higher altitudes, sodium atoms in the upper atmosphere) to glow brightly. Distortions in the image of the artificial star as it returns to Earth are measured continuously and used to deform the telescope's flexible mirror and rapidly correct for atmospheric turbulence. That sharpens images of both the artificial star and the astronomical target.[29]

Glow-mode (gas) plasma orbs can be mistaken for natural ball lightning, but they are actually produced by interfering two scalar beams over the target area. (See Chapter 11 for how they are used for battlespace targeting.)

...the energy for the plasma balls is coming from the vacuum of spacetime at the very location of the balls themselves, triggered by scalar interferometers aimed through the woodpecker grid. These kinds of balls can be used as marker beacons giving feedback for precision aiming of the howitzers [ionospheric heaters]. The energy of the marker beacon can be read back into the computers, giving precise location information for pinpoint aiming. The target area can be very small or widened out.[30]

Gas plasma orbs may serve as "guidance systems" during crop circle

25 Clifford E. Carnicom, "Orbs Require Consideration," March 14, 2004, carnicominstitute.org/wp/orbs-require-consideration/.

26 "Orb leading Chemtrail Jets," May 16, 2014. Dumbbell dumbbase, www.youtube.com/watch?v=CjDA4PpiWJI.

27 "UFO (ORB) Fires Its Gun into Chemtrail!! Twice!!" March 11, 2014. Crow777, www.youtube.com/watch?v=4byMQIk_iGQ&nohtml5=False.

28 See 2:47–4:38 of "Chemtrails Australia Gold Coast 7th Haarp Orbs, Plasma Balls, Chemtrails," Dave Brooke, January 6, 2014. www.youtube.com/watch?v=MLliG-W3JxA

29 William J. Broad, "Administration Researches Laser Weapon." *New York Times*, May 3, 2006.

30 Thomas Bearden, "Are the Russians Making 'UFOs'?" www.prahlad.org/pub/bearden/scalar_wars.htm

formation. While living in England *circa* 2005, I visited a couple of famous crop circles in the English countryside frequented by people in meditative yoga poses. I had read about the tests that revealed radioisotopes in the flattened grass and wondered if satellite computer algorithms were behind the circles.

In the Middle Ages, crop circles were obviously not attributed to algorithms but to "mowing devils." By the 1800s, it was cyclonic wind. Post-World War II aerial surveys detected buried remnants of circular buildings possibly built over naturally occurring plasma discharge vortices like big "fairy rings" produced by the Earth's electromagnetic *toroidal* forces.

Beginning in the 1970s, crop circles became increasingly complex and *technical*, much more sophisticated than slates-and-rope hoaxes might account for. But how was it being done? Much like the Apollo Moon landings, it was difficult to believe that agencies like NASA would perpetrate "high-tech hoaxes"; thus people turned to paranormal explanations. But like the creation of "sprites" thought to be natural lightning minus thunder, the answer is probably technology. Plasma can be generated and naturally occurring vortices can be directed with the precise overhead positioning of a geostationary satellite, mathematically intricate computer-generated 3D digital designs, and microwave lasers (masers).

> GPS systems enable the artists to cover vast spaces with absolute precision, while microwaves can be used to flatten large numbers of stalks at great speeds, it was claimed. An analysis of evidence in the Physics World journal reported that researchers had used magnetrons—tubes which use electricity and magnetism to generate intense heat—to mimic the physical changes in flattened stalks in some circles, which are linked to radiation.[31]

Magnetrons like those found in microwave ovens could not begin to produce the radioisotopes present in crop circles, but maser technology could do it, and easily.

Microwaves impact the nervous system and mind. In 2009, a National Aviation Reporting Center on Anomalous Phenomena (NARCAP) scientist attempted to warn pilots about possible side effects of encounters with "plasma balls":

> According to a physical theory concerning plasma ball formation, very low-frequency waves — in case generated by a previously existing plasma ball,

---

31 Nick Collins, "Crop circles 'created using GPS, lasers and microwaves'." *The Telegraph*, 1 August 2011. Also see "Physics could be behind the secrets of crop- circle artists." *Physics World*, 1 August 2011: "Matin Durrani, editor of *Physics World*, said: 'It may seem odd for a physicist such as [Richard Taylor, director of the Material Science Institute at the University of Oregon] to be studying crop circles, but then he is merely trying to act like any good scientist—examining the evidence for the design and construction of crop circles without getting carried away by the sideshow of UFOs, hoaxes, and aliens."

or by other natural causes coming from the ionosphere — if associated with high-frequency waves (such as microwaves by radar) may cause the formation of plasma vortexes in the air (Zou, 1995) whose behaviour is not possible to predict but which might constitute some danger to the conduct of any flight. It is not expected in general that very low-energy electromagnetic emission alone is able to affect the electronics aboard airplanes. What is important to consider here is that, according to the specific frequency range, it can interfere directly with brainwaves of the pilots and of passengers and often cause altered consciousness states and severe hallucinations. Of course this effect might render the conduct of any flight extremely dangerous. The physiologic effect of these waves, also in connection with pulsed magnetic fields, is now well known and demonstrated in biophysical and neurophysiologic laboratories (Persinger, 2000).[32]

"Plasma vortexes in the air" are reminiscent of the Humvee-mounted nonlethal weapon known as the *pulsed energy projectile (PEP)* that emits a short intense pulse of laser energy that creates a rapidly expanding ball of plasma that can deliver a pressure wave to stun, disable, or knock a target off their feet (or out of the air?). This EM radiation affects the nerve cells and causes excruciating pain—all from a mile away. The technology's early name was *pulsed impulsive kill laser.*

A seminal 2007 essay by targeted individual Carolyn Williams Palit alerted Internet-targeted individuals about "gas plasma generation due to the heating of chemtrails by electromagnetics":

The technical names for vertical and horizontal plasma columns are columnar focal lenses and horizontal drift plasma antennas. Various sizes of gas plasma orbs are associated with this technology. These orbs can be used as transmitters and receivers because they have great, refractory and optical properties. They also are capable of transmitting digital or analog sound. Barium, in fact, is very refractive — more refractive than glass.

What does that mean? Someone or someones are very involved in unconstitutional, domestic spying and the entrained plasma orbs carried on electromagnetic beams can be used for mind control programming. The satellites can be programmed to track and monitor various frequencies on different parts of your body. These electromagnetic beams carrying the gas plasma orbs stick due to magnetic polarity and frequency mapping and tracking

32 Massimo Teodorani, Ph.D., "Spherical Unidentified Anomalous Phenomena: Scientific Observations and Physical Hypotheses, Danger Evaluation for Aviation and Future Observational Plans." National Aviation Reporting Center on Anomalous Phenomena (NARCAP) Research Associate, October 2009. www.narcap.org/Projsphere/narcap_ProjSph_2.4_MaxTeo.pdf

to people's eyes, ears, temples, and private parts. A beam with entrained orbs carries pictures in each orb just like the different frames in a movie. It is a particle beam that is also a frequency weapon.[33]

SIGHTINGS

From strange cloud formations to orbs behaving intelligently, the sky theater is now pregnant with puzzling spectacles. From Belo Horizonte, Brazil, we hear of three small round clouds that appeared in a matter of seconds in a blue sky with two swirling clockwise as the other became a rising "cigarette puff," then took off, followed by the other two like "smoke cylinders."[34] These "clouds" sound more like vectored crafts in a plasma disguise.

It is difficult to fathom just what we are seeing in our increasingly simulated world. Sky observers point to exotic triangular crafts hiding in the plasma cloud cover, *sometimes even creating their own plasma cloud cover*. (Shades of operation #7: the detection and obscuration of exotic propulsion technology.) The TR-3B (Astra), the highly classified Aurora (SR-91), and the boomerang Northrop Grumman stealth B-2 Spirit bomber sighted since 1998 flying silent and low (500–1,200 feet) are just three possible crafts *with classified hover capability*. At his "Starburst Forum: Electrogravitics and Field Propulsion," P. LaViolette talks about the B-2's "silent field propulsion mode" witnessed over Boston:

> . . .the B-2 in flight scoops air into its engine cowlings and this air is there seeded with ions to create a multimillion volt potential which the B-2 used to power its electrokinetic ion drive. The fields induce asymmetrical electrostatic forces on the B-2 which propel it forward silently. This is an overunity drive: unbalanced forces create thrust which scoops air, which creates volt potentials.[35]

A 2012 Maryland sighting confirmed a "bright violet glow around the fighter that the B-2 charges its wing leading edge with to a multimillion volt potential."[36]

And is our plasma cloud cover cloaking an order of exotic propulsion craft beyond our own? And what of the orbs that move non-linearly *and pulsate* like organisms?[37]

---

33 Carolyn Williams Palit, "What Chemtrails Really Are." *Rense.com*, November 9, 2007.

34 Ronald Swait, July 8, 2016, Facebook.

35 P. LaViolette, "B-2 Sighting confirms field propulsion theory." *Starburstfound.org*, April 18, 2011. Also see a Sirius Disclosure interview with LaViolette, "The Underdogs of New Energy," December 27, 2013.

36 P. LaViolette, "Confirmation that the B-2 charges to high voltage its wing leading edge." *Starburstfound.org*, April 20, 2011.

37 Clifford E. Carnicom, "Orbs Require Consideration." *Carnicominstitute.org*, March 14, 2004.

What is that triangular shape hiding in the cloud?

The arching wing . . .

Something triangular moving at a fast clip
and leaving a plasma trail?

Plasma implosion?

All four photos are by Sean Gautreaux, author of *What Is In Our Skies Vol. 1 Diagrams: The Study of Cloaked Cloud Craft Above New Orleans*, 2014.

Physicist David Bohm experimented with electrons in plasma and observed *how like a biological organism plasma behaved*. In fact, the term *plasma* was coined for its resemblance to living blood cells. Because of this resemblance (including helical structures resembling DNA), "complex, self-organized plasma structures exhibit all the necessary properties to qualify them as candidates for inorganic living matter. They are autonomous, they reproduce, and they evolve."[38]

Yes, *inorganic* living matter, not carbon-based living matter, and it seems conscious, and communicates telepathically.

If plasma life forms were present in the Earth's highly ionized atmosphere 4.6 billion years ago—heated gas, chemicals, electric storms—could plasma-based life forms from space be drawn to our now-ionized atmosphere? A 2003 experiment written up in "Minimal Cell System Created in Laboratory by Self-Organization" found that the rhythmic "inhalation" of the nucleus of

38 V.N. Tsytovich, "From Plasma Crystals and Helical Structures Towards Inorganic Living Matter." *New Journal of Physics*, August 2007.

laboratory-created plasma spheres (orbs) *mimics* the breathing process of living systems and results in pulsations.[39]

These are strange, exciting, and dangerous times . . .

39 Erzilia Lozneanu and Mircea Sanduloviciu, "Minimal Cell System Created in Laboratory by Self-Organization"; published in *Chaos, Solitons & Fractals*, an interdisciplinary journal of Nonlinear Science, and Nonequilibrium and Complex Phenomena, Vol. 18, Issue 2, October 2003. Thanks to Jay Alfred ("Plasma Life Forms — Spheres, Blobs, Orbs and Subtle Bodies," *ezinearticles.com*, October 15, 2007) for a most stimulating discussion of plasma.

THE "GLOBAL VILLAGE" AS BATTLESPACE

. . .Intelligence influence matter-scientists' magic
bank telephone war investment Usury Agency
executives jetting from McDonnell Douglas to General Dynamics
over smog-shrouded metal-noised treeless cities
patrolled by radio fear with tear gas, businessman!
— Allen Ginsberg, "Pentagon Exorcism," 1967

The land shall be crisscrossed by a giant spider's web during the end of the Fourth
World and the beginning of the Fifth World.
— Hopi prophecy

The essential American soul is hard, isolate, and a killer.
— D.H. Lawrence (1885–1930)

CHAPTER FIVE

The Revolution in Military Affairs (RMA)

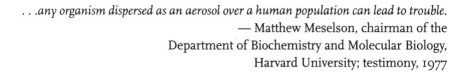

. . .any organism dispersed as an aerosol over a human population can lead to trouble.
— Matthew Meselson, chairman of the
Department of Biochemistry and Molecular Biology,
Harvard University; testimony, 1977

National socialism is nothing but applied biology.

— Rudolf Hess

In 1994, Carol Marshall's book *The Last Circle* debuted on the Internet, one of the first bold investigations into what other studies had written off as "conspiracy theory," in particular the 1980 military document "From PSYOP to Mind War: The Psychology of Victory" by Colonel Paul Vallely and Major Michael A. Aquino. Both officers had been decorated in Vietnam, but Aquino had deep ties to Nazism and was a practicing Satanist with top security clearance as a psychological warfare officer.[1] Marshall posits that the Vallely-Aquino paper virtually spawned the entire Revolution in Military Affairs (RMA) and its asymmetric approach to warfare, military operations other than war (MOOTW),[2] and viewing mainstream media as potential "Mind War operatives."[3]

In a follow-up U.S. Air Force Academy study, we learn more about why warfare must become "multidimensional":

---

1 Whittinger, John, and Bill Wallace. "Army Says Constitution Lets Satanist Hold Top Secret Job." *San Francisco Chronicle*, November 3, 1987.

2 *Peace support operations (PSO)* in the UK.

3 Vallely, Col. Paul and Maj. Michael A. Aquino. ""From PSYOP to Mind War: The Psychology of Victory." 7th Psychological Operations Group, U.S. Army Reserve, Presidio of San Francisco, 1980.

The spectrum of conflict, as portrayed in most readings, is single-dimensioned, linear, and continuous. . .[It] must be replaced by a multidimensional model, perhaps even nonlinear and discontinuous. Topological mathematicians would call it a manifold; hence, the name conflict manifold. Its primary characteristic is multidimensionality . . .[4]

Both Vallely and Aquino knew that multidimensional "Mind War" would have to include artificially produced ELF waves to impact minds:

ELF waves are not normally noticed by the unaided senses, yet their resonant effect upon the human body has been connected to both physiological disorders and emotional distortion. Infrasound vibration [up to 20Hz] can subliminally influence brain activity to align itself to delta, theta, alpha, or beta wave patterns, inclining an audience toward everything from alertness to passivity. Infrasound could be used tactically, as ELF waves endure for great distances; and it could be used in conjunction with media broadcasts as well. . .An abundance of negative condensation nuclei [atmospheric ions] in ingested air enhances alertness and exhilaration, while an excess of positive ions enhances drowsiness and depression. Calculation of a target audience's atmospheric environment will be correspondingly useful.[5]

But who other than Vallely and Aquino knew in 1980 that shifting to a "multidimensional model" of warfare would threaten the entire biosphere and the human mind?

Let's begin with a quick tour of duty through the U.S. military as it is today (and not as it is in Hollywood war films), how it and its entrenched agencies and defense contractors have militarized the United States (and NATO) for more than a half-century, and how the release of the 1,176-page "Department of Defense Law of War Manual" for the four military branches appears to be an all-in-one legal guide to superseding international human rights treaties and the U.S. Constitution.[6] In fact, the military-industrial-intelligence complex is now synonymous with the word *military* and has metastasized into a morphing Hydra with hundreds of shape-shifting heads.

In the spirit of resistance to the poisoning of our atmosphere and planet with a blizzard of chemicals, polymers, sensors, microprocessors, and genetically engineered 'bots in the name of full spectrum dominance "force multiplication," we begin with the U.S. Army Chemical Corps, renamed in 1986 the Dragon

---

4 McGarth, Lt. Col. John T., U.S. Air Force Academy. "The Conflict Manifold." *Air University Review*, May-June 1981.

5 Vallely and Aquino, "From PsyOp to Mind War."

6 Tom Carter, "The Pentagon's Law of War Manual: Part One." *World Socialist Web Site*, 3 November 2015.

Soldiers, whose regimental insignia exclaims *Elementis Regamus Proelium*, Latin for *Let us rule the battle by means of the elements*, their logo changed from the old cobalt blue benzene ring superimposed on two gold retorts crossed like swords to a gnarled tree stump and the chemical-breathing Green Dragon of the alchemists.[7] The Dragon Soldiers constitute a military Brotherhood.

Then there is the USAF Meteorological and Space Environmental Services and Air Force Weather Agency out of Offut Air Force Base in Nebraska (headquarters of Strategic Air Command) and Asheville, North Carolina, and the Air Force Combat Weather Center at Hurlburt Field, Florida and Scott Air Force Base in southern Illinois supporting Combat Weather Flights and Battlefield Weather Teams.

Policy decisions about "force multiplier" weather may rest with the USAF Meteorological and Space Environmental Services, a military/civilian phalanx of 4,100 active duty and reserve military and civilian personnel serving under the Director of Weather (AF/A30-W).

> The majority of AF weather personnel are focused on two distinct yet related functions: characterizing the past, current, and future state of the natural environment, and exploiting environmental information to provide actionable environmental impacts information directly to decision-makers.
>
> AF weather is organized in a 3-tier structure to maximize capabilities that can be accomplished in the rear area via "reachback" technology. This minimizes forward presence on the battlefield, making a "light and lean" presence consistent with the overall USAF vision for contingency operations in the 21st century.[8]

The field-operating Air Force Weather Agency (AFWA) oversees global-scale collection and production of weather. Reporting directly to the Air Force Director of Weather, AFWA plans, programs, and fields standard weather systems, and collects, analyzes, predicts, tailors, and integrates weather data, providing "timely, accurate, relevant, and consistent terrestrial and space weather products necessary to effectively plan and conduct military operations at all levels of war."[9]

> AFWA consists of a functional management headquarters; the 1st Weather Group (1 WXG) with three subordinate CONUS [continental US] operational weather squadrons (OWS); the 2nd Weather Group (2WXG), which operates three squadrons, two at Offut and one at Asheville, NC; the Air Force Combat

7 Albert J. Mauroni, *Where are the WMDs? The Reality of Chem-Bio Threats on the Home Front and Battlefront.* Maryland: Naval Institute Press, 2006.

8 "Department of Defense Weather Programs," www.ofcm.noaa.gov/fedplan/fp-fy10/pdf/3Sec3c-DOD.pdf. Thanks to Ron Angell for directing me to this site.

9 Ibid.

Weather Center at Hurlburt Field, FL, which supports the Combat Weather Flights and Battlefield Weather Teams through investigation, development, integration, exploitation, and training across new and existing systems and processes; as well as five detachments and operating locations. The 1 WXG commands three operational weather squadrons that conduct weather operations in support of Total Force Army and Air Force operations in the CONUS...[10]

These military "weather forces" are tasked with exploiting the weather for air, space, cyberspace, and ground operations that serve the four military branches and Intelligence Community. Certainly, they are concerned with the chemical and electromagnetic technology that delivers "force multiplication," but not with the decision-making.

From our mainstream media, we hear a lot about agencies like the EPA (Environmental Protection Agency), NOAA (National Oceanic and Atmospheric Administration), USGS (U.S. Geological Survey) and NASA, but not about the agencies pivotal to the creation and maintenance of our ionized atmosphere, including intelligence agencies. Here is how I characterize NASA:

- *National Aeronautics and Space Administration (NASA)* a psychological intelligence operation tasked with confusing public knowledge of covert space defense, space weapons, unmanned satellite and rocket launches, and plasma research.

- *NASA's Ames Research Center*, whose Search for Extraterrestrial Intelligence Project (SETI) is a cover story for "Human Factors" (PSY-Warfare Division), e.g., MK-ULTRA; at Sunnyvale, CA. (e.g., Project Snowbird, Project Aquarius, Project Tacit Rainbow)

More important by far to the geoengineering matter at hand are two intelligence agencies: the *National Reconnaissance Office (NRO)* and the *National Geospatial-Intelligence Agency (NGA)*.

The *National Reconnaissance Office (NRO)* flies, controls, and collects information from spy satellites of the IC and DoD, monitors exotic propulsion traffic, and coordinates energy beam weapons, all from the Pentagon basement and Dulles Airport, Virginia.

The *Department of Homeland Security (DHS)* works with the DoD and the ODNI through the NRO. As part of the Intelligence Community (IC), the NRO employs private security employees to fly American spy satellites for the DoD and ODNI. Investigative journalist Tim Shorrock in his 2008 book *Spies for Hire: The Secret World of Intelligence Outsourcing* writes that 95 percent of

---

10 Ibid.

NRO employees are defense and security contractors working for the military-industrial-intelligence complex: "With an estimated $8 billion annual budget, the largest in the IC, contractors control about $7 billion worth of business at the NRO, giving the spy satellite industry the distinction of being the most privatized part of the Intelligence Community."[11]

The career of Donald Kerr, Jr. is an excellent example of the NRO overlap of intelligence with physics. Thanks to his Cornell University BSEE in 1963, MS in Microwave Electronics in 1964, and Ph.D. in plasma physics in 1966, Kerr became director of the Los Alamos National Laboratory (1979–1985), then deputy director of Science and Technology at the Central Intelligence Agency (2001–2005), director of the NRO (July 2005–October 2007), and deputy DNI (July 11, 2007 to 2009). In fact, plasma physicist intelligence officer Kerr took the helm of the NRO just in time for Hurricanes Dennis (July 8, 2005, Category 4), Emily (July 14, 2005, Category 4), Irene (August 12, 2005, Category 3), Katrina (August 29, 2005, Category 5), and Rita (September 20, 2005, Category 5).

The National Geospatial-Intelligence Agency (NGA) (2003)[12] develops imagery and map-based intelligence (IMINT) and is horizontally integrated with the NSA, one being the eyes, the other being the ears. The NGA employs 15,400 people, many working in its four-football-fields-long headquarters south of Washington, D.C. In conjunction with the U.S. Air Force, it cuts deals with telecom corporations for dual-use tech[13].

In the upper atmosphere and space, NGA-embedded semiconductor chips and billions of photosensors are busy converting infrared radiation into electronics and sending a voltage signal to form images of objects to computer monitors hundreds of miles below. On the ground, the NGA's Mobile Integrated Geospatial-Intelligence System (MIGS) mounted on a Humvee collects information from weather disasters and street protests, then beams them up to satellites. If citizens move off the grid or out of range, an NGA Agent may arrive at the door with a census hand device to beam up the citizen's new coordinates for the Digital Point Positioning Database that maintains real-time surveillance. Like its NSA partner, the NGA makes sure everyone is plugged into the global information grid (GIG)—"horizontally integrated," even to the extent of responding to weather emergencies with its Domestic Mobile Integrated Geospatial System (DMIGS) mounted on a Humvee.

NGA's eyes are focal plane array cameras, ultra-sensitive to absorb infrared radiation instead of optical light. Semiconductor chips embedded with millions

---

11 Tim Shorrock, Spies for Hire: *The Secret World of Intelligence Outsourcing.* Simon & Schuster, 2008.

12 See James Bamford's article "The Multibillion-Dollar US Spy Agency You Haven't Heard of," *Foreign Policy,* March 20, 2017.

13 Mohana Ravindranath, "Mission Possible: A Spy Agency Builds A Tech Scout Network." *Nextgov,* October 19, 2016.

of photosensors convert the infrared radiation into electronics and send a voltage signal to form an image of the object hundreds of miles below. The NGA oversees the versatile laser radar (LiDAR) weapon for constructing 3D renderings. It oversees eye-in-the-sky drones with high-res cameras, like ARGUS-IS and its "persistent stare" feature, the equivalent of one hundred Predator drones tracking everything that moves and storing one million terabytes of data per day.

Add NORTHCOM and CIFA, DHS and Department of Justice (DOJ) fusion centers throughout the country and the Terrorist Surveillance Program (TSP) renamed PRISM,[14] and the Homeland looks more and more like a COINTELPRO police state.

The U.S. Geospatial Intelligence Foundation (USGIF) is a nonprofit corporation with above top-secret classification funded by the largest NGA and DHS contractors. USGIF GEOINT Symposiums are open windows into the highest levels of U.S. intelligence thinking.[15]

SHADOW GOVERNMENT / THE DEEP STATE

Listening posts, traffic cameras, surveillance aircraft like the U-2 Dragon Lady, telephone microphones, and satellite linkups with state of the art software like British subsidiary BAE Systems' GXP 3D mapping and GOSHAWK big picture scrutiny are on constant real-time paranoia alert for snipers, weather disasters, forest fires, snapping Homeland photos to postage-stamp clarity with MASINT (measures and signatures intelligence) seeing through walls and ceilings with infrared.

Many government agencies have been subsumed by a criminal network of graft, fraud, blackmail, corruption, contractor overcharges, health and safety violations. Agencies have become as "dual use" as the technologies they cover for during our quiet descent into a Transhumanist future on a lockdown planet. This illegal dual-use governance *apparatchik* is now termed the Deep State. Technocracy, corporatocracy, military-industrial complex—its many terms attempt to elucidate its comprehensive nature. Layers of deceit and corruption surrounding weaponized technology have erased the balance of powers defined in the Constitution. Since the National Security Act of 1947, the aftermath of Operation Paperclip, and the secretive Cold War, governance has moved from the chambers of elected officials to the back rooms and star chambers of conspirers and malefactors with no love for republics, peoples or their humanity.

---

14 The existence of PRISM was leaked by NSA contractor Edward Snowden and published by *The Guardian* and *Washington Post* on June 6, 2013. PRISM is a government codename for data collection known officially as US-984XN under the U.S. Foreign Intelligence Surveillance Court pursuant to the 1978 Foreign Intelligence Surveillance Act (FISA) footnoted earlier.

15 Shorrock, *Spies for Hire.*

The revolving door between government and corporations dedicated to new battlespace profits and careers are well oiled and spinning fast, the *Defense Science Board (DSB)* being but one example among many. The DSB serves as DoD intermediary between weapons needs and the physical sciences, including advising government officials on the science and technology of torture. Founded in 1956 to maintain a flow of war and profits for defense contractors, DSB board members hail from Aerospace Corporation, Serco, Bechtel, General Dynamics Advanced Information Systems, SAIC, Lockheed Martin, Massachusetts Institute of Technology, MITRE, Networks and Information Integration, ViaSat Inc., Loral Space & Communications, Sandia National Laboratories, and Georgia Institute of Technology. (No conflict of interest there.)

Continuing to examine the Deep State through the lens of Project Cloverleaf and geoengineering, I first want to thank the anonymous author of "Shadow Government, Structural Analysis,"[16] as well as Harald Kautz-Vella who detailed corporations and organizations according to the IP addresses they left behind when monitoring human rights organizations collecting data about Cloverleaf aerosol spraying.[17] As the anonymous author put it, "The Shadow Government is a creature of a powerful elite who need now fear being dominated by an instrument of their own creation." My sketch continues beyond the NRO and NGA to other Intelligence and Military players making liberal use of Defense Contractor players so as to avoid public scrutiny. I have placed the Executive players they are all answerable to at the top. I cannot discuss all but do discuss those most important or most interesting. Feel free to investigate and revise as you correct just how the "Octopus" is put together. (For example, I decided that the Department of Homeland Security is more like Intelligence than Executive.)

EXECUTIVE

Executive usual suspects are organized top-down as per decision-making power over the United States. It is not that government departments and agencies (DOT, DOJ, etc.) are synonymous with shadow government but that shadow players are running them.

Bilderberg Group

Council on Foreign Relations (CFR)

Trilateral Commission

---

16 truthcdm.com/shadow-government-structural-analysis/#sthash.LKQ3ZUCz.wM9QUnyy.dpbs

17 Cara St. Louis and Harald Kautz-Vella, *Dangerous Imagination, Silent Assimilation.* Self-published, 2014.

Federal Reserve System

National Security Council (NSC)

NSC-5412-2 Special Group directs black [covert] operations

PI-40 Subcommittee (40 Committee)

Joint Chiefs of Staff (JCS) Special Operations - covert directorate that implements NSC 5412 Committee directives by utilizing U.S. Special Forces Command.

U.S. Department of the Treasury (DOT)

U.S. Department of Justice (DOJ) - partially self-financed by confiscation of money and valuables from "targets of investigation."

INTELLIGENCE

General Accounting Office (GAO) - investigative arm of Congress that oversees audits.

Office of the Director of National Intelligence (ODNI) (2004) - oversees sixteen intelligence agencies and is itself overseen by the NSA.

U.S. Department of Homeland Security (DHS)

National Security Agency (NSA)

National Reconnaissance Office (NRO)

National Geospatial-Intelligence Agency (NGA)

Central Intelligence Agency (CIA) - primarily self-financing via the international drug trade and "front" businesses like InQTel for off-the-books operations, purchase of exotic munitions, and strategic bribes.

CIA's Directorate for Science and Technology (DS&T) - gathers intelligence for "Weird Desk."

NSA's Central Security Service and *CIA's Special Security Office* spy on spies and conduct special ops that cannot be entrusted to line intelligence.

Federal Bureau of Investigation, Counter Intelligence Division

Drug Enforcement Agency (DEA) conducts surveillance and interdiction of drug smuggling operations, unless exempted under "National Security" waivers.

National Security Analysis Center (NSAC) – data-mines Americans and reports to the DOJ; grew out of the FBI's Foreign Terrorist Tracking Task Force (F-Tri-F) after 9/11. (Total Information Awareness?)

Department of Energy Office of Intelligence and Counterintelligence conducts internal security checks and external security threat countermeasures through contractors.

Defense Intelligence Agency (DIA) coordinates intel gathered from Army, Navy, Marines, Air Force, Coast Guard and Special Forces; provides counter-threat measures and security at ultra-classified installations; Pentagon, VA, Fort Meade, MD.

U.S. Army Intelligence and Security Command (INSCOM) - psychological and psychotronic warfare (PSYOPS), para-psychological intelligence (PSYINT), and electromagnetic intelligence (ELMINT), Ft. Meade, MD.

U.S. Navy Office of Naval Intelligence (ONI) - intel affecting naval operations; works with

U.S. Air Force Office of Special Investigations (AFOSI) - intel affecting aerospace operations, surveillance, and coordination with NRO interdiction operations, Fort Meade, MD; works with

NASA Intelligence gathers intel relating to space flights, sabotage, astronaut and reconnaissance satellite encounters.

Air Force Special Operations Security Forces (SOSFS) - NSA/USAF joint intelligence operations unit dealing with possible threats to aerospace operations; McDill AFB, Orlando FL.

Defense Security Service (DSS) - investigates people and situations deemed a possible threat to any operation of the Department of Defense.

Naval Criminal Investigative Service (NCIS) - investigates threats to Navy/Marine operations; manages Navy security programs.

Twenty-Fifth Air Force - surveillance and interdiction of threats to the security of Air Force electronic transmissions and telemetry, and to the integrity of electronic countermeasure (ECM) warfare equipment; Lackland AFB, TX.

Federal Protective Service (FPS) - intel relating to threats against federal property and personnel.

The 2017 intelligence budget (not counting the "black budget") was $70.3 billion. *The National Security Agency (NSA)* was relatively unknown to the American people for its first few decades of existence (like the NRO). WikiLeaks and

Edward Snowden have changed all that. The NSA's dominant role in surveillance is now known, but its pivotal role under the Space Fence remains hidden. While the occasional press release about a 1999 memorandum to NSA employees banning cute big-eyed Furby stuffed owls able to record and repeat words or phrases heard[18] is amusing, the black Humvees with state-of-the-art electronics idling outside NSA headquarters like Plutonian steeds are not, nor does the chief psychologist standing by for the Associate Director for Security and Counterintelligence, ready to interrogate NSA employees in scenes reminiscent of Arthur Koestler's dystopic novel *Darkness at Noon* about Soviet Russia.

The military nature of the NSA. NSA directors are generally U.S. Navy officers, and the NSA's venture capital firm Paladin Capital Group finances high-tech start-ups that look a lot like the CIA venture capital firm In-Q-Tel. When Air Force Lieutenant General Michael Hayden became NSA director, he publicly announced Project Groundbreaker and formalized the fact that the NSA was run by the military-industrial-intelligence complex—2,690 revolving-door military contractors like Computer Sciences Corporation, Logicon (a Northrop Grumman subsidiary), Conquest Inc., SAIC, Boeing, Booz Allen Hamilton, Telcordia Technologies, etc.

The NSA has the power to bypass the Joint Chiefs of Staff and give direct commands to signals intelligence (SIGINT) military units while farming out intelligence functions: Counterterrorism under Signals Intelligence Division (SID) to Fort Gordon, Georgia; electronics intelligence (ELINT) to Buckley Air Force Base in Aurora, Colorado, where the super-secret Aerospace Data Facility answers only to the Pentagon; SIGINT to Medina Regional SIGINT Operations Center (RSOC) at Lackland Air Force Base in San Antonio, Texas; etc.

The NSA runs all U.S. intelligence agencies. The Intelligence Reform Act of December 2004 consolidated America's sixteen intelligence agencies into one Intelligence Community (IC) run out of the Office of the Director of National Intelligence (ODNI) but reporting directly to the NSA.

The NSA has the entire telecommunications industry in a vise grip. As a "dual-use" defense contractor, the telecommunications industry makes billions from NSA and military contracts and billions from consumers' obsession with dual-use cell phones, towers, Internet, etc. Verint Systems, Verizon, and the NSA are good examples of the unconstitutional revolving door favored by the military-industrial-intelligence complex:

---

18 "Furby toy or Furby spy?" *BBC*, January 13, 1999.

Skyway Global LLC, the St. Petersburg, FL company that owned the DC-9 airline busted in Mexico carrying 5.5 tons of cocaine, made its headquarters in a 79,000 sq. ft. building owned by Verint Systems (NASDAQ: VRNT), a foreign telecommunications company with a contract to wiretap the U.S. for the NSA through the communication lines of Verizon, which handles almost half of all landline and cell phone calls in the U.S. Verint's founder and CEO, Jacob "Kobi" Alexander, is a former Israeli intelligence officer who is today a fugitive from justice living in Namibia, where he has for several years been fighting extradition to the U.S. On Verint's Board of Directors is Lieutenant General Kenneth A. Minihan, former director of the NSA, which has led to speculation that the company today is a joint NSA-Mossad operation.[19]

Before becoming the thirty-first chairman of the FCC in 2013, Tom Wheeler was CEO for the wireless industry trade group Cellular Telecommunications & Internet Association (CTIA). Wheeler's job is to maintain FCC standards and assure "back doors" for whatever the military and NSA need. No conflict of interest here; move along.

Too little too late, legal maneuvers to undo or limit the NSA's gains over the past seventy years are underway. Two lawsuits challenge illegal NSA dragnet surveillance programs: the class action *Shubert v. Obama* (2006), and the Electronic Frontier Foundation's *Jewel v. NSA* (2008). Thus far, the government's invocation of the state secrets privilege has been rejected, thanks to the Foreign Intelligence Surveillance Act.[20] Oklahoma, California, Indiana, and Washington State introduced legislation in 2014 under the 4th Amendment Protection Act to cut off water, electricity, gas and all services provided by state vendors to the NSA—Washington State examples being Cray Inc., which builds NSA supercomputers, and an NSA listening station is at the Army's Yakima Training Center. In HB 2272,[21] Washington State adds prohibiting the use of unconstitutionally gathered data in state court, blocking public universities[22] from accepting NSA research monies or recruiting agents, and de-incentivizing corporations from stepping in to counter such state measures.[23]

The Department of Homeland Security (DHS) is a civilian agency answering to the Office of the Director of National Intelligence (ODNI). DHS oversees fusion centers, black projects, COG operations (including federal detention camps

---

19 Hopsicker, Daniel. "NSA links to St. Petersburg FL Drug Ring." *Mad Cow Morning News,* November 16, 2013.

20 "Federal Judge Allows EFF's NSA Mass Spying Case to Proceed." Electronic Frontier Foundation, July 8, 2013.

21 HB 2272 was "referred to the judiciary" in 2014 and by 2015 had disappeared into another HB 2272, a bill on "constitutional basic education obligation." Shades of HR 2977 . . .

22 166 schools nationwide partner with the NSA as "Centers of Academic Excellence."

23 "Washington State Bill Proposes Criminalizing Help to NSA, Turning Off Resources to Yakima Facility." Tenth Amendment Center, 2014.

often located on "closed" military bases or Bureau of Land Management lands, etc.). DHS covers domestic and international terrorism.

An example of international DHS: In Washington, D.C. on Friday, April 13, 2007, the Swedish Defence minister and DHS director Michael Chertoff signed the "Agreement Between the Government of the Kingdom of Sweden and the Government of the USA in the Area of Scientific and Technical Cooperation For the Protection of National Security,"[24] giving Sweden access to billions of NSA dollars, as similar agreements with Canada, Mexico, Australia, the UK, and Singapore have—dollars funneling into Swedish biotech corporations, institutions, universities, laboratories, etc. Such DHS "agreements" are about R&D in nuclear, biological / chemical, underwater techniques, border control, sensors and microprocessors, search and surveillance. At the end of 2011, DHS Secretary Janet Napolitano traveled to Sweden to meet with ministers of State and Justice to sign the Preventing and Combating Crime Agreement for the fluid exchange of biometric and biographic data "to bolster counterterrorism and law enforcement efforts."[25] The U.S. has similar agreements with at least twenty-two nations. Biometrics makes the global tracking and targeting of individuals easy.

DHS may have absorbed the National Applications Office (NAO, 2007–2009) previously headed up by the DHS chief intelligence officer and subject to the NRO.[26] No more warrants! the NAO lobby boasted. In 2007, the NAO was created by ODNI and Booz Allen Hamilton of McLean, Virginia to act as liaison for "disaster relief" agencies like FEMA, CIFA (Counterintelligence Field Activity), NORTHCOM, and domestic law enforcement; oversaw collection and dispersal of intelligence from all foreign and domestic listening posts, fusion centers, surveillance aircraft, and satellite image / video intelligence (IMINT). (In 2008, the shadowy Carlyle Group tucked Booz Allen Hamilton into its portfolio for $2.54 billion.)

MILITARY

Strategic Defense Initiative Organization (SDIO) under Reagan was replaced by the *Ballistic Missile Defense Organization (BMDO)* under Clinton and then renamed the Missile Defense Agency (MDA) in 2002. The SDIO and BMDO coordinated R&D and deployment of "Star Wars"

---

24 "Agreement Between the Government of the Kingdom of Sweden and the Government of the USA in the Area of Scientific and Technical Cooperation For the Protection of National Security." www.dhs.gov/xlibrary/assets/agreement_us_sweden_sciencetech_cooperation_2007-04-13.pdf.

25 "Readout of Secretary Napolitano's Meeting with Swedish Deputy Foreign Minister Frank Belfrage," Office of the Press Secretary, DHS, December 16, 2011.

26 Or parceled out to the NRO and/or the NGA.

electromagnetic pulse, optical energy weapons like the killer laser, particle beam, plasmoid, and other advanced aerospace weapons. (E.g., Project Cold Empire, Project Cobra Mist, Project Cold Witness)

U.S. Air Force Space Command (AFSC) under U.S. Strategic Command (USSTRATCOM) at Peterson AFB; coordinates the development of future technology for operations and space wars, including cyber operations and cyberwarfare; operates spy satellites for DoD agencies (NSA, NASA, NRO, DIA, etc.).

North American Aerospace Defense Command (NORAD), joint operation with Canada; aerospace warning, air sovereignty, defense, and nuclear survivable space surveillance and war command from Peterson AFB and deep inside Cheyenne Mountain, Colorado Springs, CO.

Department of Energy (DOE) - R&D of specialized nuclear weapons, particle and wave weapons including EMP, laser, particle beam and plasmoid research; high-energy invisibility "cloaking" technology, etc.

Defense Advanced Research Projects Agency (DARPA) coordinates the application of latest scientific findings to the development of new generations of weapons.

Defense Special Weapons Agency (DSWA) currently concentrates on fusion-powered, high-energy (HPM) optical energy weapons like the particle beam, X-ray laser, and EM force field weapons development and deployment.

Lawrence Livermore National Laboratories and *Sandia National Laboratories-West (SNL-W)* are involved in nuclear warhead "refinements," development of new transuranic weapons and energy applications, antimatter weapons, laser/maser applications, and teleportation experiments; at the Russian-nicknamed "City of Death," Livermore, CA.

Idaho National Laboratory (INL) houses numerous underground facilities in an immense desert installation complex larger than Rhode Island; security provided by its own secret Navy Base; involved in nuclear, high energy electromagnetic, and other research.

Los Alamos National Laboratories are the premier research labs for nuclear, subatomic particle, high magnetic field, exo-metallurgical, exobiological and other exotic technologies research; Los Alamos County, NM.

Sandia National Laboratories (SNL) and *Phillips Air Force Laboratory* are sequestered on Kirtland Air Force Base/Sandia Military Reservation;

conduct translations of theoretical and experimental nuclear and "Star Wars" weapons research done at Los Alamos and Lawrence Livermore National Laboratories into practical, working weapons; Albuquerque, NM.

Tonopah Test Range is the Sandia's DOE weapons testing facility for operational testing of "Star Wars" weapons in realistic target situations; adjacent to classified stealth and cloaked aerospace at the Groom Lake Base (USAF/DOE/CIA Area 51) and Papoose Lake Base (S-4); Nevada Test Site/Nellis AFB Range in Tonopah, NV is a thirty-level extreme security facility where classified aerospace craft are tested and flown, e.g., the Aurora hypersonic spy plane, the Black Manta (TR-3A) stealth fighter follow-on to the F-117A, the Pumpkinseed hyper-speed unmanned aerospace reconnaissance vehicle, anti-gravitational craft, etc. Haystack (Buttes) USAF Laboratory, Edwards AFB, CA.

DEFENSE CONTRACTORS

A critical component of the RMA military is its revolving door with giant lucrative defense contractor corporations not answerable to Congress. In fact, of the three components of the term "military-industrial-intelligence complex," defense contractors encompass all three. Transnational defense contractors are almost a synonym for the Intelligence Community. As Ralph Nader points out in a mind-boggling article how year after year the Pentagon a.k.a. Department of Defense goes unaudited ("its books are a mess") and is "beyond anybody's control, including that of the Secretaries of Defense, their own internal auditors, the President, tons of GAO audits publically [sic] available, and the Congress."[27] It appears that the military and its private contractors are indeed "above the law." In Afghanistan, $150 million villas were built for corporate contractors, and massive cost overruns are typical but the money keeps flowing:

> In the final analysis, the principal culprits, because they have so much to lose in profits and bonuses, are the giant defense companies like Lockheed Martin, Boeing, General Dynamics, Raytheon, Northrop Grumman and others that lobby Congress, Congressional District by Congressional District, for more, more, more military contracts, grants and subsidies. President Eisenhower sure knew what he was talking about. Remember, he warned not just about

27 Ralph Nader, "Uncontrollable—Pentagon and Corporate Contractors Too Big to Audit." *Common Dreams*, March 17, 2016.

taxpayer waste, but a Moloch eating away at our liberties and our domestic necessities."[28] (Emphasis added.)

A final list by no means comprehensive:

United Nuclear Corporation near Gallup NM. Military nuclear applications.

Defense Industry Security Command (DISCO) conducts intelligence operations within and on behalf of the civilian defense contractor corporations engaged in classified research, development, and production.

Stanford Research Institute, Inc. — intelligence contractor involved in psychotronic, para-psychological and PSY-WAR research.

RAND Corporation — CIA front involved in Intelligence projects, weapons development, and underground bases development.

Mitre Corporation / The Jason Group — Elite weapons application scientists developing cutting-edge weapons for DARPA. The Jason Group operates under the cover of the Mitre Corporation.

Edgerton, Germeshausen & Grier, Inc. (EG&G) / URS Federal Services — NSA/DOE contractor involved in "Star Wars" weapons development, fusion applications, security for Area 51 and nuclear installations, etc.

Science Applications International Corporation (SAIC) — black projects information systems contractor

Lockheed Martin — black budget aerospace; purchased Learjet in 2015.

Raytheon — classified projects; purchased Hughes Aircraft in 1997.

Booz Allen Hamilton — purchased by Carlyle Group in 2008.

Boeing Aircraft absorbed McDonnell Douglas in 1997.

General Electric — electronic warfare (EW) weapons systems; nuclear.

Reynolds Electronics Engineering — CIA and DoD contractor *Honeywell* purchased by AlliedSignal in 1999, which kept the Honeywell name.

Northrop Grumman purchased TRW in 2002.

Aerojet Rocketdyne (formerly GenCorp) formed in 2013; makes "Star Wars" rocket and missile propulsion systems, plus satellites for the NRO.

Reynolds Electronics Engineering — CIA and DoD contractor

---

28 Ibid.

Walsh Construction — CIA contractor

PSI TECH, Inc. — technical remote viewing (TRV) for military and intelligence applications; psychotronics, parapsychology, etc.[29]

Merck — Big Pharma chemical warfare

Monsanto — chemical warfare

FORWARD OPERATING BASES (FOBS)

The RMA is tailored to turning the entire Earth into a battlespace. Global basing is all about the U.S. imperial stance and the militarism—not democracy— that accompanies it.

The Pentagon owns over 29 million acres for 6,000 bases at home and in its territories, and another 3,731 sites with 1,477 Forward Operating Bases (FOBs) abroad, including 20 percent of the Japanese island of Okinawa and 25 percent of Guam—and that's not counting the bases and facilities, freight rail fleet, tactical trucks, Humvees, and law enforcement battalions in Iraq, Afghanistan, Jordan, Israel, Kuwait, Kyrgyzstan, Uzbekistan, Qatar, Ascension Auxiliary Airfield[30] off St. Helena ($337 million), Camp Ederle in Italy ($544 million), Incirlik Air Base in Turkey ($1.2 billion), Thule Air Base in Greenland ($2.8 billion), U.S. Naval Air Station at Keflavik, Iceland ($3.4 billion), and 20+ other nations, including the "Royal Air Force" military and ELINT (electronic intelligence) espionage installations in the UK to the tune of $5 billion. In total, the DoD owns over $1 trillion in assets and $1.6 trillion in liabilities (not counting the "misplaced" $2.3 trillion during Bush II).

In 2002, thanks to the Unified Command Plan, the primarily domestic U.S. Northern Command (NORTHCOM) was established. *Homeland defense command* now includes the U.S., Canada, Mexico, parts of the Caribbean, and the contiguous waters of the Atlantic and Pacific oceans up to five hundred miles off the North American coastline (including Cuba). Both Homeland Security and the North American Aerospace Defense Command are subject to NORTHCOM.

All of these installations mean profits for the military-industrial-intelligence complex. For example, Kellogg, Brown & Root have provided weapons and administration for Camp Bondsteel in Kosovo since 1999. Literally thousands of corporations depend upon hefty military contracts to support the American

---

29 See psitech.net. TRV may be brain-computer interface (BCI) technology.

30 Halfway between the equator and South Pole, Ascension Island is the transfer point for Operation Deep Freeze in the Atlantic. See Rev. Michelle Hopkins' YouTube "'HAARP' TTA [Tesla tech array] shocks the world with Ascension Isle hydroacoustic South Atlantic diamond flare!!!" January 7, 2013.

lifestyle of soldiers far from home: Sony, Danskin, Hanes Her Way, New Balance, Sara Lee Corporation, Home Depot, Procter & Gamble, Roomba Vacuum (iRobot), GlaxoSmithKline, Maytag, Sears, Samsung, NBC (General Electric), Thomasville Furniture, Lowe's, Ballpark Franks (Sara Lee), Eggo Waffles (Kelloggs), Jell-O (Kraft), Coffee Mate (Nestle), etc.

> *Serco Group* - British outsource that runs trains, hospitals, schools, missile defense systems, and border screening with a support staff of 40,000 in thirty-eight nations. Its rival is G4S (Wackenhut).

> *Kaiser Permanente* - integrated managed care consortium

Tom Corcoran of Serco provides a good example of a man who has spent his entire work life spinning in the military-corporate revolving door. He joined Serco—known as "the biggest company you've never heard of"[31]—in December 2007 as a Non-Executive Director after forty years of global business experience, particularly in senior positions in the U.S. aerospace, defense, and electronics contracting industries, including Chairman and Chief Executive Officer of Allegheny Teledyne and President and Chief Operating Officer of Lockheed Martin's Electronic and Space Sectors. He served twenty-six years in General Electric senior management positions, including Vice President and General Manager of GE's Aerospace operations. At present, he is Senior Advisor to The Carlyle Group, President of Corcoran Enterprises, LLC, a management consulting firm, and Non-Executive of Aer Lingus Ltd, L3 Communications Holdings Inc, Labarge Inc, REMEC Inc. and ARINC Inc.

Giant defense contractors like the CIA's main contractor Bechtel outsource many tasks to *private military and security companies (PMSCs)* that enjoy immunity and above-the-law status: Armor Group, Global Risk Strategies, RONCO, Control Risks Group, Erinys, Hart Security, Lifeguard, Wackenhut/G4S, MPRI, KROLL, Olive, Southern Cross, Triple Canopy, DynCorp International / Cerberus Capital Management, Blackwater / Xe / Academi, etc. Billions of dollars are spent on PMSCs handling lethal and nonlethal weapons in conflict zones but less regulated than the toy industry. Lack of oversight and immunity fit with espionage "executive actions" and outright crimes that PMSC personnel are contracted to carry out.

PMSCs are the latest Ugly American representing the bleed-over into the military and America's "Baseworld" footprint now spread over the globe.

Wackenhut Corporation / G4S Secure Solutions in Jupiter, Florida may be the granddaddy NSA/CIA/DOE cutout contractor involved in contract security operations for Top Secret Ultra and Black Budget surface and underground military reservations like Area S-4 and Sandia National Labs, plus "dirty jobs" for CIA and DIA.

---

31 "SERCO: 'The biggest company you've never heard of." *21st Century Wire*, July 7, 2013.

Blackwater, founded in 1998 by Eric Prince and other former Navy SEALs, has a 5,200-acre headquarters compound in North Carolina. By 2009, Blackwater had changed its name three times due to bad press for pimping underage children, smuggling weapons, and murdering "ragheads."

DynCorp International is more of the same in Latin America, Bosnia, Iraq, Afghanistan, Mozambique. PMSCs like Blackwater / Xe / Academi and DynCorp think nothing of destroying incriminating videos and emails, lying to the U.S. State Department, wife swapping, pedophilia, money laundering, and tax evasion, which may be why the DoD "transferred" the War on Drugs (a.k.a. international drug trade) in Latin and Central America to Blackwater / Xe / Academi.

> . . .the transition [from DoD to PMSC] will allow the government to usher billions into the War on Drugs, but to the public it will appear as if the effort is, on the periphery, nothing more than a DoD contract.[32]

In two pending lawsuits, CACI of Arlington, Virginia and L-3 / Titan of New York City—providers of "interrogators" and translators at the Abu Ghraib torture prison in Iraq—insist that since they work for the federal government during combat, they deserve sovereign immunity and "perks" like those provided by the UK giant British Aerospace / BAE Systems for the Saudi royal family as per the al-Yamamah arms deal: child prostitutes and perversions, gambling trips, yachts, sports cars, etc. (And BAE is no exception.)

The *sub-global system* gives a glimmer of how future-reaching some FOBs are. Mainstream media insist that underground and in-bottom or underwater bases are for three agendas: (1) COG (continuity of government) in case of a national emergency; (2) MX nuclear missile launch pads undetectable by satellite; and (3) nuclear test sites. All three are appropriate to a nuclear age, but now underground bases (*and* mansions *and* cities) are about possible electromagnetic, atmospheric and space events. Even underground mansion bunkers are doubling as "trophy asset" floor space with state-of-the-art filtering and shielding systems in crowded showcase cities[33] as well as secure havens from what is dropping from the sky and electromagnetic Skynet wars.

During the Cold War, the U.S. Army Corps of Engineers, tasked with excavating sprawling secret bases, originally turned to the Paperclip underground construction expert Xaver Dorsch, once the director of the Todt Org under the Nazis. By 1960, RAND Corporation[34] was busy surveying copper, iron, and

---

32 "Pentagon Outsources War on Drugs to Blackwater." *RT.com*, 19 January 2012.

33 Oliver Wainwright, "Billionaires' basements: the luxury bunkers making holes in London Streets." *The Guardian*, 9 November 2012.

34 RAND =R&D, research and development, a think tank spinoff of Douglas Aircraft known as Project RAND until 1948.

limestone mines; in Santa Barbara County, it was diatomite strata; in Arizona, under the Grand Wash and Vermilion Cliffs; near Rifle, Colorado, it was an oil shale mine in the Book Cliffs; and in southern Alaska, under the glacial ice and rock of Kenai Peninsula. In 1964, the Army Corps of Engineers chose a dozen sites,[35] and the U.S. Air Force Deep Boring Project chose sites under the Columbia Plateau, central Montana, the Sawtooth mountains, the Snake River Plain, the Denver Basin, the Wasatch/Tavaputs Plateau, the San Juan range; the Kaiparowits Plateau, Pecos and the Natanes plateau, the Aquarius mountains, Mountain Home range, south and west central Nevada, including Pyramid Lake and Carson Sink.

Ground disintegration and high explosives were used. Boring machines *melt* rock, so dirt, sand, and rock were disintegrated and no concrete had to be mixed or hauled. Bechtel's 1974 tool list was a long way from a backhoe: high-pressure water jet (continuous), low-pressure water jet (percussive), high-frequency electrical drill, thermal mechanical fragmentation, conical borer, turbine drill, explosive drill, pellet drill, ultrasonic drill, spark drill, electric arc drills, hydraulic rock hammer, subterrene, induction drilling, water cannon, electrical disintegration, microwaves, electron beam gun, jet-piercing flame, forced flame, Terra-Jetter, lasers, and plasma.

During the Reagan-Bush-Cheney reign, the Defense Nuclear Agency, U.S. Bureau of Mines, and U.S. National Committee on Tunneling Technology (USNC/TT) called a meeting to discuss deep base digging. Present were government agencies,[36] corporate and university consultants, public utilities, Local 147 of the Compressed Air and Free Air Tunnel Workers, and tunnel boring machine (TBM) corporations like the Robbins Company (Kent, Washington), Jarva, Inc. (Solon, Ohio), and Bechtel Nevada at the top of the field with lasers, plasmas, microwaves, electron beam guns, and electrical disintegration techniques. They discussed access shafts, missile nests, interconnecting passageways, adits (horizontal passageways), living quarters, electric power from iron-chlorine fuel cells, generators from waterfalls, rivers and coalmines, highways, cities, repair shops, and magnetoleviton (maglev) high-speed trains traveling at 2,000 miles per hour via magnetic fields.[37]

---

35 Oglethorpe County, Georgia (granite); Franklin County, Alabama (dolomite); Winkler and Ward Counties, Texas (dolomite limestone); Red Willow County, Nebraska (granite); Mesa County, Colorado (granite); Mesa and Montrose Counties, Colorado (salt anticline); Emery County, Utah (sandstone); Pershing County, Nevada (granite); and Mojave, Yuma, Pima, Maricopa Counties, Arizona (granite). The Corps said it would dig 4,000 feet down and 600 feet across.

36 The U.S. Geological Survey, the Bureau of Reclamation Departments of Air Force, Army, Navy, and Energy, the National Science Foundation (CIA science front established in the National Science Foundation Act of 1950), the Federal Highway Administration and Urban Mass Transportation Administration.

37 Thanks to Richard Sauder, Ph.D., *Underground Bases & Tunnels: What Is the Government Trying to Hide?* Dependable Type, 1995. Sauder's Ph.D. is in political science.

THE SUPER SOLDIER CULTURE OF VIOLENCE

U.S. Special Forces Command, Hurlburt AFB, Mary Esther, FL, and Eglin AFB under the command of Beale AFB in Marysville, CA; coordinates: U.S. Army Delta Forces (Green Berets); U.S. Navy SEALs (Black Berets), Coronado, CA; USAF Blue Light (Red Berets) Strike Force; Dolphin Society MK-ULTRA programming is at Hurlburt AFB: F3EA = find, fix, finish, exploit, and analyze. Special Forces self-finance by confiscating money and valuables from covert military operations to fund other clandestine operations.

The new American military boot print is F3EA: find, fix, finish, exploit, and analyze. It is about bypassing formal declarations of war in favor of endless "operations" under the cover of the Wars on Drugs and Terror that justify covert lightning strikes by small, well-trained, high-tech teams. In fact, "short of war" warfare looks more and more like one CIA psyop after another, more and more blurring the line between combatant and noncombatant, national sovereignty and expedience, and condemning the three branches of American governance to being little more than servants in thrall to the National Security Act of 1947.

The "asymmetric" use of capture/kill teams like SEAL Team Six and Special Operations forces (Navy SEALs, Army Rangers, Green Berets, CIA Special Activities Division, etc.) began with guerrilla warfare in Vietnam, the illegal bombing of Laos, and the grisly, illegal CIA Phoenix assassination program in Cambodia when the Pentagon's catechism became "intelligence coordination and the integration of intelligence with an action arm can have a powerful effect on even extremely large and capable armed groups."[38] "Short of war" expansion of military intelligence gathering and capture/kill programs in tandem with CIA drone-strike operations owe everything to the Phoenix assassination program.[39]

The addition of computer-based virtual reality "training" of capture/kill teams is basically advanced MK-ULTRA mind control programming, pure and simple. U.S. Special Operations Command (SOCOM) commandos are subjected to CogniSens' NeuroTracker to "improve situational awareness, multiple target tracking, and decision-making efficiency. . .[T]he brain structurally rewires itself if stimulated intensively and repeatedly."[40] This is Dr. Ewen Cameron's "psychic driving" come of age in anti-Islam jihad "training" at the Joint Forces Staff

---

38 Austin Long and William Rosenau, "The Phoenix Program and Contemporary Counterinsurgency." RAND Institute, July 2009.

39 Of the 3.8 million who died in Vietnam, 58,000 were American combat soldiers. Since the end of the Vietnam war, 200,000 Viet veterans have suicided. In 2016, suicides remain high. (Gregg Zoroya, "U.S. military suicides remain high for a 7th year." *USA Today*, May 4, 2016)

40 Katie Drummond, "Commandos Now Play Digital Brain Games As War Prep." *Wired*, December 5, 2012.

College in Norfolk, Virginia[41] and DARPA's Avatar Project "to develop interface and algorithms to enable a soldier to effectively partner with a semi-autonomous bi-pedal machine and allow it to act as the soldier's surrogate."[42]

> A 2012 lawsuit filed by veterans' groups against the CIA and DoD refers to [Dr. Ewen] Cameron's methods. The suit also states that two researchers, Dr. Louis West and Dr. Jose Delgado, working together under the early CIA MKULTRA subproject 95, utilized two protocols: brain implants ("stimoceivers") and RHIC-EDOM to program the minds of victims. RHIC-EDOM stands for Radio Hypnotic Intracerebral Control - Electronic Dissolution of Memory.[43]

Remote-operated "telepresence" is little more than the brain-computer interface (BCI) known as voice-to-skull (V2K) or synthetic telepathy.

Capture/kill candidates and "grunts" to be plugged into remote computers are now recruited primarily by poverty and secondarily by keeping the cost of higher education too high for low-income youths. The rural population of the United States is now only 16 percent of the total population, and yet 40 percent of military draftees come from rural areas.[44]

Haunted by the fact that 80 to 85 percent of American soldiers in World War II never fired on the enemy and fewer than 30 percent did so in Vietnam,[45] the military has become obsessed with the Nazi dream of the Superman super soldier, an "enhanced" neuroscience version of the ancient Spartan characterized in the dark 2007 film *300*. According to Lt. Col. Dave Grossman, author of *On Killing: The Psychological Cost of Learning to Kill in War and Society*, "Spartan warriors were subjected to a vigorous regime involving unending physical violence, severe cold, a lack of sleep, and constant sexual abuse," all of which are still part of the "pain induction" in typical MK-ULTRA programming.

The neuroscience for producing "human terminators who feel less pain, less terror and less fatigue than 'non-enhanced' soldiers and whose bodies may be augmented by powerful machines"[46] is now employed to remotely release

---

41 Noah Shachtman and Spencer Ackerman, "U.S. Military Taught Officers: Use 'Hiroshima' Tactics for 'Total War' on Islam." *Wired*, May 10, 2012.

42 Evan Ackerman, "DARPA Wants to Give Soldiers Robot Surrogates, Avatar Style." spectrum.ieee.org, February 17, 2012.

43 Jon Rappoport, "Mind control: Dr. Ewen Cameron and 'psychic driving'." *No More Fake News*, January 5, 2015. Dr. Ewen Cameron (president of the Canadian, U.S. and World Psychiatric Associations, the American Psychopathological Association, and the Society of Biological Psychiatry) employed electroshock, LSD and other drugs to create prolonged sleep during which he played tapes repeating phrases thousands of times so as to split off and produce new personalities.

44 Joel Salatin, "Ag Secretary Addresses Why We Need More Farmers: Cannon Fodder?" *Farm-to-Consumer Legal Defense Fund*, September 11, 2013.

45 Michael Hanlon, "'Super soldiers': The quest for the ultimate human killing machine." *The Independent*, 17 November 2011.

46 Lt. Col. Dave Grossman, *On Killing: The Psychological Cost of Learning to Kill in War and Society*.

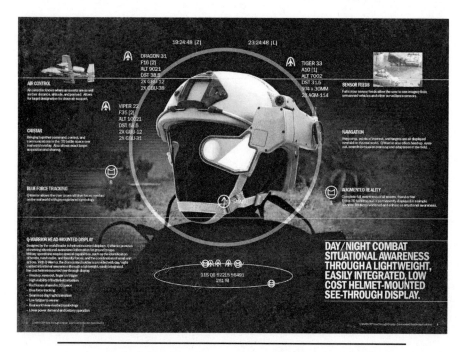

What if this graphic of Q-Warrior headgear is really about what's being done to his brain?

chemicals and zap nanobots in the brains of warriors busy at their battlespace missions. Combat helmets like Q-Warrior[47] have an augmented reality display and electronic sensors to measure the brain's electrical activity and are rigged with direct current transcranial magnetic stimulation (TMS) to enhance reasoning and learning while switching off higher levels of mentation like the conscience.

> Invented in 1985, modern-day magnetic stimulators charge up to a whopping 3,000 volts and produce peak currents of up to 8,000 amps—powers similar to those of a small nuclear reactor. That pulse of current flowing from a capacitor into a hand-held coil creates a magnetic field outside the patient's head. The field painlessly induces a current inside the brain, affecting the electrical activity that is the basis for all it does.[48]

Do we want soldiers acting without conscience? What happens when they come home and re-enter society?

Little, Brown & Co., 1995.

47 "At last, a Google Glass for the Battlefield." *Wired*, February 2014.

48 Carey Goldberg, "Zap! Scientist bombards brains with super-magnets to edifying effect." *Boston Globe*, January 14, 2003.

The Survival, Evasion, Resistance, Escape (SERE) program is Spartan programming. *Stress-inoculation torture*—waterboarding, stripping, strapping, binding, beating, anal rape, firing an unloaded gun at one's covered head, etc.— turns any human being into a killer plagued by PTSD flashbacks and nightmares, anxiety, panic attacks, depression, alcoholism, drug addiction, sleep apnea, paranoia, suicide and murder ideation.[49] Stimulants like cocaine and coca leaves, nicotine, amphetamines, crystal meth (the Nazis' favorite), and Modafinil counter fatigue and weakness, while propranolol blocks torturous memories.

Being a soldier these days is all about being little more than a guinea pig for barrages of vaccinations, popping pills, chemical and mental domination by computers and satellites, and proximity to ionized and non-ionized radiation. The malaria drug Lariam (mefloquine) produces PCP-like states of consciousness that have ended in the murders of Afghan civilians and Fort Bragg wives. Extreme panic, paranoia, and rage with out-of-body dissociative and dreamlike states in which one watches oneself perform violent acts *as if it were someone else* are among Lariam side effects.[50] As Howard Medical School psychiatrist Roger Pitman queries, "The problem is, what else are they blocking when they [administer Lariam]? Do we want a generation of veterans who return without guilt?"[51]

Of the 1.2 million men on active duty in 2012, 13,900 were raped, compared with 12,100 of 203,000 women in uniform—that's thirty-eight men and thirty-three women per day. Given that male survivors of rape report abuse less than female survivors, and all combatants who report abuse are systematically punished either by ambush or officer neglect, male rape may be much greater, particularly as male rape is often viewed as an initiation rite in cultures of violence.[52]

Beyond rape, there is the increase in suicides and murder-suicides. At Joint Base Lewis-McChord (JBLM) south of Seattle, eleven "suspicious deaths" took place in 2011 alone: suicides, murders, setting fire to wives after serial deployments, waterboarding of a daughter. The Afghan kill team that went on a three-month kill spree against Afghani civilians was from JBLM.[53]

In 2011, a federal lawsuit was filed in Virginia to expose and hold the military

---

49 Valtin, "More Evidence SERE Training Caused PTSD in Some Soldiers." *Invictus*, May 28, 2012.

50 Martha Rosenberg, "A Nightmare Drug, Military Suicides and Killing." *CounterPunch*, April 8, 2014.

51 Ibid.

52 Bill Briggs, "Male rape survivors tackle military assault in tough-guy culture." *NBCNews.com*, May 16, 2013. Also see Michael Kasdan's "What's Unusual About Sayerville's Locker Room Sexual Assaults? Nothing." *The Good Men Project*, October 14, 2014; and Cooper Fleishman's "In Yale Fraternity Pledging, Rape Is a Laughing Matter," *The Good Men Project*, October 15, 2010.

53 Winston Ross, "Army Base on the Brink." *The Daily Beast*, September 9, 2011.

accountable for this culture of violence. Two men and fifteen women, veterans and active duty, claim:

> ...that Defense Secretary Robert M. Gates and his predecessor Donald H. Rumsfeld 'ran institutions in which perpetrators were promoted and where military personnel openly mocked and flouted the modest congressionally mandated institutional reforms'. . .[T]he two defense secretaries failed 'to take reasonable steps to prevent plaintiffs from being repeatedly raped, sexually assaulted and sexually harassed by federal military personnel.'[54]

And what was the story behind the NATO helicopter crash on May 11, 2011 in Afghanistan that killed everyone on board, including U.S. Navy SEAL Team Six, the Naval Special Warfare Development Group that had purportedly killed al-Qaeda leader Osama bin Laden in Pakistan? And four months after the crash, December 22, 2011, the suicide of SEAL Team Six's Commander Job W. Price?

Are we all simply cannon fodder now?

"MY ELF WEAPON"

In 1995, when *Time* magazine announced the RMA, who could have foreseen that all of human society (and anything and anyone living), foreign and domestic, was being redefined as a potential battlespace? Is this what former Navy reservist Aaron Alexis unwittingly triggered at the Washington Navy Yard on Monday, September 16, 2013, when he shot and killed twelve people? On his shotgun was carved "My ELF weapon." Officials may not have known what the carved words meant, but the Navy certainly did: the Office of Naval Research (ONR) had spent decades researching the biological and neurological effects of ELF waves:

- Put a person to sleep
- Make a person tired or depressed
- Create a feeling of fear
- Create a zombie state
- Create a violent state
- Create a sexually aggressive state
- Change cellular chemistry

---

54 Ashley Parker, "Lawsuit Says Military Is Rife With Sexual Abuse." *New York Times*, February 15, 2011.

- Change hormone levels

- Inhibit or enhance mRNA synthesis/processes

- Control the DNA transaction process

- Control biological spin and proton coupling constants in DNA, RNA & RNA transferases

The month before the Navy Yard debacle, Alexis had told Rhode Island police that "he was hearing voices of three people who had been sent to follow him and keep him awake and were using 'some sort of microwave machine' to send vibrations into his body, preventing him from falling asleep."[55] Perhaps while being driven mad by sleep deprivation, Alexis had been involved in testing the Army's Burke Pulsar that fits into an M4 rifle like a standard suppressor—two wide antennas, a piezoelectric generator, and a blast shield. The claim is it's for use against electronics, sort of a mini-EMP, but who knows?[56]

"Nonlethal" weapons wire everywhere wireless for war. Many components of weapons systems double as consumer products that have been tested for decades on nonconsensual citizens, like the video game helmets loaded with TMS and neuro-feedback that "detect the player's thoughts, emotions and expressions, then translate them into their character in the game."[57] The question is, can video game TMS helmets be trusted as long as the military views them as dual use? *Caveat emptor.*

The United States has always been wired for war. Two million miles of power lines spew a minimum of 500mG (milligauss), with 4mG being enough to produce leukemia in small children. Those living under and around power lines are also subject to a constant 60Hz entrainment created by the magnetic field, thanks to the wiring in their homes. Freelance journalist Jim Stone points out the covert dual use behind how the nation has been purposefully wired:

. . .the longest possible antenna needed for ANY cell activity in the U.S. is around 20 inches, for the lowest possible cell-related frequencies. More common would be 9-inch antennas. Why then are the antennas which adorn cell towers up to eight feet long? You can easily make a high-powered coil antenna suitable for low MHz frequencies with that much room to work with . . .[58]

55 Sari Horwitz, Steve Vogel and Michael Laris, "Officials: Navy Yard shooter carved odd messages into his gun before carnage." *Washington Post*, September 18, 2013.

56 "US Army begins testing a weapon overwhelming enemy electronics." *Pravda.Ru*, April 23, 2015.

57 Vriti Saraf, "Unlocking the Power of the Mind." *media.www.theticker.org*, October 6, 2008.

58 Jim Stone, "Mind control via electronic manipulation," www.jimstonefreelance.com/cells.html, June 14, 2012.

Air warfare has evolved from bullets to bombs to missiles to EMPs (electromagnetic pulse HPMs or high-power microwaves) and DEWs (directed energy weapons) aboard UCAVs (unmanned combat aerial vehicles) able to plant a DE payload within fifty feet of a target once Special Ops use their laser pointers or GPS receivers to designate the targets.

Precision-strike warfare began in the 1991 Persian Gulf War under former CIA director and then-President George H.W. Bush. Battlefield surveillance by satellites and instantaneous targeting capability made large armies and traditional battlefields passé. Precision intelligence, precision weapons, and tactical teams like those described earlier are now the primary strengths in asymmetric electromagnetic warfare. At the end of World War II, 12 million troops were on active duty; today, 1.4 million are active frontline and 1.1 million active reserves.

War is no longer a matter of strength in numbers; it is tactical and a matter of electromagnetic signatures, frequencies, and pulses. Asymmetric warfare with full spectrum dominance may sound good on paper—no risk to pilots, greater accuracy and stealth, no scattered flechettes to murder half the population—but the present high-tech invisible business of war is absolutely inimical to the biosphere and lives of all creatures on Earth.

In space, there are now satellite-disabling lasers and beam weapons ranging from HPMs (high-power microwaves) and acoustic beams (high power, very low frequency) to pulsed energy projectiles (PEPs), chemical lasers that generate localized high-pressure plasma, and CEIR (computer-enhanced infrared) heat-seeking lasers. HPMs in air defense suppression or cyberwarfare can knock out electronics, scramble computer memories, and enter false targets to create chaos and lower defenses.

While hypersonic spy satellites scree from four hundred miles up, EA-6B jets jam radars and radios, F-16CJs destroy anti-aircraft installations, and F-15C fighters loose Sidewinder missiles. Hypersonic diamond-shaped stealth UAVs covered with ceramic tiles can accelerate to Mach 8 (5,720 miles per hour) if not brought down by a mobile microwave gun like the Ranets-E.[59] Global Hawk drones fly at 65,000 feet as JSTARS (joint surveillance and target attack radar system) with moving-target indicator radar feeds data to command centers. The F-15E Strike Eagle and F-35 Joint Strike Fighter are perfect for front-line offense, and cruise missiles like the ALCM (air-launched cruise missile) Tomahawk, Joint Air-to-Surface Standoff Missile (JASSM), or Britain's Storm Shadow can take in an HPM payload and deliver it.

Besides the TR-3A Black Mantra and Aurora SR-91 peeking through chemical cloud cover (mentioned in Chapter 4), Boeing's RC-135 Rivet Joint recon jet intercepts communications and pulses Silent Sound Spread Spectrum

59 Tatyana Rusakova, "New microwave radiation gun to take down drones." *Russia & India Report,* 26 July 2015.

(S-quad or S4) over AM, FM, HDTV, and military bands. S4 is a dual-use hypersonic sound weapon, patent #5,159,703. After collecting, analyzing, and replicating emotional EEG patterns (*excitation potential signatures*) then stored on computers as low-amplitude emotion signature clusters, these clusters are then piggybacked on S4 carrier frequencies to trigger some emotions at the expense of others.[60] S4 made its war theater debut during the first Gulf War (August 1990 – February 1991). After Saddam Hussein's command-and-control electronics in Kuwait were knocked out, emotion signature clusters of fear, anxiety, and despair were broadcast on 100MHz FM radio channels as subliminals to bogus military orders and patriotic and religious music.[61] Silent Sound, Inc. was the first to sell CDs of excitation potential signatures, after which Israel's eXaudios created Magnify, a telephone software that decodes and analyzes excitation potential signatures in the voice.[62]

Laser-guided bombs and Joint Direct Attack Munitions (JDAMs) can perform precision strikes night or day, in rain, cloud, snow, or sandstorm. JDAM guidance kits make dumb bombs smarter, even when launched from 10,000 meters up and 10.5 miles from target. The Navy's Tomahawks are smart and the Army's Apache Longbow helicopter gunships with their radar-guided Hellfire missiles even smarter and more lethal. B-2s, B-52s, and B-1s carry multiple JDAMs programmed for different targets. The seven-foot, fifty-pound ShadowHawk is a 150-gram drone helicopter loaded with a 50,000-volt stun gun, a 12-gauge shotgun, and a 40mm grenade launcher.

The ATL[63] or advanced tactical laser (L3 Communications/Brashear and HYTEC Inc.) fires through a rotating turret extending from Boeing's C-130H beam control system. The ATL aircraft can direct a low-power laser to find, track, and hit ground targets. By setting off a storm of electrons, cobalt ionization bombs can disrupt infrared and detection laser systems. (The Russians bypass radar limitations by utilizing infrared beams, though pulsed neutron beams are another matter.)

Reusable payloads are a bonus. For example, the PGS (prompt global strike) program, code-named FALCON (force application and launch from CONUS), is a reusable HCV (hypersonic cruise vehicle) that launches into space, delivers its payload, then returns to Earth. FALCON's global reach includes unmanned reusable HCVs, SLVs (small launch vehicles), and CAVs (combat aerial vehicles) that can launch satellites on short notice, carry nuclear or kinetic payloads like

---

60 Jason Jeffrey, "Electronic Mind Control: Brain Zapping, Part One." *New Dawn*, March-April 2000.

61 Judy Wall, "Aerial Mind-Control: The Threat to Civil Liberties." *Nexus* magazine, October-November 1998. Also see "High Tech Psychological Warfare Arrives in the Middle East." ITV News Bureau, Ltd., Dhahran, Saudi Arabia, March 23, 1991.

62 Eric Bland, "Computer Software Decodes Emotions Over the Phone." *Discovery News*, May 8, 2010.

63 The Ice Age *atl atl* and Nahuatl *atlatl* both mean "spear-thrower." The Nahuatl term *xiuhcoatl* refers to a "fire serpent" *atlatl*, possibly a prehistory electromagnetic weapon.

the tungsten *kinetic orbital bombardment projectiles* called "rods from God" or robust nuclear earth penetrators that penetrate deeply buried or particularly hardened underground bases. Deployment can be from missiles, MSPs [military space planes], or space-based platforms like satellites. A single Minuteman III ICBM will lift three CAVs 7,000 miles or more, each for a different target, weapon, or purpose.

The Nautilus antimissile laser, previously known as the MIAAD-182 or midrange infrared advanced antipersonnel disabling system, is a lightning-fast, high-intensity infrared laser pulse that destroys missiles in flight. On the ground it can permanently blind combat troops by entering the aqueous front shell of the eyeball and damaging the thin membrane of the retina in the rear. The Nautilus is the ultimate Buck Rogers death ray.

Abandoning analog for digital has multiplied the velocity and scope of military data transmissions (and eavesdropping). The Battlefield Optical Surveillance System (BOSS) mounted on trucks and Humvees scans and deciphers the landscape with lasers and sensors; once it recognizes a watch-list retina, it "paints" the target with a laser beam and the target dies. Boomerang uses an array of microphones to decipher the speed and direction of a shot and recognizes not just where the shot came from but what kind of weapon. The Large Area Coverage Optical Search-while-Track and Engage (LACOSTE) is imaging technology for day or night tactical surveillance of all moving vehicles and humans in urban areas.[64]

Ships like the 97,000-ton nuclear-powered *U.S.S. Harry S. Truman* are armed with an electromagnetic rail gun capable of firing a projectile 230 miles with a Mach 7 muzzle velocity and Mach 5 impact velocity—sheer kinetic energy, not conventional explosives—as its point droppers in the Strike Intelligence Analysis Center examine digital real-time black-and-white feed from Predator drones at 20,000 feet in search of Hellfire missile targets.

Air defense radar, C4 computers, CBW (chemical / biological weapons) storage and production units, underground weapons bunkers— Americans have given the store, and more, to their military.

While all weapons (including nonlethals) are designed to maim or kill, the small bombs that stay around long after the battle to maim curious children are particularly pernicious. *Cluster* and *carpet bombs* open in mid-air and disperse hundreds of small sub-munitions over an area as large as several football fields. B-52s deliver carpet bombs to "soften" enemies before Special Forces teams enter, the hope being that after the dust and smoke clear, they will be walking in circles and mumbling, blood oozing from their ears and noses. The United States has refused to sign an international ban on these small bombs. Israel prefers DIMEs (dense inert metal explosive), carbon fiber casings packed with tungsten

---

64 Dave Hodges, "Will We Be the New Viet Cong? Or, Is It Too Late?" *The Common Sense Show*, February 17, 2014.

powder that produce an intense explosion in a small space, then dissolve human tissue. And don't forget the IEDs (improvised explosive devices). Some drones have been specifically created to sniff out these bombs and defuse them.

Thermobaric weapons like the fuel-air bomb use heat and pressure to kill, the first stage being to disperse a gas or chemical agent in an enclosure the gas will fill, the second being ignition. What kills is the pressure wave and subsequent vacuum that ruptures ears, lungs, and organs. Fuels for solid fuel-air explosives (FAEs) like ethylene oxide and propylene oxide are highly toxic and if undetonated will burn personnel to death. Optimized to create blast over heat, its deceptive code name is *light anti-structure munition*.

SWORDS (Special Weapons Observation Reconnaissance Detection Systems) is a land-based robot standing a meter high whose soldier operator is a half-mile away in VR (virtual reality) goggles. Unlike Talon and Packbot robots that disarm bombs, SWORDS carries a machine gun with an electronic aim. QinetiQ Group, a joint venture of the UK Ministry of Defense and the shadowy Carlyle Group, owns SWORDS designer Foster-Miller of Waltham, Massachusetts. (In Chapter 6, we will examine DARPA's Systems of Neuromorphic Adaptive Plastic Scalable Electronics (SyNAPSE) program modeling artificial brains after the human brain for robots.)

MAV [micro air vehicle] explosive robotic insects silently fly and hop and crawl into buildings while their human pilots sit in climate-controlled trailers outside Las Vegas. Reconnaissance MAVs like WASPs can be fitted with a C4 mission-specific payload for functional defeat—say a cloud of metal-coated fibers that destroy computers or shut down electronic power. WASP and A160 Hummingbird drones weigh less than a pound but pack a 1.8 gigapixel camera with SkyGrabbers capturing the feed. Project Anubis[65] WASPs carry sensors, data links, and munitions payloads. Flying at 20,000+ feet, WASPs can follow sixty-five targets at a time and learn by mimicry and observation, thanks to their human pilot. Quad-rotor iDrones look like flying saucers, and SWARMS (scalable sWarms of autonomous robotics and mobile sensors) move in a pack. "Drone-swarm" tech is all about the "hive mind"—micro-drones working together *as one* to make decisions and complete their assignments, like the Perdix drone (wingspan 30 cm, Mach 0.6):

> Perdix are not pre-programmed synchronized individuals, they are a collective organism, sharing one distributed brain for decision-making and adapting to each other like swarms in nature," said William Roper, director of the Strategic Capabilities Office at the US Department of Defense.[66]

---

65 The canine-headed Egyptian god Anubis weighed the hearts of the dead.

66 Thomas McMullan, "US military drops swarm of self-thinking drones from jets." *Msn.com*, January 11, 2017.

Drone-swarm tech—including self-aware killer robots[67]—comes under the Autonomous Real-time Ground Ubiquitous Surveillance Imaging System or ARGUS-IS, named after Argus Panoptes, the hundred-eyed giant.

We finally arrive at Cerberus, the unmanned mobile integrated tower with mounted sensors that guards the gates of Forward Operating Bases, similar to Raytheon's G-BOSS (ground-based operational surveillance system). "Networked, multiple Cerberus towers form a mesh network, communicating autonomously with each other to act as a virtual fence."[68] *A virtual fence.* As *Defense Industry Daily* admits about the three-headed mythical dog, "Cerberus permitted souls to enter the realm of the dead, but allowed none of them to leave."

Is this our fate with the Space Fence?

In the next chapter, "Mastering the Human Domain," I go deeply into *network-centric warfare (NCW)*, the cornerstone of the Revolution in Military Affairs that has converted the American military into something very different than it once was.

---

67 Nafeez Ahmed, "The Pentagon is building a 'self-aware' killer robot army fueled by social media." *InsurgeIntelligence*, May 12, 2016.

68 "Cerberus: Standing Guard Over US Military's Forward Bases." *Defense Industry Daily*, June 23, 2011.

CHAPTER SIX

Mastering the Human Domain

I can assert that the vast majority of the computer systems currently in use, the huge systems that span our planet—military ones, for example—are beyond our understanding. I do not merely mean that there is no one left who can grasp their working, but that the time when we could do that is past and gone. It is no longer possible to understand them.

— Joseph Weizenbaum, *Computer Power and Human Reason,*
"From Judgment to Calculation," 1984

Quarry considered VIXAL. He pictured it as a kind of glowing celestial digital cloud, occasionally swarming to earth. It might be anywhere—in some sweltering, potholed industrial zone stinking of aviation fuel and resounding to the throb of cicadas beside an international airport in Southeast Asia or Latin America; or in a cool and leafy business park in the soft, clear rain of New England or the Rhineland; or occupying a rarely visited and darkened floor of a brand-new office block in the City of London or Mumbai or São Paolo; or even roosting undetected on a hundred thousand home computers. It was all around us, he thought, in the very air we breathed. He looked up at the hidden camera and gave the slightest bow of obeisance.

"Leave them," he said.

— Robert Harris, *The Fear Index,* 2012

Any discussion of computers and their role in a planetary lockdown must begin with *secrets.*

Off the island Antikythera in the Aegean on Maundy Thursday 1900, sponge divers found a small lump of corroded bronze and wood with gear wheels inside dated between 85 and 50 B.C. It was a geared computer once used to calculate the past, present, and future positions of the heavens.

Futurist Arthur C. Clarke (1917–2008) made it clear that magic and technology are at least kissing cousins, if not closer. The modern cryptic relationship between machines and their encoded secrets harks back to soldier-playwright Sophocles

(496–406 B.C.) who introduced the *mechane* to Greek drama by lowering it onto the stage to provide the *deus ex machina* for the plot's supernatural intervention. Electrical computer ciphers are equally magical in that they appear at the stroke of a key. The year the Bomb shattered matter (and, some think, the United States), its father Vannevar Bush (1890–1974)[1] had a dream in which a *deus ex machina* called *memex* was perched on a doctor's desk, calling up patients' files and case histories. In the July 1945 *Atlantic Monthly* article "As we may think," Vannevar explained how one day soon whole encyclopedias with associative connections would appear magically on the memex screen, and professionals and laymen alike would turn to it as to a library or oracle. The memex would be self-teaching and relieve people of the need for memory or recall.

Concurrent with Vannevar's dream was the arrival of the U.S. Army's EINAC (Electronic Numerical Integrator and Computer) with 18,000 vacuum tubes and miles of wiring, followed by England's Mark I (1948), EDSAC (1949), and the United States' EDVAC (1951), UNIVAC I (1951), and ILLIAC I (1952). Thus was the cyber revolution set in motion.

For centuries, high-degree Freemasons and wealthy *cognoscenti* have quietly used gear models for computation and forecasting the future, at least since the bronze head answered yes or no to Gerbert d'Aurillac (920–1003), the Benedictine monk-professor at the University of Rheims elected Pope Sylvester II. From the cybernetics journal *Computers and Automation* (October 1954):

> We must suppose that Pope Sylvester II, Gerbert d'Aurillac, was possessed of extraordinary knowledge and the most remarkable mechanical skill and inventiveness. This speaking head must have been fashioned "under a certain conjunction of stars occurring at the exact moment when all the planets were starting on their courses." Neither the past nor the present nor the future entered into it, since this invention apparently far exceeded in its scope its rival, the perverse "mirror on the wall" of the Queen, the precursor of our modern electronic brain. Naturally, it was widely asserted that Gerbert was only able to produce such a machine because he was in league with the Devil and had sworn allegiance to him.

From British intelligence agent Lord Francis Bacon's seventeenth-century ciphered binary code to the encrypted Dayton Witch with its cipher book (once the property of the cybernetics department at Brunel University in High Wycombe just up the road from Sir Francis Dashwood's "magickal" Hellfire estate in Oxfordshire), families with wealth and standing have had access to intellectual, computational, and magical help. Perhaps the U.S. State Department still has a Black Chamber (Cipher) Bureau at 131 East 37th Street in New York City, just a few blocks from industrialist J.P. Morgan's mansion-cum-museum.

1 He was also a co-founder of Raytheon in 1924.

Why not? The electromechanical rotor cipher machine known as the Enigma was recently the topic of *The Imitation Game*, a film about Alan Turing, the father of modern computation.

Computers, their ciphers and binary code go hand in hand with cryptography and cybernetics such as were practiced by visionary millionaire George Fabyan (1867–1936), patron of acoustics and perpetual motion. In fact, the National Security Agency—secretly founded thirty-six years (6 X 6) after Fabyan's death—honored him with a plaque at the Riverside Acoustic Laboratory on his three-hundred-acre Fox River Valley estate forty miles west of Chicago. The plague read *To the memory of George Fabyan from a grateful government.*

The fact is that high-speed data processing has availed the elite of their dream of social and eugenic engineering. In 1946, the Cybernetics Group morphed into the Feedback Mechanisms in Biology and the Social Sciences, and later into the World Federation of Mental Health. Sandwiched among cybernetics, biology, and mental health was the discovery of LSD, the U.S. Navy's Project Chatter, and the 149 subprojects of the CIA's drugs-hypnosis-pain induction program with an acronym for public dissemination almost as nifty as HAARP: MK-ULTRA. In the same timeframe, the U.S. Air Force privatized its research and development arm as RAND Corporation to protect military projects and cyphers from congressional curiosity. After the Manhattan Project, polymath John von Neumann (1903–1957) pushed for neural nets (the conceptual forerunner of the Internet) and mutual assured destruction (MAD) war-game strategies.

From ferrite-core memory and transistors to hard disks, linked modular mainframes, Sketchpad, and vacuum tubes to solid-state logic, the military-industrial-intelligence complex zeroed in on cybernetics. In 1961 the silicon crystal chip miniaturized the memory storage that transistors had been handling, packet switching and digital expanded the bandwidth and laid out the redundant, multilevel web for teletype data and voice.

After the Cuban Missile Crisis of 1962, Secretary of Defense McNamara called for automated intelligence production. The Defense Intelligence Agency (DIA) already had an Automatic Data Processing Center on a rented IBM 360/30 mainframe that could index, store, and retrieve intelligence like U-2 photos of Soviet military installations. The RAND packet-switching idea had a go, born out of a need for an alternate Cold War command-and-control (C2) network. The first linkup of Pentagon computers talking to each other in a closed network of ARPANET known as COINS[2] (Intelligence Community Computer Communication Network) was launched on December 31, 1966 from the DIA Security Office. Streams of digital data were broken into short bursts, followed by the early email software known as SMTP (simple mail transfer protocol).

Industry and the State Department were already scouring the Earth for what the new cybernetic weapon would need: lead and cadmium for circuit boards,

---

2 COINS is now the acronym for counterinsurgency.

lead oxide and barium for monitor cathode ray tubes, mercury for switches and flat screens, brominated flame retardants, and most recently niobium and titanium for superconduction. Resources, dictators and war are practically synonymous.

From the beginning, the National Security State had plans for Vannevar Bush's democratic dream. Computers as multiprocessors were churned out: the 360 IBM series with the internal military-intelligence compatibility, Seymour Cray's 6600, 7600, and Cray I, Intel's microprocessor mesh Paragon, and Xerox PARC mouse-windows and ALTO in 1971.

And then came the supercomputer with its customized units called blades housing multiple nodes (CPUs, GPUs). Imagine 10 quadrillion calculations per second (10 petaflops). . .But then something happened in the 1980s—the theft of a piece of computer software—that secretly ricocheted through world events even more powerfully than the Enigma.

THE PROMIS BACKDOOR

Beyond embedded journalists, news blackouts, false flag events, blacklisted and disappeared Internet domains—the plotline of America's "free press"—there are now ISP-filtering programs subject to Homeland Security guidelines that sift through emails and toss some into a black hole. Insiders and the NSA-approved, however, can get around such protections of networks by means of the various hybrids of the PROMIS backdoor.

The 1980s theft of the Prosecutor's Management Information System (PROMIS) software handed over the golden key that would grant most of the world to a handful of criminals.[3] In fact, this one crime may have been the final deal with the devil that consigned the United States to its present shameful descent into moral turpitude.[4]

PROMIS began as a COBOL-based program designed to track multiple offenders through multiple databases like those of the DOJ, CIA, U.S. Attorney, IRS, etc. Its creator was a former NSA analyst named William Hamilton. About the time that the October Surprise Iranian hostage drama was stealing the election for former California governor Ronald Reagan and former CIA director George H.W. Bush in 1980, Hamilton was moving his Inslaw Inc. from non-profit to for-profit status. His intention was to keep the upgraded version of PROMIS that Inslaw had paid for and earmark a public domain version funded by a Law Enforcement Assistance Administration (LEAA) grant for

---

3 See *The Octopus: The Secret Government and Death of Danny Casolaro* by Kenn Thomas and Jim Keith (Feral House, 1996).

4 Moral turpitude is a legal concept referring to conduct considered contrary to community standards of justice, honesty or good morals. (*West's Encyclopedia of American Law*)

the government. With 570,000 lines of code, PROMIS was able to integrate innumerable databases without any reprogramming and thus turn mere data into information.

With Reagan in the White House, his California cronies at the DOJ offered Inslaw a $9.6 million contract to install public-domain PROMIS in prosecutors' offices, though it was really the enhanced PROMIS that the good-old-boy network had set its sights on.

In February 1983, the chief of Israeli antiterrorism intelligence was sent to Inslaw under an alias to see for himself the DEC VAX enhanced version. He recognized immediately that this software would revolutionize Israeli intelligence and crush the Palestine *Intifada*. Enhanced PROMIS could extrapolate nuclear submarine routes and destinations, track assets, trustees, and judges. Not only that, but the conspirators had a CIA genius named Michael Riconosciuto who could enhance the enhanced version one step further, once it was in their possession.

To install public domain PROMIS in ninety-four U.S. Attorney offices as per contract, Inslaw had to utilize its enhanced PROMIS. The DOJ made its move, demanding temporary possession of enhanced PROMIS as collateral to ensure that all installations were completed and that only Inslaw money had gone into the enhancements. Naïvely, Hamilton agreed. The rest is history: the DOJ delayed payments on the $9.6 million and drove Inslaw into bankruptcy. With Edwin Meese III as Attorney General, the bankruptcy system was little more than a political patronage system, anyway.

The enhanced PROMIS was then passed to the brilliant multivalent computer and chemical genius Riconosciuto, son of CIA Agent Marshall Riconosciuto.[5] Recruited at sixteen, Michael had studied with Nobel Prize-winning physicist and co-inventor of the laser Arthur Shallo. Michael was moved from Indio to Silver Springs to Miami as he worked to insert a chip that would broadcast the contents of whatever database was present to collection satellites and monitoring vans like the Google Street View van, using a digital spread spectrum to make the signal look like computer noise. This Trojan horse would grant key-club access to the backdoor of any person or institution that purchased PROMIS software—as long as the backdoor could be kept secret.

Meanwhile, the drama between Hamilton and the conspirators at DOJ continued. A quiet offer to buy out Inslaw was proffered by the investment banking firm Allen & Co., British publisher (*Daily Mirror*) Robert Maxwell, the Arkansas corporation Systematics, and Arkansas lawyer (and Clinton family friend) Webb Hubbell. Hamilton refused and filed a $50 million lawsuit in bankruptcy court against the DOJ on June 9, 1986.

---

5 Marshall Riconosciuto was close friends with Fred Crisman, the first person Clay Shaw called when he heard that attorney Jim Garrison was implicating him in the Kennedy assassination. Strangely, Crisman was also involved in the UFO incident at Maury Island, Washington in June 1947.

Bankruptcy Judge George F. Bason, Jr. ruled that the DOJ had indeed stolen PROMIS through trickery, fraud, and deceit, and awarded Inslaw $6.8 million. He was unable to bring perjury charges against government officials but recommended to the House Judiciary Committee that it conduct a full investigation of the DOJ. The DOJ's appeal failed, but the Washington, D.C. Circuit Court of Appeals reversed everything on a technicality.

Under then-President George H.W. Bush (1989–1993), Inslaw's petition to the Supreme Court in October 1991 was scorned. When the IRS lawyer requested that Inslaw be liquidated in such a way that the U.S. Trustee program (AG Meese's feeding trough between the DOJ and IRS) could name the trustee who would convert the assets, oversee the auction, and retain the appraisers, Judge Bason refused.

Under then-President William Jefferson Clinton (1993–2001), the Court of Federal Claims whitewashed the DOJ's destruction of Inslaw and theft of PROMIS on July 31, 1997. Judge Christine Miller sent a 186-page advisory opinion to Congress claiming that Inslaw's complaint had no merit—a somber message to software developers seeking to do business with Attorney Generals and their DOJ. For his integrity, Judge Bason lost his bench seat to the IRS lawyer.

Throughout three administrations, the mainstream Mockingbird media obediently covered up the Inslaw affair, enhanced PROMIS being a master tool of inference extraction able to track and eavesdrop like nothing else. Once enhanced PROMIS was being sold domestically and abroad so as to steal data from individuals, government agencies, banks, and corporations everywhere, intelligence-connected Barry Kumnick[6] turned PROMIS into an artificial intelligence (AI) tool called SMART (Special Management Artificial Reasoning Tool) that revolutionized surveillance. The DOJ promised Kumnick $25 million, then forced him into bankruptcy as it had Hamilton. (Unlike Hamilton, Kumnick settled for a high security clearance and work at military contractors Systematics and Northrop.) Five Eyes / Echelon and the FBI's Carnivore / Data Collection System 1000 were promptly armed with SMART, as was closed circuit satellite high-definition (HD) television. With SMART, Five Eyes / Echelon intercepts for UKUSA agencies became breathtaking.

The next modification to Hamilton's PROMIS was Brainstorm, a behavioral recognition software, followed by the facial recognition software Flexible Research System (FRS); then Semantic Web, which looks not just for link words and embedded code but for what it means that this particular person is following this particular thread.

Then came quantum modification. The Department of Defense paid Simulex, Inc. to develop Sentient World Simulation (SWS), a synthetic mirror of the real world with automated continuous calibration with respect to current real-world

---

6 Barry's father Frank Kumnick played a role in the FBI-US Marshal siege against Randy Weaver (August 21–31, 1992).

information. The SEAS (Synthetic Environment for Analysis and Simulations) software platform drives SWS to devour as many as five million nodes of breaking news census data, shifting economic indicators, real world weather patterns, and social media data, then feeds it proprietary military intelligence and fictitious events to gauge their destabilizing impact. Research into how to maintain public cognitive dissonance and learned helplessness (psychologist Martin Seligman) help SEAS deduce human behavior.

Coupled with Semantic Web, SWS became the testing environment by which military and intelligence could foresee what adversaries, neutrals, and allies were planning to do and thus prevent, alter, or accommodate their future behaviors. The SWS mirror world is like *Where's Waldo?* the children's book for upcoming spies, replete with big and little institutions, terrains, streets, homes, individuals, even crowds. Feed SWS what you plan to do—take over private water wells in California, unleash weather warfare on the poor in New Orleans or Haiti, phase out the border between Canada and the U.S., back Israel's Palestinian policies—and SWS software prophesies the reaction. *Crisis management.*

Needless to say, military-industrial-intelligence players like DARPA, Eli Lilly, Lockheed Martin, and Homeland Security all employ SWS. Simulate a crisis, then either run it in the real world or not as a controlled false flag or real event. During Noble Resolve 07, the JFCOM-J9 (Joint Innovation and Experimental Directorate of the U.S. Joint Forces Command) worked with Homeland Security and multinational forces to run real-time, round-the-clock simulations for dozens of nations. Reactions under stress may require that several solutions be at hand, but generally JFCOM-J9 is confident after six decades of slow-boiling-frog cognitive dissonance that SWS predictions will hold in the U.S. as well as in other nations (like Ukraine and Greece) whose cultures have been carefully subjected to SWS scrutiny.

As military contractors, telecom corporations comply with the NSA's mandate that their equipment include a backdoor so that the NSA's TAO hackers (Tailored Access Operations), the "premier hacking ninja squad" with "a catalog of all the commercial equipment that carries NSA backdoors," can then intercept the online orders and bug them:

Storage products from Western Digital, Seagate, Maxtor and Samsung have backdoors in their firmware, firewalls from Juniper Networks have been compromised, plus networking equipment from Cisco and Huawei, and even unspecified products from Dell. . .*Spiegel* notes that the [Snowden] documents do not provide any evidence that the manufacturers mentioned had any idea about this NSA activity. Every company spokesperson contacted by *Spiegel* reporters denied having any knowledge of the situation, though Dell officials said instead that the company "respects and complies with the laws of all countries in which it operates."[7]

---

7 Lily Hay Newman, "The NSA Actually Intercepted Packages to Put Backdoors in Electronics." *Gizmodo*, December 29, 2013.

In 2014, China removed high-end servers made by IBM and Microsoft and replaced them with local brands.[8] Had they finally discovered the PROMIS backdoor?

Now we are ready for a look at supercomputers, quantum computers, and artificial intelligence (AI) as we wonder if they too have secret backdoors.

SUPERCOMPUTERS

Supercomputers now digest data 24/7 and map every square inch of planet Earth while programs like LifeLog electronically bind every human being to the Smart Grid. The military-intelligence penchant for mythical names fits well with intelligence machines like the supercomputer in Brussels 666,[9] a partner in crime with the U.S. Naval Research Laboratory's BEAST (Battle Engagement Area Simulator/ Tracker), a real-time space battle management simulator seven times faster than a Cray Y-MP for functionally equivalent optimization of 3D code. (BEAST can model 32,000 objects from inputs provided by sixty-four satellite sensors.) The U.S. Army Research Laboratory's Excalibur—named after King Arthur's magical sword—is also well named for technological advantage on the battlefield.

Then there is the Big Mac supercomputer at Virginia Tech—home of the mid-latitude-to-polar SuperDARN network discussed in Chapter 7—peaking out at 17.6 teraflops, second only (at the time of this writing) to Japan's Earth Simulator with 35.6 trillion calculations per second. The ASCI-Q at Los Alamos National Laboratory, built by Hewlett-Packard, weighs in at 13.8 teraflops, and Lawrence Livermore National Lab's Sequoia at 16 petaflops. IBM's Blue Gene processes 10^{14} operations per second, compared to 10^{16} per second of the human neocortex.

Moving toward quantum computers and artificial intelligence (AI), IBM's Neuromorphic System is reverse engineered from the human brain based on a neurosynaptic computer chip called IBM TrueNorth developed by Cornell University and DARPA's Systems of Neuromorphic Adaptive Plastic Scalable Electronics (SyNAPSE) program. TrueNorth is capable of "deep learning" (16 million neurons, 4 billion synapses) and utilizes low electric power (2.5 watts)— just like the human brain.

> A single TrueNorth processor consists of 5.4 billion transistors wired together
> to create an array of 1 million digital neurons that communicate with one
> another via 256 million electrical synapses. It consumes 70 milliwatts of power
> running in real time and delivers 46 giga synaptic operations per second . . .[10]

---

8 Tyler Durden, "First Cisco and Microsoft, Now IBM: China Orders Banks To Remove High-End IBM Servers." *Zero Hedge*, May 27, 2014.

9 Gematria 666 is the pure male power without mercy of Mars, the Egyptian Crocodile Eater of Souls.

10 Don Johnston, "Lawrence Livermore and IBM collaborate to build new brain-inspired supercomputer." *PhysOrg*, March 31, 2016.

SyNAPSE's intent is to "develop electronic neuromorphic machine technology that scales to biological levels"[11]—in other words, to reverse engineer a brain by building parallel processing chips one square micron in size and arraying them in a basketball-sized carbon sphere suspended in a gallium aluminum alloy (liquid metal for maximum conductivity) in a powerful wireless router "tank" communicating with millions of sensors already released around the planet and linked to the Internet.

> These sensors gather input from cameras, microphones, pressure and temperature gauges, robots, and natural systems—deserts, glaciers, lakes, rivers, oceans, and rain forests.[12]

"Natural systems" no doubt include human beings and animals, as well, but I guess that's better left unsaid. As function follows form, SyNAPSE's "neuromorphic, brain-imitating hardware autonomously gives rise to intelligence" by mirroring the human brain's 30 billion neurons and 100 trillion synapses, then surpassing its 1,000 trillion operations per second.

Meanwhile, McGill University in Montréal, Canada has developed a "biological supercomputer" powered by adenosine triphosphate protein strings ("molecular units of currency") and as small as a book, but with the mathematical capabilities of giant supercomputers—and it doesn't overheat.[13]

QUANTUM COMPUTERS

Because the D-Wave is so good at specific problems, [Google] thinks some classical/quantum combination may prove ideal. . .[M]aybe the "neocortex" of future AIs will be comprised of a quantum chip, whereas the rest will remain classically driven.[14]

Classical computers use bits of information in 1s and 0s, 1s being positive charges on a capacitor, 0s being an absence of charge. A quantum computer uses *qubits*—strings of ions held in place by an electrical field and manipulated by laser pulses—and can simultaneously mix 1s and 0s in a quantum state called a *superposition* on a single atom or electron that can be in two places at once or spin clockwise and anticlockwise at the same time. In fact, "measuring a qubit

---

11 Wikipedia.

12 James Barrat, *Our Final Invention: Artificial Intelligence and the End of the Human Era*. New York: St. Martin's Press, 2013.

13 "Bio Breakthrough: Scientists Unveil First Ever Biological Supercomputer." *Sputnik News*, February 28, 2016.

14 Jason Dorrier, "Google Buys Quantum Computer for Artificial Intelligence Lab at NASA." *SingularityHUB*, June 5, 2013.

knocks it out of superposition and thereby destroys the information it holds."[15]

Still, it is the qubit that makes "the weirdest feature of quantum mechanics"—*quantum entanglement*—possible:

> . . .this property enables distinct quantum systems to become intimately correlated so that an action performed on one has an effect on the other, even for systems that are too far apart to physically interact.[16]

This is *quantum teleportation,* "a reliable and efficient way to transfer quantum information [measured as qubits] across a network. . .for a future quantum Internet, with secure communications and a distributed computational power that greatly exceeds that of the classical Internet."[17]

What is most attractive about quantum computers to the Global Security State is their ability to quickly factor large numbers—the mainstay of electronic surveillance and data security—and equally quickly sift through masses of unsorted data to find one person or one event. These two applications alone would make the quantum computer and its superpositioning a game-changer, but add in quantum teleportation and it becomes irresistible.

The first quantum computer comes from D-Wave Systems in Burnaby, British Columbia, Canada, and its first customer was aerospace giant Lockheed Martin in 2011 for its University of Southern California Quantum Computation Center. The Quantum Artificial Intelligence Laboratory (QuAIL) run by the Google-NASA-Moffatt Field Universities Space Research Center consortium has a D-Wave Two, and D-Wave's 2048 adiabatic quantum computer is at NASA Ames and Google X, which is connected to CERN through the UC Berkeley hub for the ESnet5 fiber optic network with a speed of nearly 100 gigabits per second on a 300 GHz band for "research in high energy physics, climate science and genomics.[18] *ESnet5 is all about terabytes and terahertz transmitters.*[19] The NSA's new $2 billion Utah Data Center now operates a 512-qubit chip code-named "Vesuvius" and bankrolled by Goldman Sachs. Vesuvius can execute more than

15 Amy Nordrum, "Quantum Computer Comes Closer to Cracking RSA Encrypton." *IEEE Spectrum,* 3 March 2016. The RSA algorithm is the most widely used encryption method.

16 Stefano Pirandola and Samuel L. Braunstein, "Physics: Unite to build a quantum internet." *Nature.com,* 12 April 2016.

17 Ibid.

18 "ESnet5: Shadow Internet with 100 Gbps Connectivity Speed." *CIOReview,* January 30, 2015. Email from Anthony Patch, June 9, 2016: "The latest network is the go-between Helix Nebula, which began in Europe and is now linked up to the 2048, with Berkeley as its link. Since the founding of the World Wide Web at CERN, Berkeley has been the hub controlling data flow and distribution throughout the U.S. and Europe. All data streams from CERN are routed through UC Berkeley, specifically the Lawrence Berkeley Lab up on the hill, where the Advanced Light Source building housing the Synchrotron particle accelerator is located."

19 105 gigabits (0.1 terabit) 275–450 GHz. Terahertz transmitters are 10X faster than 5G networks: ultra-high speed wireless.

100,000,000,000,000,000,000,000,000,000,000,000,000 computations at once, which would take millions of years on a standard desktop.

D-Wave's site (*dwavesys.com*) boasts of the amazing data crunching and searches its "exotic tool" can achieve: genomic analysis, "looking for bad guys in large amounts of data," space exploration, building AIs, simultaneous comparisons of multiple solutions in the wake of, say, a "natural" disaster—and a 512-qubit quantum computer's "central hubs" that can unlock *any* encrypted file.

> A longtime goal among cryptologists has been to perfect the "quantum Internet"—which in the most basic way possible, uses the main principle of quantum mechanics to transfer communications from one point to another. . .like a hub-and-spoke network in which all messages anywhere in the network get routed from a main node—a central hub.[20]

Who needs a PROMIS backdoor when there is D-Wave?

But it is the Mephistophelean nature of D-Wave's qubit chip that truly sets it apart from a mere supercomputer. Eric Ladizinsky, co-founder and chief scientist of D-Wave who built the 512-qubit quantum computer, equates it with the Manhattan Project and *magic*:

> Quantum computers are not made of simple transistors and logic gates like the CPU on your PC. They don't even function in ways that seem rational to a typical computing engineer. Almost magically, quantum computers take *logarithmic* problems and transform them into "flat" computations whose answers seem to appear from an alternate dimension.

> For example, a mathematical problem that might have 2 to the power of *n* possible solutions — where *n* is a large number like 1024 — might take a traditional computer longer than the age of the universe to solve. A quantum computer, on the other hand, might solve the same problem in *mere minutes* because it quite literally operates across multiple dimensions simultaneously.[21]

The environment of the very small, niobium D-Wave chip cloistered in its ten-foot black cabinet must be kept colder than deep space at −273.13°C (just above absolute zero[22]). *It is the cold that makes the chip behave as a superconductor.* Also, to function optimally the D-Wave processor requires an extremely low magnetic environment—50,000X lower than the Earth's ambient magnetic field.

---

20 Stephen Lam, "The US government has been running a quantum Internet for over two years." *RT*, May 6, 2013.

21 Mike Adams, "Skynet rising: Google acquires 512-qubit quantum computer; NSA surveillance to be turned over to AI machines," *NaturalNews.com*, June 20, 2013.

22 By international agreement, absolute zero is −273.15° on the Celsius scale (−459.67° on the Fahrenheit scale).

[Note: Interestingly enough, *Lucifer* (Large Binocular Telescope Near-infrared Utility with Carnera and Integral Field Unit for Extragalactic Research) in Arizona is chilled to -213°C (-351°F) for near-infrared observations.]

Unlike the supercomputer, the D-Wave quantum computer is indeed artificial intelligence. Its *binary classification* is its ability to categorize and label vast amounts of complex input data (text, images, videos, phone calls, etc.); its *quantum unsupervised feature learning (QUFL)* is its ability to learn on its own, creating and optimizing its own programs; its *temporal QUFL* enables it to predict the future based on information it learns through binary classification and QUFL; and its *artificial Intelligence via Quantum Neural Network processes* means it can completely reconstruct the human brain's cognitive processes and teach itself how to make better decisions and better predict the future.

So it is no surprise that In-Q-Tel, the private investment arm of the CIA, is a major investor in D-Wave,[23] nor is it surprising that government agencies and defense contractors in the business of collecting mass surveillance and tracking data are buying up D-Waves as fast as they're built.

ARTIFICIAL INTELLIGENCE (AI)

Every aspect of learning or any other feature of intelligence can be so precisely described that a machine can be made to simulate it.

— Proposal for the Dartmouth Conference, 1956

Artificial intelligence (AI) systems on a less sophisticated scale than D-Wave are up and running everywhere, from Internet apps like facial and voice recognition and profiling, to translating one language to another, predicting hedge fund capital movements,[24] even building AI algorithms that can build AI algorithms.[25] Airlines, banks, traffic flow, hospitals, insurance, utilities, telephone exchanges, factories, military, Internet—in fact, what *isn't* being run by AI? A new algorithm now learns handwriting as fast as a human child.[26]

SpaceX founder Elon Musk has put together OpenAI, a $1 billion fund "to assist humans in staying at least one step ahead of technology." At MIT, Musk

---

23 Michael Brooks, "What could the NSA do with a quantum computer?" *New Statesman*, 27 June 2013.

24 Richard Craib, "Rogue Machine Intelligence and A New Kind of Hedge Fund." Numerai *Medium.com*, June 21, 2016.

25 Cade Metz, "Building AI Is Hard—So Facebook Is Building AI That Builds AI." *Wired*, May 6, 2016. Other social media—Twitter, LinkedIn, Uber, etc.—will soon be using Facebook's Flow tool "to help engineers build, test, and execute machine learning algorithms on a massive scale . . ."

26 Peter Dockrill, "Scientists have developed an algorithm that learns as fast as humans." *Science Alert*, 11 December 2015.

clearly stated, "With artificial intelligence, we are summoning the demon."[27] His solution? To found Neuralink that will manufacture micron-sized devices that link human brains with computers for what he calls "consensual telepathy":

> Artificial intelligence and machine learning will create computers so sophisticated and godlike that humans will need to implant "neural laces" in their brains to keep up, Musk said in a tech conference last year.
>
> "There are a bunch of concepts in your head that then your brain has to try to compress into this incredibly low data rate called speech or typing," Musk said in the latest interview. "If you have two brain interfaces, you could actually do an uncompressed direct conceptual communication with another person."[28]

Theoretical physicist Stephen Hawking admitted on BBC that a type of system so advanced that it could re-design itself at an ever-increasing rate would eventually exponentially outpace human beings. He then signed an open letter (with one thousand AI experts) in protest of the military AI arms race.[29]

On January 21, 2016—too little, too late—USAF General Paul J. Selva told the Brookings Institute:

> "There are ethical implications. There are implications that I call the 'Terminator conundrum,'" Selva said. "What happens when that thing can inflict mortal harm and is empowered by artificial intelligence? How are we gong to deal with that? How do we know with certainty what it's going to do? Those are the problem sets I think we're going to deal with in the technology sector."[30]

A month later, Deputy Defense Secretary Bob Work admitted before the Atlantic Council that the Aegis combat system commanding missile warships is already being run by an AI:

> "Machines [are] taking over from humans key decisions of when to launch both offensive and defensive missiles. . .In the next decade it's going to become clear when and where we will delegate authority to machines. We will delegate [some] authority to machines," Work explained.[31]

---

27 Brianna Blaschke, "Elon Musk Funds $1B Proposal To Stop Human Destruction From 'Demon' of Artificial Intelligence." *Activist Post*, May 20, 2016.

28 "Elon Musk on mission to link human brains with computers in four years: report." *Reuters*, April 21, 2017.

29 Claire Burnish, "Stephen Hawking Warns Humanity: Leave Earth Before the Ruling Class Destroys It." *Antimedia.org*, January 22, 2016.

30 Bianca Spinosa, "As AI advances, military leaders mull the 'Terminator conundrum'." *Defense Systems*, January 25, 2016.

31 "Artificial Intelligence Now Decides Targets on Aegis Ships." *Sputnik News*, March 5, 2016.

Lockheed Martin's "semiautonomous" Long Range Anti-Ship Missile is already up and running. "Semiautonomous" means that a human operator can select the target, but the missile will then fly hundreds of miles out of contact with its human controller and attack the target, whatever the human operator's second thoughts.[32]

The 2016 report "Autonomous Weapons and Operational Risk" by Paul Scharre of the Center for a New American Security (CNAS)[33] warns against letting AIs make killing decisions, pushing instead for "centaur warfighting," integrated decision-making between humans and computers.[34] Some military brass go so far as to insist that "real" intelligence is what humans can do that machines cannot, while others view AI as a "Third Offset"—computer-based high-tech that will offset a smaller military and "create a new class of 'Iron Man'-style fighters."[35]

NETWORK-CENTRIC WARFARE AND JADE 2

Each human being is an emitter of radio waves, a living broadcasting station of exceedingly low power. The stomach will send out not only infrared heat waves, but the entire spectrum of light, ultraviolet rays, x-rays, radio waves and so on. Of course all these radiations are fantastically weak and the radio waves are among the weakest. But the fifty-foot aerial of the Naval Research Laboratory in Washington, D.C., the most accurately constructed aerial in existence, could pick up radio signals coming from your stomach more than four miles away.

— John Pfeiffer, *The Changing Universe*, 1956

The military doctrine of full spectrum dominance supporting C4 (command, control, communications, and cyberwarfare) has recast the whole of earthly life into a digitalized "battlespace" even as supercomputers, quantum computers, AI systems, and *network-centric warfare (NCW)* come online. A perpetually ionized antenna atmosphere has opened the floodgates to a network-centric cyberwarfare that makes the entire biosphere and near-earth into a 24/7-ready

32 John Markoff, "Pentagon Turns to Silicon Valley for Edge in Artificial Intelligence." *New York Times*, May 11, 2016.

33 The CNAS think tank began in 2007 in Washington, D.C., the year after the infamous Project for the New American Century (PNAC) ceased operation and morphed into the Foreign Policy Initiative.

34 John Markoff, "Report Cites Dangers of Autonomous Weapons." *New York Times*, February 28, 2016.

35 Markoff.

battlespace. In fact, U.S. Cyber Command (USCYBERCOM) is now its own agency and no longer under U.S. Strategic Command. NSA Director Admiral Michael S. Rogers is also commander of USCYBERCOM and chief of Central Security Service (cryptology).

In his 2009 article "Network-Centric Warfare,"[36] Tom Burghardt stresses that the Revolution in Military Affairs (RMA) is really about electronic full spectrum dominance warfare—what *Air Force Magazine* calls "compressing the kill chain." In the 2009 version of NCW, highly classified nano-sized sensors included:

- Tagging tracking and locating gear (TTL), homing beacons to guide drone strikes

- Dynamic optical tags (DOTs), small active retro-reflecting optical tags for two-way data exchange

- Radar responsive tags that include battlefield situational awareness, unattended ground sensors, data relay, vehicle tracking, search and recovery, precision targeting, special operations, drug interdiction, etc.

Now, self-aware artificial intelligence composed of neural networks of quantum processors is communicating with "situational awareness" nano-sensors unleashed by the trillions into the environment, NCW's "operational art" being dependent upon "low-density, high-demand and vulnerable air- and space-based sensing and communications systems."[37]

> For example, aerosols would be sprayed over enemy troops, or chemicals would be clandestinely introduced into their food supply. Then biosensors flying overhead, says Thomas Baines at Argonne National Laboratory in Illinois, would "track their movement from their breath or sweat," so they could be targeted for attack. . .With microprocessors making smaller weapons systems. . .aircraft carriers and manned bombers may become obsolete . . .[38]

Remember: "enemy troops" now include civilians, given that NCW and "situational awareness" are weaponized for the new battlespace of cities, neighborhoods, and rural areas—in other words, *everywhere*.

Meanwhile, all around us AI networks are crunching sensitive data (voice stress, financial profiles, social networking patterns, etc.), remote-controlling nanobots and RFIDs, webcams and websites, cell phones, microphones, electric grids, airplanes,

---

36 Tom Burghardt, "Network-Centric Warfare." *Global Research*, June 11, 2009.

37 Col. Alan D. Campen, "Look Closely At Network-Centric Warfare." *SIGNAL*, Armed Forces Communications and Electronics Association (AFCEA), January 2004.

38 Douglas Waller, "Onward Cyber Soldiers." *Time* magazine, August 21, 1995.

vehicles, drones, weather systems, "exotic weapons," etc. Space Fence geopolitical control depends entirely upon interlocking AI networks surveilling everything and everyone in what the military calls the *Human Terrain System (HTS)*.

NCW and "compressing the kill chain" includes mind control and "perception management," as plugging the entire world into a wireless grid in which each dimension is connected *orthogonally* to every other dimension points directly to RMA "multidimensionality," as do the all-too-lethal nonlethal (less-than-lethal / electronic) weapons, including your cell phone:

> The conventional and nuclear weaponry dimensions might be continuous; within each of these dimensions, there is a smooth gradation in size and efficacy of weapons employed. A third dimension, targeting, might then be added; this would likely be a discontinuous dimension with a discrete point for each type of target, namely industrial, military, etc. A fourth dimension might contain the issues involved, a political dimension. A fifth could be the cost or economic dimension, etc.. . .Interactions between factors would be represented by a hypersurface in this multidimensional space. . .A model of conflict as a multidimensional manifold, with the idiosyncrasies and pathology of a mathematical hypersurface, increases our predictive power tremendously.[39]

The RMA is structured on the understanding that the most powerful weapon systems are those that can subdue populations by using subtle energy fields and tricking the body and mind into reacting to incoming signals as normal and natural—signals that can just as well be used to create total disorientation or remotely trigger mystery illnesses, heart failure, respiratory distress, etc. This is DARPA's Combat Zones That See (CTS) and military operations in urban terrain (MOUT) graduating from small CCTV cameras to "nano-swarms":

> . . .a "swarm" of near-microscopic nano-devices that function in consort [sic] with each other to produce either a visual image or an audio image of whatever is in the vicinity of the "swarm." The more nano-devices that are present, the more fidelity or resolution they can resolve and transmit. These little particles can be made "sticky" so that they attach themselves to almost anything: a wall, a tree, a vehicle, even a person's clothing. It may even be that they can be made selectively sticky, so that they only will stick to something when told to do so.[40]

Key to understanding the *turning point* that the May 2015 "military exercise" known as JADE Helm represents (Joint Assistant for Deployment and Execution at the helm) is to grasp the fact that *we the people are already NCW wired* into AI and, by proxy, the Space Fence. As I stressed in Chapter 3, "The Nano Assault,"

---

39 Ibid.

40 Robert G. Willscroft, Ph.D., "Combat Zones that 'See' Everything." *Defense Watch*, July 14, 2003.

nano-scale "situational awareness" sensors have already been loaded into the environment and, by proxy, our lungs, bloodstreams, and brains. Burghardt's dynamic optical tag (DOT) sounds remarkably like *quantum dots*. Then there is neural or smart dust, nanospheres, carbon nanotube transistors and integrated circuits, electrochemical energy-storage nanos with anode-cathode nanowires inside polymer core shell separators, etc. All are nanotechnology telemetric sensors released into our environment for us to wear, breathe, and ingest, each sensor programmed to gather data that is then remotely accessed, transmitted, and stored by AIs.

Berghardt called attention to the civilian version of Radar Responsive tags made in 2009 by Gentag:

> According to Gentag, "the civilian version. . .is a lower power technology suitable for commercial civilian applications, including use in cell phones and wide area tracking." Conveniently, "Mobile reader infrastructure can be set up anywhere (including aircraft) or can be fixed and overlaid with existing infrastructure (e.g. cell phone towers)."[41]

Virginia-based Inkode developed tiny, low-power metal fibers that embed themselves "in paper, plastic and other materials that radio frequency waves can penetrate." Does "and other materials" include lungs, blood, and skin? The fibers reflect radio waves back to the reader in a "resonant signature," and the reader of data flowing in from our sensor-armed environment is an AI. Queralt in Wallingford, Connecticut invented "an integrated behavioral learning engine" to "learn an individual's or asset's habits over time":

> The core of Queralt's system is the behavioral engine that includes a database, a rules engine and various algorithms. Information acquired by reading a tag on an asset or individual, as well as those of other objects or individuals with which that asset or person may come into contact, and information from sensors (such as temperature) situated in the area being monitored, are fed into the engine. The engine then logs and processes the data to create baselines or behavioral patterns. As baselines are created, rules can be programmed into the engine; if a tag read or sensor metric comes in that contradicts the baseline and/or rules, an alert can be issued . . .[42]

Feeling secure yet?

May 2015 was the beginning not of just an NCW rollout but of a complete makeover of what military war games and exercises are about. In the United States, it was called JADE Helm (i.e., JADE at the helm); in Estonia, Hedgehog 15;

---

41 Tom Burghardt, "Network-centric warfare." *Global Research,* June 11, 2009.

42 Beth Bacheldor, "Queralt Developing Behavior-Monitoring RFID Software." *RFID Journal,* April 23, 2009.

in Lithuania, Operation Lightning Strike; in Norway, Dynamic Mongoose; in the Mediterranean Sea, Joint Sea 2015-I was run by the Russians and Chinese.[43] All exercises are "simulations" of *network-centric operations* (NCOs, i.e., cyberwarfare) dependent upon the *global information grid (GIG)*, once a key component of the Project for the New American Century (PNAC) / Foreign Policy Initiative.

According to Raytheon's BBN Technologies / DARPA Abstract in the 2001 Final Technical Report "Joint Assistant for Development and Execution (JADE)," JADE 2 is AI quantum computing technology capable of utilizing vast stores of NSA-collected data to produce a *Human Terrain Analysis (HTA)* tool that will guarantee mastery over the *Human Domain*.

> JADE (Joint Assistant for Deployment and Execution) is a knowledge-based mixed-initiative system that supports force deployment planning and management. JADE uses case-based and generative planning methods to support the development of large-scale, complex deployment plans in minimal time. JADE incorporates the technology of three tools:
>
> Prodigy-Analogy, a combined case-based and generative planner developed by Carnegie Mellon University;
>
> ForMAT (Force Management and Analysis Tool) that supports case-based force deployment planning developed by [Raytheon] BBN Technologies [and MITRE; supports NSA knowledge acquisition]; and
>
> PARKA, a highly indexed knowledge-based management system developed by the University of Maryland.
>
> With JADE, a military planner can build a preliminary force deployment plan, including the Time Phased Force Deployment Data (TPFDD) in less than one hour. This speed in plan construction is possible because JADE supports the rapid retrieval and reuse of previous plan elements for use in the development of new plans. In addition, JADE employs an easy to use map-oriented drag and drop interface where force modules (FMs) from previous plans (cases) whose force capabilities and composition match the requirements of the current situation can be dragged from the case library and dropped onto a geographic destination. Plan modification and/or adaption is supported through remindings, e.g. each time that a force module is created or is copied into a plan (TPFDD) the user is automatically reminded of the need for geographical changes.[44]

---

43 Other NCW "war games" may not have been made public. For example, in October 2015 on Wake Island "a tandem defense capability in the Pacific unmatched in the world today by successfully combining U.S. Aegis BMD and THAAD multi-layered defense in a live-fire exercise against both cruise and ballistic missiles" ("Guam — A Fiesta," Missile Defense Advocacy Alliance, November 30, 2015).

44 "Joint Publication 3-35: Deployment and Redeployment Operations," Joint Chiefs of Staff, 31 January 2013. The primary AI "military planner" in 2001 was the force management module for ForMAT and PARKA; now, JADE 2 is served by the Prodigy module.

JADE 2 was publicly announced, but its purpose remains obscured beneath cryptic military acronyms and references to "simulations."[45] The truth is that JADE 2 is a network-centric software-based AI program "at the helm" that can reconfigure its own network topography to optimize data transfers. Not only is it *aware* but it is self-adaptive, self-reactive and self-modeling for predictive forced deployment. It is preemptive in that it can interpret commanders' intent and potential behaviors by means of macro-cognition mind mapping.

JADE 2 first collects masses of data from remote sensors and computers at data-dump places like fusion centers and the Intelligence Community Comprehensive National Cybersecurity Initiative Data Center in Bluffdale, Utah, then moves it into an HTA module to develop an HTS to determine the behavioral parameter norms (and therefore *Human Domain Deviations*) for individuals, groups, dense populations, etc. In seconds, JADE 2's Prodigy Logic Module can generate holographic simulations of battle plans, pre-crime plans, and "kill chains." The CBR (case-based reasoning module) then examines the mission statement and comes up with an ACOA (adaptive course of action), *all within seconds.*

After sucking up sensor data, JADE 2 thinks, plans, and executes, its ultimate mission being to transform C4 into total mastery of the Human Domain. With network-centric HTA and activity-based Intelligence (ABI) tools, mastery of the Human Domain under the Geospatial Intelligence (GeoINT) neural net is imminent.

> JADE 2 has the ability to use vast amounts of data being collected on the Human Domain to develop an HTS [Human Terrain System] for geographic locations to identify and eliminate targets flagged on a GIG [global information grid] in Network Concentric Warfare.[46]

It is not JADE 2 that "casts a neural net" but the nanosensors and microprocessors swimming in our bloodstreams and brains and ionized atmosphere. All the mapping of our brains, emotions, and behaviors goes into developing HTSs, but Mastering the Human Domain is the military objective.

Finally, several months after the JADE 2 exercises throughout the American Southwest and elsewhere had run their course in the mainstream media, a brouhaha erupted in Washington, D.C. regarding "the controversial battlefield anthropology program known as the Human Terrain System."[47] Apparently, the

---

45 I am indebted to system and network engineer DJ Walsh's Level9News YouTubes. Please watch them all, beginning with "

www.youtube.com/watch?v=FiKBPmq37Yo, and her July 26, 2016 interview on John B. Wells' show Caravan to Midnight, "Pentagon's HSCOI Program Using Social Media as Targeting Telemetry."

46 DJ Walsh, "The History of JADE II: Planetary Conquest By a Global A.I. Warfare System." Level9News YouTube, May 19, 2015.

47 Tom Vanden Brook, "$725M program Army 'killed' found alive, growing." *USA Today*, March 10, 2016.

Army had said that the HTS was terminated in 2014, but then Rep. Duncan Hunter (R-CA) on the Armed Services Committee discovered it was still very much alive, being funded, and expanding. Given that HTS was initially presented to Congress as a "battlefield anthropology program" and not a biometrics AI sweep of behavioral parameters, it is no wonder that Congressman Hunter was incensed that from the beginning, the Army had misled Congress about the HTS and was even then conjuring up a cover story:

> . . .documents obtained by *USA Today* show that Army officials had simply changed the program's emphasis from deploying social scientists with troops to providing information to commanders from a group of experts at Ft. Leavenworth, Kansas. They also began referring to the program as the Global Cultural Knowledge Network.

"Cultural knowledge," huh? (Are congressmen watching DJ Walsh's YouTubes?)

CHAPTER 7

The "Star Wars" Space Fence Rises Again

▼

Let the future tell the truth, and evaluate each one according to his work and accomplishments. The present is theirs; the future, for which I have really worked, is mine.

— Nikola Tesla (1856–1943)

In this era of science fiction that is no longer fiction—when the military uses the Moon for laser target practice[1] and earth-penetrating tomography (EPT) provokes deep plate shifts and accelerates the inner core[2]—it becomes imperative to doubt just how "natural" our Earth events are. Certainly there are solar flares and sunspots, but what about the impact of tweaking the auroral electrojet and blasting the ionosphere with microwaves? Plausible deniability hangs like a shroud over military operations and experiments, preventing the public from discerning what is "natural" from what is anthropogenic.

Thus far, we have taken a look at the various components of what it takes to keep our lower atmosphere ionized and primed for optimal Space Fence operation. I have attempted to alert the reader to the presence of phenomena not covered in their high school or college physics: æther, plasma, and scalar waves, all of which play a pivotal role in the atmospheric physics of geoengineering still denied the public. We examined the secret presence of nanotechnology already invading every aspect of our lives, including the air we're forced to breathe, the foods we eat, and the water we drink. As signs and wonders of a science hidden from the public appear in the global skies, we must consider that the Earth and human society have been weaponized in the name of the Revolution in Military Affairs, at the cold heart of which lies the shadowy Deep State of the military-industrial-intelligence complex ever in service to the planetary lockdown of the

---

1 "NASA 'shoots moon' in search of water ice Friday." AP, October 9, 2009.

2 The Earth's inner core is directly related to the Earth's magnetic field via dynamo action in the liquid outer core.

AI-run Space Fence as the *modus operandi* by which earthly civilization will finally be offered up to a Transhumanist future.

The planetary technology comprising the Space Fence system is mind-boggling. It is this Rubik's cube that this book begins deciphering. In this chapter, we concentrate on how the Space Fence is being primed aerially and in space; in Chapter 8, we'll concentrate on the ground-based infrastructure.

Into our thin lower atmosphere, chemical nanoparticles, microprocessors and sensors are being injected by jets even as sounding rockets spew chemical compounds, microprocessors and sensors into the upper atmosphere to set the scene for *Space Situational Awareness.* One by one, up go the rockets from Wallops, Canaveral, Poker Flat, and Vandenberg to release aluminum sulfate, lithium (red), barium (green / bluish purple),[3] and Mylar components that ignite in the upper atmosphere and space. Centrifugal force then whirls the thermite brew into a Saturn-like ring around the equator while above Earth a man-made aurora blossoms to be seen as far south as Missouri, South Carolina, Arkansas, Oklahoma, and Texas.[4]

In 2012, Japanese researchers at the National Astronomical Observatory foresaw another Maunder Minimum or Little Ice Age and that the Sun's two magnetic poles could become four.[5] NASA spun the opposite: that the minimum was actually a maximum.

> Something unexpected is happening on the sun. 2013 is supposed to be the year of Solar Max, the peak of the 11-year sunspot cycle. Yet 2013 has arrived and solar activity is relatively low. Sunspot numbers are well below their values in 2011, and strong solar flares have been infrequent for many months.
>
> The quiet has led some observers to wonder if forecasters missed the mark. Solar physicist Dean Pesnell of the Goddard Space Flight Center has a different explanation: "This *is* solar maximum," he suggests. "But it looks different from what we expected because it is double peaked."[6]

"Something unexpected"? Double peaked? Four magnetic poles? Another Carrington Event?[7] Cosmic weather reports are now as anxiety-provoking as regional weather reports.

3 T. Neil Davis, "Lithium Red Sky, Article #312." Alaska Science Forum, April 16, 1979.

4 "Northern Lights over St. Cloud, Minnesota," March 17, 2015: www.telegraph.co.uk/news/worldnews/northamerica/usa/10351617/Spectacular-Northern-Lights-display-filmed-over-US-Midwest.html.

5 Seiji Tanaka, "Sun may soon have four poles, say researchers." *The Asahi Shimbun*, April 20, 2012. The Little Ice Age was seventy cold years in the seventeenth century when England's Thames River froze and cherry blossoms bloomed late in Kyoto.

6 "Solar Cycle Update: Twin Peaks?" *NASA Science News*, March 1, 2013.

7 The Solar Superstorm of 1859 when a solar flare or CME hit the Earth's magnetosphere.

Then in 2013, a veritable theater of meteors and asteroids erupted on the scene—or were they plasma connected with sounding rocket thermal events? On February 15, 2013, a meteor burned across the Russian sky only to disintegrate above Chelyabinsk[8] less than a day before Asteroid 2012 DA14 made the closest recorded pass in history. Exploding over Russia's Ural Mountains, the Chelyabinsk meteor struck just sixty miles from nuclear and chemical weapons disposal facilities. Some compared it with the 1908 Tunguska event discussed in my *Chemtrails, HAARP* book. A month later, a meteor exploded "like a fireball" over Cape Town, South Africa. More meteors plummeted over Texas, Florida, San Francisco, New York City, Cuba, and Australia.

Strangely, the timing of the "meteor" events were just before the Near-Earth Object (NEO) threat mitigation conference in Spain in early May 2013, sponsored by Space Situational Awareness of the European Space Agency and the Spanish corporation Elecnor Deimos Space.[9] On May 31, Asteroid 1998 QE2 buzzed the Earth, followed on September 29 by purported meteorite pieces dropping on the small Yucatán town of Ichmul:

> The falling object was accompanied by a strong thundering noise and a loud blast. . .flashing blue hazes and a power outage. . .police started to play with the gathered pieces and formed humanoid figures whose images have caused wonder and excitement among locals and foreigners . . .[10]

NASA chief Charles Bolden recommended prayer,[11] but Nazi engineer Wernher von Braun (1912–1977) may have been right about the U.S. utilizing fear to cow the public—first, of the Russians, terrorists, and nations of concern, then of asteroids, UFOs and extraterrestrials.[12] The ionized sky theater was a perfect platform for the latter.

The Space Fence as represented by the mainstream media, however, is just a tabulator of orbiting "space debris" (about 200,000 objects) that threatens our satellites, and about keeping tabs on space events like the Chinese "kinetic kill" of their own satellite Fengyun 1C with an antisatellite on January 11, 2007, the debris of which damaged a small Russian "Ball Lens in Space" (laser-ranging retro-reflector) satellite.[13]

8 "Meteor explodes, rains down on central Russia; 1100 injured." *New Zealand Herald*, February 16, 2013.

9 Lee Rannals, "Scientists Gathering To Discuss Asteroid Threat Mitigation." *redOrbit.com*, May 8, 2013.

10 "Huge fireball explosion creates power outage in Yucatán, Mexico." *Strange Sounds*, 30 September 2013.

11 Irene Klotz, "Large asteroid heading to Earth? Pray, says NASA." *Reuters*, March 19, 2013.

12 Reported by Dr. Carol Rosin, President of the Institute for Cooperation in Space (ICIS), who testified at the Disclosure Project Press Conference, National Press Club, Washington, DC, on May 9, 2001 regarding Dr. Wernher von Braun's claim that a succession of "false flag wars" would follow the end of the communism vs. capitalism Cold War (1945–1991).

13 Karl Tate, "Russian Satellite Crash with Chinese ASAT Debris Explained." *Space.com*, March 8, 2013.

The Space Fence nuncio arrived on cats' feet with Air Force Global Strike Command first launching an unarmed LGM-30G Minuteman III ICBM from Vandenberg Air Force Base (August 19, 2015), then three months later the Navy launching a nuclear-tipped Trident II (D5) missile toward Kwajalein Atoll from a ballistic submarine, smearing the sky with a blue-green plume:

> The Navy's fleet of 14 ballistic submarines can each carry 24 Trident missiles, each tipped with 14 independently targetable thermonuclear warheads. . .The test on Saturday featured the launch of a missile outfitted with a dummy warhead toward the Kwajalein Atoll, a missile test site that's part of the Marshall Islands in the western Pacific. . .The U.S. military's nuclear weapons strategy rests on a triad of delivery systems—bombers, submarines and land-based missiles. . .The submarine missile test came late Saturday after Defense Secretary Ashton Carter addressed a defense forum at the Ronald Reagan Presidential Library in Simi Valley about the U.S. "adapting our operational posture and contingency plans" to deter Russia's "aggression."[14]

Nice touch, that tip of the hat to Ronald Reagan whose administration initiated the "Star Wars" program now culminating in the latest addition to the ground-based system upon which the Space Fence depends going up on the Kwajalein Atoll at the old Ronald Reagan Ballistic Missile Test Site.

The Space Fence rises from "Star Wars"

> The Naval Space Surveillance System field stations comprise a bi-static radar that points straight up into space and produces a "fence" of electromagnetic energy. The system can detect basketball-sized objects in orbit around the Earth out to an effective range of 15,000 nautical miles. Over 5 million satellite detections or observations are collected by the surveillance sensor each month. Data collected by the Fence is transmitted to a computer center at Dahlgren [VA], where it is used to constantly update a database of spacecraft orbital elements. This information is reported to the fleet and Fleet Marine Forces to alert them when particular satellites of interest are overhead. The Navy's space surveillance system is one of about 20 sensors that together comprise the nation's worldwide Space Surveillance Network directed by U.S. Strategic Command in Omaha, Nebraska.[15]

The Space Fence actually began with the Navy Space Surveillance System (NAVSPASUR) in 1957, just after the Soviets launched the Sputnik satellite. Designed to track both transmitting satellites and those that were quiet,

---

14 W.J. Hennigan, "Navy launches second test missile off Southern California coast." *Los Angeles Times*, November 9, 2015.

15 "Navy Space Surveillance System [NAVSPASUR]." *GlobalSecurity.org*, no date.

NAVSPASUR's ground base consisted of a nine-radar array "fence" (217MHz each) from Georgia to Southern California at the 33rd parallel north: two transmitters at Gila River, Arizona (pre-recalibration frequency 219.97MHz) and Jordan Lake, Alabama (pre-recalibration frequency 216.99MHz); a more powerful addition at Lake Kickapoo, Texas (768kW radiated power, pre-recalibration frequency 216.983MHz);[16] and six receiving stations, four of which are still operating in San Diego, California, Elephant Butte, New Mexico, Red River, Arkansas, and Hawkinsville, Georgia.

The 1983 Strategic Defense Initiative (SDI), known familiarly as "Star Wars," was presented as a multi-layered outer space defense system based on "non-chemical kinetic and directed energy weapons"—kinetic kill and speed of light weapons, neutral particle beams, ground-based lasers, electrons using fighting mirrors and hyper-velocity guns—against invading ballistic missiles divided into flight-orbit stages of booster, late booster, mid-orbit, and last-stage.

The plan in the 1980s was that a space-based constellation of forty platforms would deploy 1,500 kinetic interceptors. But what happened was that the initial stage alone—Brilliant Pebbles, a satellite constellation of 4,600 kinetic interceptors (KE ASAT) in low Earth orbit, each weighing 100 pounds (45 kg), and their associated tracking systems—would cost $125 billion, and that wasn't counting the next stage deployment of even larger platforms, including laser and particle beam weapons like the Mid-Infrared Advanced Chemical Laser (MIRACL). It became evident that "Star Wars" was premature and that a more sophisticated ground-based system would have to be developed to support space-based platforms.

Along came Bernard Eastlund and his 1987 HAARP patent, leading to a decade of HAARP experiments that solved the problem of keeping the lower atmosphere ionized to sandwich between near-earth orbit space platforms and a conductive ground-based infrastructure. HAARP fulfilled every military hope and more: it altered the relationship between the ionosphere and the troposphere while Project Cloverleaf provided jet deliveries of conductive nanoparticles around the globe as smaller and mobile ionospheric heaters were built, and radar installations, towers, and phased-array installations proliferated.

On October 1, 2004, NAVSPASUR was passed from the U.S. Navy to the U.S. Air Force 20th Space Control Squadron and renamed the AN/FPS-133 Air Force Space Surveillance System (SSS / the VHF Fence), a key component of the Space Surveillance Network (SSN).

In August 2013—one year before HAARP's shutdown—the AFSSS ceased operation so it could be recalibrated to the frequencies and pulses of the global infrastructure of ionospheric heaters, radar installations, towers, NexRads, wind farms, fracking wells, etc.

---

16 All Space Fence recalibrated frequencies are classified and non-trackable. They are evolving into S-band (2–4 GHz with wavelengths of 15–7.5cm) on multiple exchange frequencies. Related bands are E / F (NATO).

In 2014, the Lockheed Martin SATCOM Technologies team (Lockheed Martin, Raytheon, AMEC, AT&T, and General Dynamics) began building a six-acre array system on the Kwajalein Atoll 2,100 nautical miles southwest of Honolulu[17] that would replace the AFSSS with an S-band (2.2–2.3GHz)[18] ground-based radar system of four hundred or so units in service to continuous *space situational awareness.*

> "The ground-based receive array is an elegant merger of a huge physical structure built with the precision of a complex scientific or medical instrument," said Mike DiBiase, a vice president and general manager of General Dynamics Mission Systems. "The SATCOM Technologies-built array has the sensitivity to locate, identify and track objects as small as a softball, hundreds of miles above the Earth's surface."[19]

A scaled-down version of the Lockheed Martin Kwajalein Atoll next-generation space surveillance system opened in 2016 in New Jersey as a "test site."[20]

As part of the Space Situational Awareness Group of the U.S. Air Force, the Space Based Space Surveillance (SBSS) system detects and tracks space objects in orbit around the Earth while the previously classified Geosynchronous Space Situational Awareness Program (GSSAP) satellites are loaded with dedicated SSN electro-optical sensors in communication with Air Force Satellite Control Network (AFSCN) ground stations like Schriever Air Force Base in conjunction with the 50th Space Wing of Air Force Space Command (AFSPC) in Colorado Springs. (The present incarnation of GSSAP gives a whole new meaning to "neighborhood watch.")

> GSSAP satellites will support Joint Functional Component Command for Space (JFCC SPACE) tasking to collect space situational awareness data . . .[21]

Broadly speaking, AFSPC has four missions: (1) space forces support; (2) space control; (3) force enhancement (weather, communications, intelligence, missile warning, navigation); and (4) force application. Translated, this is C4.

---

17 "Construction Underway for U.S. Air Force Space Fence Radar in the Marshall Islands." Lockheed Martin press release, March 2015. (2014 net sales for Lockheed Martin were $45.6 billion.)

18 Shorter S-band wavelength means greater detection capability for small satellites and debris—data feeding into the Joint Space Operations Center Mission System at Vandenberg Air Force Base monitoring space weather and foreign launches.

19 "General Dynamics Completes U.S. Air Force Space Fence Radar Array Ground Structure." *PRNewswire*, April 7, 2016.

20 Mike Gruss, "Lockheed Martin opens Space Fence test site in New Jersey." *SpaceNews*, March 28, 2016.

21 "Geosynchronous Space Situational Awareness Program (GSSAP)," Air Force Space Command Public Affairs Office, April 2015.

The 50th Space Wing satellite operators of the 1st Space Operations Squadron uplink C4 calculations for weapons command from MacDill Air Force Base (Patriot missile and Iron Dome) and are in touch with the Kwajalein Atoll installation that feeds data to the Joint Space Operations Center at Vandenberg Air Force Base and with Eglin Air Force Base Site C-6 radar station whose AN/FPS-85 phased-array radar runs the radar / computer processing.

It is important to remember that the U.S. Air Force[22] was tutored by Paperclip Nazi scientists like Hubertus Strughold, M.D., who conducted pilot stress tests and experiments in radiobiology and human radiation at the School of Aviation Medicine (SAM) near Randolph Air Force Base in San Antonio, Texas. Today, SAM, the Human Effects Center of Excellence, and the Air Force Research Laboratory continue to research nonlethal weapons like lasers, masers, microwave hearing, synthetic telepathy / voice-to-skull (V2K), brain-machine interface (BMI), etc. In fact, AFSPC at Peterson Air Force Base may be the military hub of artificial telepathy operations: "It's the 'mission control' center where rocket scientists, AFRL, HAARP, spy satellites, radar dishes, microwave towers, beam weapons, human experimentation and spooky intelligence agencies like NSA, NRO and DIA all come together."[23]

Now let's move on to the control over the poles that plays heavily in Space Fence operation. The Nazis studied the poles but not necessarily for the mythical reasons disseminated after the Nuremberg show trials.

MAGNETIC NORTH & SOUTH

[NASA Jet Propulsion Laboratory senior research scientist Surendra Adhikari:]
The pole used to be heading along ~75 degrees west longitude, toward Canada,
during the 20th century. It is now heading along the central meridian, i.e. 0
degrees longitude, toward the UK. So this shift in direction would roughly be
about 75 degrees to the east, from Canada to the UK.[24]

In November 2013, The European Space Agency (ESA) launched three 9-meter SWARM satellites to monitor and map the magnetosphere from 300–530 km

---

22 The latest U.S. Air Force Strategic Master Plan was released on May 21, 2015 in order to translate "the United States Air Force's 30-year strategy, *America's Air Force: A Call to the Future*, into comprehensive guidance, goals, and objectives." www.af.mil/Portals/1/documents/Force%20 Management/Strategic_Master_Plan.pdf?timestamp=1434024300378.

23 "Focus On: The U.S. Air Force Space Command." Artificial Telepathy blog, June 22, 2009.

24 Jacqueline Howard and Chris D'Angelo, "Climate Change May Be Causing Earth's Poles To Shift." *Huffington Post*, April 11, 2016.

(186–330 miles) above the planet. It was SWARM that scientifically confirmed the North Pole drift which Inuit Indian shamans had discerned years before, and SWARM that provides data regarding the South Atlantic Anomaly, an indicator of a possible geomagnetic pole reversal.[25]

The North and South Poles are moving, as is our magnetosphere. If the Earth's axis is shifting, the passive sounding "climate change" (melting ice sheets, loss of water mass in Eurasia, etc.) makes what is going on at the poles worthy of headlines. It may even be time to reconsider the Earth axis deviating considerably from its current position in the only extant map of an ancient sky we have, the Dendera Zodiac on display at the Musée du Louvre in Paris.[26]

Given the state of the art fifth-generation EISCAT 3D ionospheric heater with its 100,000 simultaneous antennas, and four previous generation heaters are located in the Arctic region—EISCAT in Norway, HAARP Gakona, HIPAS Fairbanks in Alaska, and Brookhaven National Laboratory on Long Island, New York—and myriad "space trails" of chemical and metal nanoparticles being spewed from rockets launched in the Northern Hemisphere (like Poker Flat Research Range and NASA's Wallops Flight Facility), it is safe to say that the poles remain a big military concern and play heavily into Space Fence "space situational awareness."

If you were to believe retired U.S. Army Colonel David Hunt, it's Russian saber-rattling that U.S. interest in the Arctic is about, given that Russia shares the Arctic Sea with NATO nations under a USUK thumb (Greenland, Canada, Sweden, Norway, and Finland):

> Recently, Russia submitted a claim to the United States for 1.2 million square kilometers of Arctic sea shelf, including the North Pole. The territory could hold about 5 billion tons of oil and gas resources. Above its northern coastline, they've also asserted ownership of the emerging Northern Sea Route, the Arctic's fastest-growing shipping route.
>
> This comes after Russia spent years militarizing the region unabated and without challenge. Not only have the Russians placed a flag via submarine on the seabed of the North Pole and rehabilitated a Soviet-era military base, they've also launched a full-alert combat readiness exercise with 38,000 troops, 110 aircraft, 41 ships, and 15 submarines. They've added a 6,000-soldier permanent military force in the Arctic's northwest Murmansk region, equipped with new radar and guidance system capabilities and coastal defense missile systems. Russia currently has a fleet of six nuclear-powered icebreakers and at least a

---

25 F. Javier Pavon-Carrasco and Angelo De Santis, "The South Atlantic Anomaly: The Key for a Possible Geomagnetic Reversal." *Earth Science*, 20 April 2016.

26 According to mythographer John Lamb Lash, the Dendera Zodiac is a "time-scanning dial" recording the mythic pattern of humanity's experience over a cycle of 26,000 years (a Platonic year) due to end in 2216 CE. Cf. "The Dendera Revelation: Our Moment in the Mythic Order of the Ages," *Metahistory.org*.

dozen diesel-powered icebreakers, and three more nuclear-powered icebreakers will be added by decade's end.[27]

But that's old-school propaganda for an old-school public. We've entered the Space Age and planet Earth is being weaponized while new commercial possibilities are opening up.

Plasma. Our magnetosphere is basically a cloud of electrified gas called plasma that surrounds and protects our Earth. It was once assumed that this plasma was coming from solar winds, but now we know it is spewing from the poles, though solar winds energize the magnetotail, as well. Particles from the poles zoom up as far as sixty Earth radii (400,000km) into the magnetotail, then zip back along the plasma sheet and bounce back and forth along the magnetic field lines (mirroring).

In 1981, a RIMS instrument (retarding ion mass spectrometer) on a Dynamics Explorer satellite discovered gases flowing out into space from both magnetic poles. After measuring the polar "fountains" of ions of hydrogen, helium, oxygen, and nitrogen, the solar wind theory was abandoned: the Earth generates all it needs to shield itself from solar electromagnetic radiation. Later, the TIDE instrument (thermal ion dynamics experiment) on the *Polar* spacecraft measured the "fountain" from twice the distance while neutralizing the spacecraft's plasma sheath so it could read just how much low-energy plasma was flowing out of the atmosphere and into the magnetosphere.

> "The really incredible thing," [plasma physicist Rick Chappell] said, "is that if you do a very careful model of the magnetic field and electric field, and then put in 10eV particles, they go into the plasma sheet [extending outward from the magnetic equator] and are energized at least a thousandfold."[28]

A century before the discovery of plasma fountains issuing from the poles, ether scientist John Worrell Keely of Philadelphia laid out three polar forces that make up the governing conditions of "the controlling medium of the universe" he and others of his time called *ether:* magnetism, electricity, and gravity. Magnetism provides polar attraction and gravity polar propulsion, gravity being "nothing more than a concordant attractive sympathetic stream flowing towards the neutral center of the earth." Keely's patron Clara Bloomfield-Moore put it this way:

The great polar stream with its exhaustless supply of energy, places at our

27 David Hunt, "America Still Making The Wrong Moves In Arctic Chess Game With Russia." *The Daily Caller*, August 28, 2015.

28 "Earth weaves its own invisible cloak — Polar fountains fill magnetosphere with ions." *NASA Science*, 1997. Also see "Plasmas Can't Hide From Neutralized TIDE." *NASA Science*, November 20, 1996.

disposal a force. . .We have but to hook our machinery on to the machinery of nature and we have a force, the conditions of which when once set up remain forever, perpetual molecular action the result."[29]

Because these polar fountains play an immense role in maintaining the strength of the magnetosphere, it seems obvious that "natural" planetary changes like the occasional magnetopause collapse or geomagnetic pole reversal (every 780,000 years) are connected to the military and its "team" of plasma physicists and geoengineers whose "experiments" run the risk of imminent danger to both the ionosphere and magnetosphere.

When the U.S. Naval Research Laboratory (NRL) announced that HAARP had produced "a sustained high density plasma cloud in Earth's upper atmosphere. . .to be used for reflection [artificial mirrors] of HF radar and communication signals,"[30] did it not occur to physicists that "plasma clouds or balls of plasma" are now also being created in our lower ionized atmosphere?

Ice-free Arctic. Also in 2013 (the infamous year of asteroids and meteors, remember?), a U.S. Navy research project determined that the Arctic Ocean might be ice-free by summer 2016, "opening the door to vast reserves of fossil fuel, and eventually freeing up a shipping lane between Europe and Asia."[31] The Navy study admits to including measurements of "chemical and biological processes," which makes this projected date more feasible than 2100. In fact, a 2016 *Arctic News* article echoes the same projection: on May 9, 2016, Arctic Sea ice was 1.1 million square kilometers less than it was on May 9, 2012. The now erratic, wavy jet stream is pulling warm air along its path, and cold air is moving south out of the Arctic.[32] And now that the ice is being melted, 15,600 kilometers of fiber-optic cable is being laid for a 24-terabit connection between Tokyo and London.[33] Wiring the top of the world for wireless Space Fence transmissions.

Methane and Project Lucy. But with all the melting of glaciers, excavating for oil and digging for fiber-optic cable, what of methane gas releases? Methane gas in the Arctic issues from ancient permafrost carbon stored beneath the shallow sea of the East Siberian Arctic Shelf. The concern is that rapid warming could release as much as 500 billion tons of carbon as methane. Escaping through conduits known as *taliks*, methane bubbles to the surface and adds yet more heat pressure to the already-heating atmosphere.

---

29 Theo Paijmans, *Free Energy Pioneer: John Worrell Keely.* IllumiNet Press, 1998.

30 "NRL Scientists Produce Densest Artificial Ionospheric Plasma Clouds Using HAARP," NRL News Release, February 25, 2013.

31 David Schmalz, "NPS researchers predict summer ice might disappear by 2016, 84 years ahead of schedule." *Monterey County Weekly*, November 27, 2013.

32 Sam Carana, "Arctic Sea Ice gone by September 2016?" *Arctic News*, May 13, 2016.

33 Amy Nordrum, "How to Thread a Fiber-Optic Cable Through the Arctic." *IEEE Spectrum*, 26 January 2015.

In 2015, a SoCalGas natural gas well blowout in California released over 100,000 tons of greenhouse gas methane—the largest methane leak in U.S. history. It spread out into the densely populated San Fernando Valley and was laced with volatile organic compounds like ethane, benzene, butanes, and pentanes.[34] After the panic died down, the issue became greedy Big Oil and its fracking and bottom-line production, processing, pipeline, and storage infrastructure at the expense of public health.

Given the propaganda and lies in our corporate-controlled media, plus the number of misled and embedded scientists who buy into the military-industrial-intelligence complex domination of peer reviews, university and lab grant monites, careers, etc., it is difficult to weigh the "mass extinction" threat factor of methane releases. Papers like Malcolm P.R. Light's "The Non-Disclosed Extreme Arctic Methane Threat" that blame "human emissions" for global warming and methane increasing in the atmosphere 3X what it was two hundred years ago (methane's warming potential being 100X that of carbon dioxide), is weakened by the fact that "military emissions" are absent and so Light ends up attributing causality to mere "pollution clouds" symptoms of geoengineering: "The Gulf Stream. . .is warming up more than usual due to global warming. Specifically, pollution clouds pouring eastward from the coast of Canada and the United States are the main culprit in heating up the Gulf Stream."[35]

In his 2014 article "HAARP and Project Lucy in the Sky with Diamonds," researcher Jim Lee references the 2012 Total gas leak and huge methane release in the North Sea that convinced the Arctic Methane Group (AMEG) that geoengineering would be a necessary evil. In the context of the Arctic Natural Gas Extraction Liquefaction & Sales (ANGELS) plan for drilling under the ice to extract gas/oil, Lee explains how ionospheric heaters can be used to keep methane at bay:

> Project Lucy [would] involve three radars focusing their beams on methane clouds and turning those methane clouds into diamond dust, something formerly left to the science-fiction world of Alchemy. Apparently the methane molecule and a diamond are very similar, and they believe with a 13.56 MHz frequency they can break methane down and turn it into diamond dust, which will reflect sunlight and slow global warming (aka SRM, or Solar Radiation Management). Further, the director of HAARP says they can form noctilucent clouds above the HAARP IRI [ionospheric research instrument] using three

---

34 "Study: California Blowout Led to Largest U.S. Methane Release Ever." UC Davis, February 25, 2016.

35 Malcolm P.R. Light, "Act now on methane." *Arctic News*, December 21, 2013. See Light's paper "The Non-Disclosed Extreme Arctic Methane Threat," December 22, 2013, sites.google.com/site/runawayglobalwarming.

radars, and if their plan works, the heaters at HIPAS (Alaska), Arecibo (Puerto Rico), EISCAT (Tromsø), and Sura (Russia) "could immediately attack the atmospheric methane as well."[36]

Under Project Lucy, the Big Oil-military alliance put together a *mobile* microwave transmission system of ionospheric heaters mounted on submarines, aircraft, boats, and oil drilling rigs purportedly to zap low-altitude methane clouds to break their first C-H bonds as soon as they erupt from the Arctic Ocean.[37]

It is also possible that Project Lucy is a "dual use" technology and not just about zapping methane, given turning methane clouds into diamond dust in an era when new diamond mines, like gold mines, are becoming scarce, and military contractors like Lockheed Martin want to develop liquid methane for stealth aircraft like the Mach 6 Aurora replacement of the CIA's SR-71 "Blackbird."[38] In short, methane is a moneymaker in a disaster capitalism era. Extract it, store it, convert it into propane and other gases with UV light, sequester it as hydrates, and sell it.

Lucy may actually be about mobile ionospheric heaters interfering rays over target areas as *scalar weapons*, and not necessarily just in the Arctic Circle.

Operation Deep Freeze. Antarctica at the South Pole is much the same but different, too. Antarctica hides a landmass under its glaciers and so is not an open sea like the Arctic Pole. It is the world's coldest, driest, and highest continent and is therefore more strategic, which is why its absent landlords row over the complicated treaty they signed when they cut Antarctica up like a pie.

Not everything in Antarctica is military. Similar to the Svalbard Global Seed Vault on the island of Spitsbergen, Norway, dedicated to preserving natural seeds in this genetic modification era, the Protecting Ice Memory project launched in 2015 seeks to preserve ice cores—the geological records of the times in which they formed—in an Antarctic snow cave at -54 degrees Celsius (-65 degrees Fahrenheit) in this era of melting glaciers.[39]

Operation Deep Freeze is decidedly military in that it is under the jurisdiction of Pacific Air Forces at Joint Base Pearl Harbor-Hickam, Hawaii with LC-130 Hercules support from the New York Air National Guard, sealift support from the U.S. Coast Guard and Military Sealift Command, engineering and aviation services from U.S. Navy Space and Naval Warfare Systems Command, and cargo

36 Jim Lee, "HAARP and Project Lucy In the Sky with Diamonds," April 1, 2014, climateviewer.com/2014/04/01/haarp-lucy-sky-diamonds/.

37 Also see Malcolm R. Light, "Radio and Laser Frequency and Harmonic Test Ranges for the Lucy and HAARP Experiments and their Application to Atmospheric Methane Destruction." *Arctic News,* October 7, 2012.

38 www.fas.org/irp/mystery/aurora.htm

39 Catch Team, "Thanks to global warming, we now need an ice library in Antarctica." *New Scientist,* 6 August 2016.

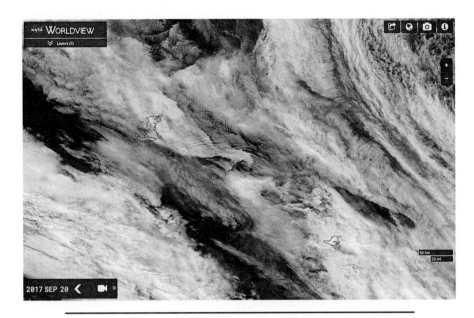

Auckland Island, New Zealand, September 20, 2017, Photo by V. Susan Ferguson metaphysicalmusing.com/ :
"Fascinating stuff. Insane scalar physics."

Auckland Island is the island between Antarctica and NZ that when it blows, it sets the Ring of Fire off. The
pattern is usually the east coast of NZ, then Fukushima-ish, then Mexico, West Coast US, Alaska, and then
often the Bering Straits, which usually ends in the Ring lighting up. When this area quakes (and they have
been giant quakes), we know a big one is coming. — Rose Paige *(The Con Trail)* in New Zealand

handling from the U.S. Navy. The three U.S./Five Eyes outposts providing "air
and maritime cargo and passenger transport throughout the Antarctic Joint
Operations Area"[40] are based in New Zealand: Harewood at Christchurch
Airport, Tangimoana at Manawatu, and Waihopai at Marlborough, all connected
to McMurdo Station (77°51'S, 166°40'E) and Amundson-Scott South Pole
Station (90°0'S, 0°0'E).

Deep Freeze "explorations science"[41] translates to military objectives run by
U.S. intelligence (Raytheon, etc.) and in service to the Space Fence.

Two days in a car with a Raytheon employee who worked in the American
sector of Antarctica, coupled with the Feral House 2005 book *Big Dead Place:
Inside the Strange and Menacing World of Antarctica* by Nicholas Johnson and
Eirik Sønneland, convinced me of how truly bizarre operations in the coldest
place in the world can be. In the years since the HAARP experiments began and
ended, the number of U.S. personnel in Antarctica and the number of C-130

40 Lt. Col. Edward Vaughn, "Operation Deep Freeze." *Armed With Science*, U.S. Department of
Defense, no date.

41 The Antarctic Treaty System (1961) now has fifty-three members and has been set aside "as a
scientific preserve, establishes freedom of scientific investigation and bans military activity on that
continent." (Wikipedia) Only twelve nations signed off on the treaty, all in 1959.

airlifts per year increased dramatically. Raytheon—once the owner of HAARP patents—commands the only way on and off Antarctica and makes all the supply runs for the American sector, often in C-17 Globemaster IIIs out of Joint Base Lewis-McChord, Washington State. Raytheon handles strategic inter-theater airlift, tactical deep field support, aeromedical evacuation support, search and rescue response, and "supply runs" for drugs, Johnnie Walker Black, and San Miguel beer for the cooks, waitresses, maids and workers in the American sector forced to live inside and underground nonstop.

And what was really going on in January 2014 when the icebreaker *Polar Star* rescued 120 people aboard the Russian icebreaker *MV Akademik Shokalskiy* stranded in Commonwealth Bay, as well as people from the Chinese icebreaker *Xue Long* (*Snow Dragon*) that had come to its rescue? On the Russian ship was a team of scientists researching climate change and "how one of the biggest icebergs has altered the system by trapping ice."[42]

The biggest clue, however, is the number of barges arriving in the hundreds loaded with equipment, two hundred of which remain in Antarctica to serve as a floating antenna/phased array farm (towers, heavy equipment, antenna gear, communications electronics, and mobile SBX units) that can be moved in water and on ice as Magnetic South moves. Twelve of them work in tandem with the twelve Starfire optical laser systems discussed later in this chapter.

For optimal Space Fence operations, Deep Freeze engineers are tasked with maintaining a delicate balance between Magnetic South and Magnetic North by means of ongoing fine-tuning, calibrating and experimentation. With the help of ionospheric heaters in the Northern Hemisphere, the bipolar *maser* outflow can be increased in the polar regions.[43] Increase the ion flow along the magnetic lines of force and it will be mirrored back toward Earth; increase the charge potential in the ionosphere and "fountain" up along the coherent inner core of Birkeland currents into the Sun's electromagnetic circuit. *Increase the charge potential of the Sun and voltage can be induced to increase solar activity*, as per Tesla in 1901. Christopher Fontenot (*amicrowavedplanet.com*) added this in an email:

> As Birkeland currents interface at the Poles, they ionize atoms and create the 'funnel' of energy that cause the Earth's electrojets. This interaction perturbs Alfvén waves ["whistlers"] in the ionosphere. Alfvén waves are pilot waves. Imagine a drill bit tip and all magnetic lines of force spiraling out from that pilot wave. Alfvén waves occur in both longitudinal and transverse forms, which means they either have no frequency (longitudinal) or have an oscillation frequency (transverse).[44]

42 "'Stuck in our own experiment': Leader of trapped team insists polar ice is melting." *Fox News*, December 30, 2013; Rod McGuirk, "U.S. icebreaker to rescue 2 ships in Antarctica." AP, January 5, 2014.

43 K. Papadopoulos et al., "HF-driven currents in the polar ionosphere." *Geophysical Research Letters*, 21 June 2011.

44 Email, May 3, 2017.

During the March 2015 saber-rattling over Ukraine,[45] nothing was said about Magnetic North. The truth is that the struggle over who will control Magnetic North and the plasma energy pouring from the poles is far from resolved.

SATELLITES

When Schumann first estimated the resonant frequency of the Schumann cavity, he used EM radiation pressures as a primary factor. Therefore, atmospheric pressures are a product of that EM radiation on the upper layers of the Ionosphere and convergence of air masses is created by that EM potential. Some refer to these anomalies as Earth spots. Essentially, they are an interface between the Ionosphere and Earth, which in turn creates a channel for discharge of EM energy. This phenomenon has been the study of weather modification scientists. The ability to induce these 'channels' via satellite, air-, sea-, and ground-based facilities is the mission of Full Spectrum Dominance.

— *Philip Francis*[46]

Satellites are invaluable to Space Fence operation.

A revolutionary technological shift occurred at the beginning of the third millennium C.E. that the public recognized only in terms of their televisions: analog systems were replaced by digital phased arrays. Individual transmitters and receivers (T/R modules) were recalibrated, and wideband performance was thus upgraded for dual-use military / civilian operations. The shift to the digital means a computational power and ability to store *on the scale of DNA*[47] and represents a *massive* expansion of "eyes in the sky" and new weapons platforms for C4: "Multiple radar, communications, and electronic warfare functions can be served by a single antenna having such an architecture."[48]

Satellites now deliver television, computer, and iPhone signals. The secretive National Reconnaissance Office (NRO) was the sole launcher of satellites until 1997 when the U.S. Commerce Department issued licenses to nine American corporations for eleven classes of satellites with a range of reconnaissance (surveillance) powers. Today, satellites owned by spooky military contractors

---

45 "Russia Launches Massive Arctic Military Drills." AP, March 16, 2015.

46 See www.everythingselectric.com/electric-universe-london-pf/ for more on Philip Francis, a self-professed "dyslexic auto-didact."

47 Andrea Leontiou, "World's shift from analog to digital is nearly complete." *NBCNews.com*, February 10, 2011.

48 David Jen, et al., "Distributed Phased Arrays and Wireless Beamforming Networks." *International Journal of Distributed Sensor Networks*, 5:283-302, 2009.

like Boeing, BAE Systems, Booz Allen Hamilton, Northrop Grumman, Lockheed Martin, L-3 Communications, and Science Applications International Corporation (SAIC) are vacuuming up images and communiqués from planet Earth inhabitants and feeding them to the NRO—now passing responsibility for launch codes and orbital parameters to the Space Fence—NASA, the National Security Agency (NSA), the National Geospatial-Intelligence Agency (NGA), U.S. Geological Survey (USGS), and the Pentagon's Defense Counterintelligence and Human Intelligence Center (DCHC).

> Currently the NRO has two types of spacecraft operating in these orbits: a series of communications satellites known as Quasar or the Satellite Data System (SDS) and a fleet of SIGINT [signals intelligence] birds known as Trumpet.[49]

Trumpets replaced the Jumpseats of the 1970s, and orbits are still either Molniya (highly elliptical) or geostationary. Commercial *and* military satellites may be armed with NASA's TWINS magnetosphere research instruments and SBIRS (space-based infrared system) HEO (highly elliptical orbit) missile defense sensors.

Satellites have come a long way since the Soviet Sputnik in 1957 and the early 2000-watt geostationary (synchronous orbit) communications satellites tracking at a purported altitude of 22,300 miles. Coaxial video lines delivering television network programs were abandoned in the 1960s and 1970s, and cable systems of HBO, C-SPAN, and other providers began using satellite. The tax-paying public was slow to realize that high-tech satellites meant not just entertainment and news but increasing surveillance at home and that the airwaves were no longer theirs. Free TVRO systems with dishes homing in on satellite frequencies meant a television in every home and the usual profit schemes like scrambling cable signals with Macom's DES algorithm to force viewers to buy equipment to descramble the signals.[50]

The U.S. Navy launched Seasat in 1978 and Geosat in 1985 purportedly to bounce satellite radar off ocean surfaces to measure the topography and gravity of sea surface and floor. But Navy survey ships had already mapped the oceans with sonar and probed the deep chasms cutting across the Atlantic, Pacific, and Indian Ocean basins, so it is probable that Seasat and Geosat were measuring and recording much more than topography and gravity.

By the close of 2010, satellite surveillance had moved further toward the Space Fence imminence. Three satellites were launched that cast a 3.2G LTE wireless broadband "net" over the Earth. First, the American corporation LightSquared launched SkyTerra 1, a seventy-two-foot L-band (1–2GHz) reflector-based antenna

49 William Graham, "Atlas V launches NROL-35 out of Vandenberg." NASA Space Flight, December 12, 2014.

50 Thanks to Gary Bourgois for this history: "Satellite System Security," *Full Disclosure*, No. 28.

with five hundred spot beams able to focus 11,900 watts of power anywhere in North America—the largest, most powerful commercial antenna reflector ever put in orbit.[51] Ten days later, the UK's Avanti Communications, in concert with the European Space Agency, launched Hylas 1 with a 2.6-ton antenna[52] to provide two-way coverage across Europe, and the French corporation Eutelsat Communications launched KA-SAT with eighty-two spot beams.[53] (In August 2012, Hylas 2 extended coverage to the Middle East and Africa.[54])

In 2014, Google bought SkyBox Imaging, SkyBox being a refrigerator-sized spy satellite on the cheap, rather like a roving CCTV that captures high resolution:

> Just one week after Google announced they'd purchased SkyBox, the US Department of Commerce lifted restrictions on high resolution, allowing commercial satellites to trade in what's been called "manholes and mailboxes" imagery. . .Clive Evans, lead imagery investigator with LGC Forensics: "When you reach this sort of frequency you can begin to add in what we call 'pattern of life' analysis. This means looking at activity in terms of movement—not just identification."[55]

Google's 2016 DigitalGlobe WorldView-4 now trumps WorldView-3's thirty-one-centimeter resolution images and offers "short-wave infrared resolution that sees through dust, smog and smoke [and chemical trails] as well as things on Earth invisible to the naked eye."[56] In sync with NASA Goddard's Deep Space Climate Observatory satellite,[57] WorldView-4 will take Al Gore's dream of "a clearer view of our world" up to four hundred miles and 17,000 miles per hour, orbiting every ninety minutes—thanks to Lockheed Martin and DigitalGlobe, whose largest customer is the spooky U.S. National Geospatial-Intelligence Agency (NGA) providing GEOINT. (It was the NGA that tracked Osama bin Ladin to Abbottabad, Pakistan.[58]) The U.S. Congress has kindly

51 Marin Perez, "LightSquared launches satellite for 4G LTE network." *IntoMobile*, November 15, 2010.

52 Vlad Savov, "Hylas 1 satellite blasting off today, will rain down broadband from above." *engadget*, November 26, 2010.

53 en.wikipedia.org/wiki/KA-SAT

54 www.avantiplc.com/fleet-coverage/coverage

55 James Vincent, "Skybox: Google Maps goes real-time—but would you want a spy in the sky staring into your letter box?" *The Independent*, 21 June 2014.

56 "U.S. government OKs DigitalGlobe to sell high-resolution earth images." *Denver Post*, June 11, 2014.

57 Ari Phillips, "A Sneak Peek At NASA's New Satellite That Has Been 16 Years in The Making." *ThinkProgress.org*, February 4, 2015.

58 Nicholas West, "U.S. Defense Contractor To Blanket Earth With New Surveillance Technology." *Activist Post*, July 17, 2016.

given DigitalGlobe its blessing to sell its high-res (and tomographic) images to mining, oil, gas, etc.

As the aerial eyes and ears platforms for the C4 Smart Grid below, satellites serve the specially engineered and recalibrated antennas on airplanes and jets, helicopters and drones—like the advanced flying psyop warfare station EC-130 Command Solo that targets minds and bodies, and the MC-12W twin turboprop capturing full-motion video and SIGINT. Next-generation KH-12s (Keyholes) vacuum up real-time enhanced infrared imaging with three-inch resolution so NetTrack software can "stitch together information from a variety of sensors (synthetic aperture radar, optical, video, acoustic, moving target indicators, etc.), and hand off to the right platform when appropriate."[59]

State-of-the-art satellite technology is imaging targets on Earth in real time, and not just as the sky's eyes and ears. Besides ending privacy, satellites mean a whole new dimension to the Vietnam-era euphemism "winning hearts and minds," including everything from propaganda to "no-touch" torture and mind control. Deep-space tracking antennas can capture Voyager signals, which at 10^{-16} are weaker by *20 billion times* than an electronic digital watch. In other words, radio transmission of a cubic centimeter of brain matter is well within the detection range of satellites armed with SLF/ELF reception gear and antennas like the ELF satellite array, given that *human thought broadcasts on the ELF band.* Such arrays work perfectly as a Very Large Array (VLA), given that a VLA is 100X as sensitive as Ohio State University's 1977 radio telescope (2×10^{-22} W m-2 per channel)—very high resolution for detecting brain activity hundreds of miles below.

> . . .the development time for this technology places the capability to detect brainwaves as far back as the early 1970s. Given an average lifespan for a satellite as 5 years, with an initial deployment during 1970, the satellite technology would be in its 8th generation.[60]

By the early 1990s, properly equipped geosynchronous orbit satellites were able to read minds, influence behavior, and detect human speech underground or behind walls unprotected by lead. Beams from high above the Earth are able to "interfere" and lock onto human targets and knock them down. Subliminals can be broadcast into the brain, including signals ordering the target to do something criminal, sexual, or violent. Up to now, ordering up a commercial or government satellite to track an individual target took big resources and embedded contacts; now, all one needs is the individual's *signature frequency*

59 Nathan Hodge, "Darpa's Simple Plan to Track Targets Everywhere." *Wired*, May 21, 2009.

60 Deep Thought, "Can A Satellite Read Your Thoughts? — Physics Revealed." [This three-part article was at the Freedom From Covert Harassment and Surveillance (FFCHS) site that has been taken down. See "FFCHS - An Introduction," www.youtube.com/watch?v=PtrYF7pGZlM.]

obtained by microprocessors comparing incoming signals with computerized images or signatures of what the target should look like. A signatured target never escapes.[61] Satellite technology as it stands now spells the arrival of the perfect high-tech crime and clean getaway.

Now that radar can lock onto targets through the cloud cover that inhibits electro-optical sensors with ViSAR (video synthetic aperture radar),[62] electro-optical satellites that can collect data from across the electromagnetic spectrum are all the rage. Electro-optical light waveforms are highly efficient with 1,000X more data capacity than radio frequency. For example, the OptiSAR constellation of sixteen satellites on two orbital planes: eight satellites in polar Sun-synchronous orbit, the other eight in a medium-inclination orbit 20–45 degrees relative to the equator. Each SAR satellite carries two sensors, one L-band (low resolution), one X-band (high resolution).[63] The only weakness electro-optical light waves are known to have is they are easily disrupted by atmospherics like weather.

Laser Light Communications was the first Optical Satellite Service (OSS) provider with its Global Hybrid Satellite-Terrestrial All Optical Network (HALO). Under the Defense Information Systems Agency (DISA)[64] and in collaboration with Raytheon, HALO's constellation of twelve satellites in medium earth orbit (10,000 km) uses high-powered laser to coordinate with terrestrial and undersea fiber-optic networks as one global surveillance network in alliance with the Space Fence.[65] Not only are waves of light from one "Point of Presence" to another employed to overtake the controls of an aircraft,[66] but LEDs are replacing streetlights so that waves of light can be steered to remotely track a human being:[67]

> The network that Laser Light brings to the relationship is a MEO [medium earth orbit] constellation, which gives it global coverage instead of regional coverage. The MEO system, known as HALO, also allows it to pick up data at one Point of Presence [POP] and, after a couple of hops, deliver it to another POP somewhere else in the world," said [Managing Director of Laser Light Robert] Brumley. "When you put this together we have a large amount of data capacity—the

---

61 Thanks to John Fleming's "The Shocking Menace of Satellite Surveillance," July 14, 2001.

62 "Pentagon's DARPA Develops Radar for Air Attacks Through the Clouds." *Sputnik International*, March 24, 2015.

63 Peter B. de Selding, "UrtheCast wants 8-satellite system in addition to proposed 16 satellite radar/optical constellation." *SpaceNews.com*, March 31, 2016.

64 DISA manages the Global Information Grid (GIG).

65 Linked to Boeing's Uninterruptible Auto-pilot system designed to take control of a commercial aircraft.

66 And is linked to Boeing's Uninterruptible Auto-pilot system.

67 "J," www.laserlightcomms/newsroom.php.

equivalent of terrestrial—and, at the same time, we have global reach, which a regional FSS [Fixed Satellite Service] by its very nature does not."[68]

While the Australian telecom Singtel Optus Pty Limited combines geostationary satellite and terrestrial networks to provide broadcast, mobile and Internet to Australia, New Zealand, and Antarctica (and Operation Deep Freeze), Laser Light's HALO integrates the existing infrastructure of regional Fixed Satellite Service (FSS) providers in a SpaceCable system:

> Laser Light plans to have 100 POPs [Points of Presence] around the world. In partnering with regional telecom companies, the optical, hybrid satellite/terrestrial network promises to deliver data transmission rates up to 100 times greater than conventional high-frequency satellite systems with the ability to distribute that data globally...While the new system, which Laser Light has coined SpaceCable, is complicated, the tradeoff is mutually beneficial: the network allows Optus to deliver more data at faster rates to more people while Laser Light gains access to more undersea cable and fiber, thus enabling a more reliable network in the face of atmospherics [weather]...Laser Light's business model calls for a minimum of 65 percent of traffic traveling over the SpaceCable and 35 percent or less over terrestrial.[69]

Satellite connection with the global fiber optic grid means enabling a laser optical surveillance grid or web over the entire Earth. In 2013, China launched the quantum optical fiber communication network project called the Beijing-Shanghai Line, the world's first wide-area fiber optic quantum private network. Joining it will be the first quantum communications satellite. Because quantum technology is indivisible and cannot be cloned, it will be used for secure transmissions.[70]

Satellites are getting smaller and smaller. The 4" CubeSats (1–10 kg) mentioned in Chapter 4, along with pico satellites (0.1–1 kg) and femto satellites (0.1 kg) have come a long way since the TV-sized "microsats" designed by artificial intelligence (AI) were launched in 2006.[71] Corporations like Media Development Investment Fund are using nano- and microsatellites to draw every person on Earth into the telecom Web. Free unrestricted Internet access! Broadcast of data over wide radio waves (datacasting)! Meanwhile, radar beams down everywhere.

---

68 "The Speed of Light: Laser Light and Optus Explain Optical Communications Partnership to Via Satellite Magazine." *Via Satellite* press release, May 4, 2015.

69 Ibid.

70 Huang Jin, "China developed the world's first quantum satellite ready to be launched this July." *People's Daily Online*, March 8, 2016.

71 "'Borg' Computer Collective Designs NASA Space Antenna." NASA, February 16, 2006.

Using a technique known as User Datagram Protocol (UDP) multitasking, which is the sharing of data between users on a network, Outernet will beam information to users. Much like how you receive a signal on your television and flick through channels, Outernet will broadcast the Internet to you and allow you to flick through certain websites. . .If everything goes to plan, the Outernet project aims to ask NASA for permission to test the technology on the International Space Station.[72]

Miniaturized satellites mean less expense, more takers, more space garbage, more "programming" straight from space into the brain. Everyone, not just 60 percent of the planet,[73] is to be plugged in.

Space-based platforms like satellites can be armed with orbital bombardment missiles or kinetics, or interceptor ASATs (anti-satellite weapons) like Brilliant Pebbles and Brilliant Eyes for use against comsats, early warning sats, navsats, recon sats, military and civilian satellites, a single satellite or satellite constellation, space-based laser systems and space stations. "Parasitic satellites" contain nanometer-sized components that utilize microelectronics to attach to host satellites until remotely commanded to interfere with or destroy it.[74] A "tractor beam" device called the Magnetic Field Architecture (MFA) creates a specialized magnetic field that can pull satellites into a fleet that moves together or "draw as well as repel satellites at the same time, meaning it will hold a satellite at a distance and won't allow it to move away or toward the capture device."[75]

Intelligence, surveillance, reconnaissance, remote sensing, electro-optical assault, orbital bombardment, ASAT—it's easy to see why satellite resilience and security, along with anti-jamming and anti-radiation, is crucial to maintaining these pivotal Space Fence players.

SOUNDING ROCKETS AND "DUSTY PLASMA"

Hurricanes are arriving seasonally and then being broken up. Notice that all the typhoons and hurricanes have been in the Pacific this year [2014]. As soon as Operation Deep Freeze and Ascension Island are up and running, we'll see the same thing in the Atlantic. It's ongoing fine-tuning

72 "Forget the Internet — soon there will be the OUTERNET: Company plans to beam free Wi-fi to every person on Earth from space." *Daily Mail*, 5 February 2014.

73 Adam Chandler, "How Much of the World Has Regular Internet Access?" *The Atlantic*, September 22, 2015.

74 Cheng Ho, "China Eyes Anti-Satellite System: Space platforms have become crucial in today's conflicts." *Space Daily*, January 8, 2001.

75 Fiona MacDonald, "NASA has partnered with a hoverboard company to build an IRL [in real life] 'tractor beam.'" *Science Alert*, 3 September 2015.

and experimentation. They know it works, but the system has to be precise: the timing, the magnetosphere ringing perfectly, which is why all the rocket firings from California are happening. Vandenberg AFB has fired over 250 sounding rockets in 18 months to disperse aluminum oxides in the upper atmosphere and space as part of the new Space Fence.

— *Billy Hayes, "The HAARP Man," 2014*

Rockets have been blasting off from Earth for over half a century now, either delivering satellites into orbit or providing thrust into space for landing modules and instruments of exploration, such as the Proton-M rocket launched from the Baikonur Cosmodrome in Kazakhstan, carrying the Trace Gas Orbiter that will analyze the composition of the Mars atmosphere[76]—rockets with state-of-the-art thrusters like the propellant-free EmDrive microwave thruster.[77]

But *sounding rockets* are also being launched for suborbital experiments from launch pads at Wallops Island, Virginia (37°51.456'N 75°30.594'W); Poker Flat Research Range (65°06'56.16"N 147°26'56.58"W) Cape Canaveral Air Force Station, Florida (28°29'20"N 80°34'40"W); and Vandenberg Air Force Base, Lompoc, California (34°43'57"N 120°34'05"W)—rockets not regulated by the Clean Air Act of 1963 (amended in 1970, 1977, and 1990). What the public is not told is that whatever the experiment, what is spewing from these rocket engines—like that from jet engines and supplementary systems—is secretly formulated to beef up lower atmosphere and ionospheric densities.

In 1972, physicist Wilmot N. Hess, director of the Environmental Research Laboratories of National Oceanic and Atmospheric Administration (NOAA), announced the modification of the near-earth space environment at a Society of Engineering Science conference in Tel Aviv:

> In the last few years experimenters have artificially modified the space environment. We can now produce artificial aurorae. We can change the population of the Van Allen radiation belt. We can artificially modify the ionosphere from the ground, and our other ideas about artificial experiments for the future stretch as far as trying to copy the sweeping action being carried on naturally by Jupiter's moons . . .[78]

---

76 "Russian-European interplanet station ExoMars-2016 launched from Baikonur," *INTERFAX*, March 14, 2016. "

77 Beverley Mitchell, "NASA Confirms 'Impossible' Propellant-free Microwave Thruster for Spacecraft Works!" *inhabitat*, August 30, 2014: "The NASA team's report stated: "Test results indicate that the RF resonant cavity thruster design, which is unique as an electric propulsion device, is producing a force that is not attributable to any classical electromagnetic phenomenon and therefore is potentially demonstrating an interaction with the quantum vacuum virtual plasma."

78 *Astronautics and Aeronautics, 1972: Chronology of Science, Technology, and Policy.* National

From Jupiter's moons to retro-engineering the rings of Saturn . . .

HAARP's 1990 Executive Summary "Joint Services Program Plans and Activities" of the U.S. Air Force Geophysics Laboratory and U.S. Navy Office of Naval Research stresses over and over again the need for "chemical releases" for "space-based efforts," with "particle beams and accelerators aboard rockets (e.g., EXCEDE and CHARGE IV), and shuttle- or satellite-borne RF transmitters (e.g., WISP and ACTIVE)."[79]

> The heart of the program will be the development of a unique ionospheric heating capability to conduct the pioneering experiments required to adequately assess the potential for exploiting ionospheric enhancement technology for DOD (Dept. of Defense) purposes. . .for investigating the creation, maintenance, and control of a large number and wide variety of ionospheric processes that, if exploited, could provide significant operational capabilities and advantages over conventional C3 systems.[80]

In 2009, native Californian president and co-founder of the Agriculture Defense Coalition Rosalind Peterson (*agriculturedefensecoalition.org*) called public attention to the Charged Aerosol Release Experiment (CARE) sponsored by the Naval Research Laboratory and Department of Defense Space Test Program, laying out yet more artificial modification of the space environment.[81] To create artificial noctilucent (polar mesospheric) clouds in near-earth orbit, rocket exhaust particles must be able to trigger an "artificial dust cloud." Supposedly, the dust particles are not released until the sounding rocket is 55+ miles (90+ km) above Earth, after which they "settle back down to a lower altitude."[82]

Since the successful HAARP experiments and resurrection of the SDI Space Fence, seeding the heated regions of the ionosphere with "dusty plasma" has been all the rage. Dusty plasma is basically smart dust in space. As military and private rocket launches multiply to maintain a fleet of 2,400–4,000 surveillance/communications satellites, each with a five-year lifespan, sounding rockets are adding dusty plasma to the ring of conductive metal particulates settling around the equator to facilitate high-speed global WiFi coverage from space, like NASA's Orbiting Rainbows Project that manipulates and controls orbiting engineered dust clouds with radio frequency, optics, and microwaves

---

Aeronautics and Space Administration [NASA], 1974. Thanks to activist Max Bliss for drawing my attention to this document.

79 www.viewzone.com/haarp.exec.html.

80 Ibid.

81 Rosalind Peterson, "U.S. Navy to Conduct Massive Atmospheric Experimental Tests." *NewsWithViews.com*, September 9, 2009. Peterson was a Keynote Speaker at the 60th UN DPI/NGO Conference on Climate Change in New York on September 5–7, 2007.

82 Clara Moskowitz, "NASA Rocket to Create Clouds Tuesday." *Space.com*, September 14, 2009.

. . .to enable a new vision of space system architecture with applications to ultra-lightweight space optics and, ultimately, in-situ space system fabrication. . .A cloud of highly reflective particles of micron size acting coherently in a specific electromagnetic band, just like an aerosol in suspension in the atmosphere, would reflect the Sun's light much like a rainbow.[83]

Reading between the lines of mainstream launch accounts makes it clear that contributing to space environment modification is now the primary task of suborbital sounding rockets like NASA's Dynamo Project launch from Wallops Flight Facility on July 4, 2013 to study the electrical current (dynamo) of the ionosphere as space scientist Robert Pfaff of NASA's Goddard Space Flight echoed Hess' justification forty years earlier, this time by referencing dynamos on Jupiter, Saturn, Uranus and Neptune.[84]

In *Chemtrails, HAARP*, I discussed how draining the Van Allen Belts responsible for our radiation shielding has always been a high priority for a functional Space Age—as important as studying the planetary dynamo circuit. In 1996, the HiVOLT (High Voltage Orbiting Long Tether) dynamo experiment failed—or did it? NASA's story says it did:

The space tether experiment, a joint venture of the US and Italy, called for a scientific payload—a large, spherical satellite—to be deployed from the US space shuttle at the end of a conducting cable *(tether)* 20 km (12.5 miles) long. The idea was to let the shuttle drag the tether across the Earth›s magnetic field, producing one part of a dynamo circuit. The return current, from the shuttle to the payload, would flow in the Earth›s ionosphere, which also conducted electricity, even though not as well as the wire. . .the [experiment] on February 25, 1996, began as planned, unrolling mile after mile of tether while the observed dynamo current grew at the predicted rate. The deployment was almost complete when the unexpected happened: the tether suddenly broke and its end whipped way into space in great wavy wiggles. The satellite payload at the far end of the tether remained linked by radio and was tracked for a while, but the tether experiment itself was over.[85]

HiVOLT was originally Russian physicist V.V. Danilov's idea for draining the Van Allen radiation belts around the Earth—our planetary shield now viewed by Space Age scientists as keeping human beings from straying beyond their lower earth orbit (LEO). A charged tether would change the pitch angle of charged particles and thus dissolve the offending inner Van Allen belts.

83 Marco Quadrelli, "Orbiting Rainbows: Optical Manipulation of Aerosols and the Beginnings of Future Space Construction." NASA/JPL, 2012.

84 Tariq Malik, "NASA's Fourth of July Fireworks: 2 Rockets Launching Today." *Space.com*, July 4, 2013.

85 www-istp.gsfc.nasa.gov/Education/wtether.html.

So the tether broke after brief but direct "contact with the ionosphere," and was swept up in Earth's orbit. Dutchsinse (Michael Janitch) in his June 22, 2015 YouTube revisit of the incident ("6/22/15 – NASA 'accidentally' DRAINS the Van Allen Belts – STS-75 'Tether Incident'") notes that global spikes in temperature, earthquakes, and volcanic eruptions began in earnest after the 1996 "tether incident,"[86] and maintains that the tether is still orbiting as a lightning rod that is sending Van Allen Belt radiation (electrons) into the Earth's core.

So are rogue tethers serving to drain the Van Allen Belts or "change the population of the Van Allen radiation belt"? I assume not (or not enough), given the 2014 video "Orion: Trial By Fire" of a NASA Orion Deep Space Mission engineer still bemoaning getting past those pesky radiation belts.[87]

On September 16, 2015, CARE II was launched on a NASA Black Brant XI sounding rocket from Andoya, Norway—sponsored, of course, by the U.S. Naval Research Laboratory and the Department of Defense Space Test Program.

> After entering the ionosphere, 37 small rockets were fired simultaneously to inject 68 kg of dust comprised of aluminum oxide particulates, accompanied by 133 kg of molecules such as carbon dioxide, water vapor, and hydrogen.[88]

The engendering of "'dirty plasma' with high-speed pickup ions" was closely tracked by ground receivers and radar at the Finnish Meteorological Institute, the Sodankyla Geophysical Observatory, the U.S. Naval Research Laboratory Plasma Physics Division, EISCAT, the Institute of Atmospheric Physics (Germany), the Institute of Space Physics (Sweden), and others in Oslo and at the University Center in Svalbard, plus the UK's QinetiQ.[89] Dusty plasma generates waves for the purpose of scattering radar signals for remote sensing—in other words, Space Fence surveillance.

Not a month later, NASA launched a sounding rocket from Wallops loaded with "vapor tracers," yet more chemicals to modify the space environment:

> NASA explains that it has actually been injecting various vapor tracers into the atmosphere since the 1950s—these trails help scientists understand "the naturally occurring flows of ionized and neutral particles" in the upper atmosphere by injecting color tracers and tracking the flow across the sky. Tonight, NASA says it's *ejecting four different payloads of a mix of barium*

---

86 Note the pulsating plasma orbs that surround the broken tether in NASA film footage. NASA insisted they were "ice crystals."

87 www.nasa.gov/press/2014/october/nasa-premieres-trial-by-fire-video-on-orion-s-flight-test.

88 Daniel Parry, "NRL Rocket Experiment Tests Effects of Dusty Plasma on the Ionosphere." *NRL* news release, October 6, 2015.

89 Affiliated with MI6 the same way InQTel is affiliated with the CIA.

and strontium, creating a cloud with a mixture of blue-green and red color.⁹⁰ [Emphasis added.]

The vapor tracers released include trimethylaluminum, which reacts with oxygen to produce carbon dioxide, water vapor, and aluminum oxide, plus lithium and barium mixed with thermite in a payload canister. Thermite means *heat* and is a mix of iron oxide and nano-aluminum to vaporize the metal into iron and aluminum oxide.

Thus charged particles are being released into the upper atmosphere to induce artificial plasma plumes that exactly correlate with lower atmosphere artificial plasma cloud cover.

In the Japanese language, "dusty plasma" means "star crumbs":

> Nanoparticles, also called cosmic dust (star crumbs), are formed by gas expelled from celestial bodies. Within this cosmic dust, the nanoparticles which are formed first greatly influence the evolution or alteration of nanometer-size solid matter into becoming a planet. . .The predominant component in this material is oxidized aluminum. . .Considered to be *an originating material of light,* which first appears at 13μm [micrometers] around a celestial body that forms oxidized material.⁹¹ [Emphasis added.]

If nucleated aluminum and silica oxides were the first nanoparticles to govern the evolution of cosmic dust ("dusty plasma"), then the 2008 children's film *The Golden Compass* about fountains of glowing "dust" in the northern polar regions forbidden to commoners by the evil Magisterium world order wants us to think kindly upon the nanoparticles being spewed to modify the atmosphere and near-earth orbit. Dusty plasma—a continuation of the conductive metal nanoparticles being dumped by jets and zapped in our lower atmosphere—is slowly accreting a Saturn-like conductive "fence" around the equator. It is the ancient alchemical formula *As above, so below* all over again as the electro-chemical processes in the ionosphere and lower atmosphere are woven into a plasma mesh or grid of domination.

LIGHTNING, SPRITES AND STARFIRE

LASER — light amplification by stimulated emission of radiation
MASER — a microwave laser

90 Kelsey Campbell-Dollaghan, "A NASA Experiment Is Going to Light Up the Sky With Beautifully Colored Clouds Tonight." *Gizmodo,* October 7, 2015.

91 "Successful Launch of the Sounding Rocket S-520-30 Experiment using a microgravity environment to reproduce star 'crumbs.'" Hokkaido University press release, September 28, 2015.

Transmitting and steering electricity through the air over distance requires conductive density. Thus, keeping the lower atmosphere energized with metal nanoparticles constantly charged and colliding has given the green light to a vast array of laser operations:

> . . .lasers work by passing energy through a "lasing medium," which causes electrons in the medium to reach a specific excited state and to interact with electromagnetic waves so as to give the wave that extra, exciting energy. This interaction, called stimulated emission, is reliable enough that with the excitation of certain media, we can create beams of light specific enough to do everything from slice raw carbon to transferring data across continents. . .light guns are broken down according to the type of wave emitted. . .the more useful rubric [being] the way we power up, or "pump," the lasing medium.[92]

Whether powered by gases (carbon dioxide, fluorine, deuterium fluoride), chemicals, solid fibers, or wafer-thin diodes, lasers are about getting the energy into the electrons—"jacked up atoms just itching to offload their energy to the right passing wave"—then sending the electrons into the target. Firing a two-nanosecond pulse from a 5TW nitrogen/helium laser needs a certain chemical metals-plasma gas density in order to create the essential *laser-induced plasma channels (LIPCs)*. Initially, two interfering beams—a lesser (femtosecond) laser and a greater (nanosecond) dress beam laser—were needed for LIPCs, but now only one beam is necessary.

> The idea of using laser to channel electricity through the air, which is normally not conductive, was first proposed in the 1970s and further explored through the 1990s. The research was based on the idea that by superheating a very narrow column of air, it would be possible to create a straight path along which an electric charge could flow. . .The highly focused laser beams superheated a narrow line of air molecules, stripping off their outer electrons and producing a filament of charged plasma. The higher-than-normal concentration of free electrons in the plasma overcame the atmosphere's natural insulator properties, making it much more conductive.[93]

Consider the role lasers play in manipulating the natural roll-up of atmospheric weather moving east-northeast off the Pacific Ocean. A relay system of sixteen stationary *Starfire optical lasers* near or on military installations

92 Graham Templeton, "A deeper look into lasers, particle beams, and the future of war." *ExtremeTech.com*, April 25, 2013. See the YouTube "Lightning in the lab: Femtosecond laser generating plasma in air" by David Sheludko, March 29, 2009.

93 Pavel Polynkin, "Laser 'Lightning rods' channel electricity through thin air." *Physics*, August 19, 2014. Also see Ludger Woste et al., "LIDAR-Monitoring of the Air with Femtosecond Plasma Channels." *Advances in Atomic, Molecular, and Optical Physics*, 53:413-441; 01/2006.

or air bases in the middle of nowhere—from the original Starfire Optical Range at Kirtland Air Force Base in Albuquerque, New Mexico, northwest to Tonopah, Nevada and further north to Boise, Idaho, with eight at Eglin Air Force Base in Florida—is utilized to manipulate the jet stream for weather deliveries. As coastal air pressure is bundled into a north-south LIPC "log" off the West Coast, moisture from the South Pacific is folded into it and shunted north to the jet stream loop over Vancouver Island, which then piggybacks the additional moisture east for weather operations in the Midwest, Texas, Florida, the Atlantic coast, etc. With differently shaped laser beams—such as the Teramobile laser system[94] provides—the LIPC waveguide can be made into an S-shape to "hit an obstacle and reconstitute itself on the other side."[95]

The laser's ability to deliver a precision spark of electricity *even through obstacles* makes it a formidable, versatile optical energy weapon. LIPCs can as likely carve a conductive path for a sprite (.0001-second discharge) in the upper atmosphere as provide a high-voltage beam for a "lightning strike" (SDI "Star Wars" technology) at "anything that conducts electricity better than the air or ground surrounding it."[96] Indeed, Starfire is key to *space situational awareness*—"To maintain space situational awareness, the Air Force conducts research in laser guided star adaptive optics, beam control, and space object identification."[97]

NASA research into natural plasma effects like lightning in the Global Electric Circuit (GEC) has been in large part about cloning lightning as an optical weapon system.[98] Naturally triggered by cosmic rays and accelerated solar wind particles, two thousand bolts of lightning flash every minute, some striking people. Microwave dishes were set up on a tower monitoring station thirteen miles from the Lake Maracaibo lakebed into which the Catatumbo River of Venezuela empties, the world's largest single generator of tropospheric ozone,[99] in order to study Catatumbo lightning, particularly how "the electrical

---

94 "The Teramobile system is the first mobile laser yielding 5 terawatts (TW) and 100 fs (10-13 s) pulses, with 350 mJ pulse energy at 10 Hz repetition rate. It concentrates the state-of-the-art laser technology in a 20' standard freight container, allowing field measurement campaigns." teramobile.org/teramobile.html

95 Lizzie Wade, "Laser beams make lightning tunnels." *Science* magazine, 19 June 2015.

96 Clay Dillow, "Army Demonstrates a Weapon That Shoots Laser-Guided Lightning Bolts." *Popular Science*, June 28, 2012.

97 "Starfire Optical Range at Kirtland AFB, New Mexico." Kirtland AFB Fact Sheet, March 9, 2012. The sister site of the Directed Energy Directorate is the Air Force Maui Optical and Supercomputing Site (AMOS) at the crest of the dormant volcano Haleakala. The Maui Space Surveillance System (MSSS) includes an Image Information Processing Center and Supercomputing facility and combines satellite tracking and laser projection.

98 For example, Sotirios A. Mallios and Victor P. Pasko, "Charge transfer to the ionosphere and to the ground during thunderstorms." *Journal of Geophysical Research Atmospheres*, November 2011.

99 Billy Hayes: "We also installed microwave dishes on the same tower for UPI/API communications links. The Company had interest in cloning the natural events there as optional weapons systems."

properties of the air are somehow altered as the incoming charged particles from the solar wind collide with the atmosphere."[100] Could cloud-to-ground and intra-cloud lightning discharges between the ionosphere and Earth be controlled and created, perhaps in conjunction with *sprites* and *whistlers (Alfvén waves)*?

Sprites are super-fast releases of electrical energy in lightning form (plasma) so intense that *they appear to straddle spacetime and reach into antimatter* to produce plasmic heat and energy 10X more powerful than lightning (albeit electrically weaker), with discharges of over a billion volts. Like lightning, sprites send waves throughout the Schumann cavity that spike our AC global circuit known as the Schumann resonance. In short, sprites are planetary bell-ringers[101] in service to keeping our atmosphere charged, whereas the very low frequency (VLF) radio waves (1 to 30 kH) we call whistlers or Alfvén (magnetohydrodynamic) waves follow from lightning and sometimes produce audible whistles, hisses and booms.

Are we getting a picture of how the Earth's atmospheric layers are a natural electrical system, and how weaponizing it would be pivotal to full spectrum dominance of planet Earth? In *Chemtrails, HAARP*, we studied how ionospheric heaters like HAARP stir up the ionosphere so that ions spin down the Earth's magnetic lines of force to turn our lower atmosphere into a transceiver antenna. No doubt ionizing the atmosphere impacts the *Birkeland currents* connecting our poles to the Sun and the electrojet boundaries between Earth and the Sun.[102] Because both Birkeland currents and Alfvén waves are concerned with solar activity, we have to ask: In the military quest for full spectrum dominance, is electromagnetic manipulation of these elements connected to plans to harness the plasma power of the Sun, as Tesla indicated would be possible?

Through sprites and their relationship to antimatter, NASA recognized how plasma develops an electrical field by utilizing a few particles of antimatter, which pointed to the need for a magnetic field for maneuvering antimatter and aligning instruments like HAARP necessary for pinning a magnetic flux field in place once it was known where a plasmatic effect was happening.

Antimatter will be taken up again in the CERN chapter. Suffice it to say that it is far from science fiction, having been intensely studied since the 1930s. In fact, matter and antimatter are the yang and yin of what we call reality. Every type of subatomic particle has an antimatter counterpart, and yet if matter and antimatter collide, they annihilate each other in an immense burst of energy.

One of Gakona HAARP's objectives was to explore how antimatter could be farmed. By utilizing high-voltage pulses from the Starfire Optical Range, NASA began farming antimatter along with sprites and whispers around

---

100 "High speed solar winds increase lightning strikes on Earth." Institute of Physics, 15 May 2014.

101 See NOVA YouTube "Extreme Rare Lightning: Sprites (Edge of Space)," June 17, 2015.

102 One above the magnetic equator (the equatorial electrojet) and those near the Northern and Southern Polar Circles (the auroral electrojets).

1998, concentrating on sprite currents connecting the Earth and ionosphere. It was quickly discovered that the magnetosphere had to be maneuvered and sprite energy steered over certain geographic (geomagnetic) areas *prepared by specific chemical signatures.* Sprites are initiated by specialized laser and maser systems like Starfire or the mobile Sandia Transportable Triggered Lightning Instrumentation Facility (SATTLIF) usually parked at the Sandia Lightning Simulator (SLS) but able on a moment's notice to be transported by flatbed truck, aircraft, or offshore oil platform.[103] SATTLIF's helium masers create triple-path discharges.

As a crucial component of the Directed Energy Directorate, Starfire is in service to the Space Fence. Initially, Starfire was ground-based, but now that we've gone digital and tetra frequency allows components to be as small as a pack of cigarettes, Starfire is particularly C4 versatile. Tetra frequency means less weight, less size, less power, all of which means it is not as likely to be destroyed by EMPs.[104] A tetra hertz is so compact that an EMP goes right through it without making contact with the circuits, and "less power" is no problem, given that we no longer live in the power age. *This is the digital age.*

Starfire can transmit fiber-optic or direct light communications by putting them on a mirror or shooting them into a fiber-optic line and sprinkling them all over the Earth to be received all at one time, like a scanner in outer space that can send massive amounts of information instantaneously to multiple ground stations. To our limited vision, Starfire's 20 million joules of power look as small as a pinhead, but as it flashes up through the atmosphere, it is actually sharp, bright, and *gigantic.* Fire it at radio frequency and it becomes invisible; fire it at microwave frequency and it becomes a maser, the *real* weapon of destruction.

Starfire can target individual populations or individuals, track and locate them, measure how fast they're moving, how big they are, and read their biological signatures all the way down to their genetic code. (Tetra frequency.) And when Starfire's reflective power returns as it must, it brings with it all the signals, frequencies, distances, and signatures of everything it's picked up on the way out and back. It is a super-spy tool.

Even the pencil-thin laser available on the open market can hit the moon. Point it at the cockpit of an aircraft and it can cause "flash blindness." In 2014, the Civil Aviation Authority reported 168 laser-pilot incidents at London Heathrow alone. On February 15, 2016, Flight VSO25 with 252 passengers and

---

103 Sandia National Laboratories is operated and managed by Sandia Corporation, a wholly owned subsidiary of Lockheed Martin Corporation. Sandia Labs and Kirtland Air Force Base are practically synonyms.

104 The electromagnetic pulse was first detected during a nuclear test at Johnston Island in 1962. Thus it is an electric field and a magnetic field released immediately after a nuclear explosion. It will burn out electronic circuitry, communications systems, computers, etc. See *The Language of Nuclear War: An Intelligent Citizen's Dictionary* (Harper & Row, 1987).

fifteen crew had to return to Heathrow after a laser beam was pointed at the cockpit and the First Officer felt unwell.[105]

The military is energetically developing tactical weapons-grade lasers like the focused high-energy laser (HEL), which, when fired at a drone, rocket or mortar, will heat it until it disintegrates, just like in a *Star Wars* film. Israeli state-owned Rafael Advanced Defense Systems spokesman Amit Zimmer brags that lasers mean an unlimited magazine.[106] Israel's considerable arsenal includes the *Iron Beam*, the *Iron Dome*, the *Arrow* system (intercepts missiles in space), and *David's Sling*.[107]

Then there is Lawrence Livermore National Laboratory's "Death Star"—the *High-Repetition-Rate Advanced Petawatt Laser System (HAPLS)*—producing one petawatt (one quadrillion watts) of power in extremely fast pulses (thirty femtoseconds per pulse, or 0.03 trillionths of a second per pulse),[108] and Osaka University's 100-meter two-petawatts laser that "instantaneously concentrated energy equivalent to 1,000X the world's electricity consumption and entered the record books as the most powerful laser beam ever emitted."[109]

In tandem with endless war, lasers are also leading the way for remarkable medical advances—and yet harnessing lasers to genetics and the creation of life sounds exceedingly like yet another weapon, like the billion kilowatts of the Prague Asterix Laser System that produce a 7,600°F shockwave that directed at plasma containing *formamide*, an ancient primordial substance, will trigger chemical reactions leading to the formation of the four RNA nucleobases adenine, guanine, cytosine, and uracil.[110]

The Cold War may have ended a quarter of a century ago, but a "space situational awareness" race has taken its place. Ultimately, the buildup of "space surveillance assets" under the new and upgraded Space Fence SSDS program that intends to "indirectly enable a range of decisive responses that will enable counterspace threats ineffective"[111] is all about the new battlespace encompassing the Earth and the heavens. Near-Earth conductivity is being amped up with the same metal nanoparticles that have transformed the troposphere into an antenna farm. Whether for upper atmosphere wars or lower atmosphere

---

105 "'Laser Incident' Forces Jet To Return To London." *Sky News*, 15 February 2016.

106 Magazine: an ammunition storage and feeding device within or attached to a repeating firearm.

107 "Israel says it is close to developing 'Star Wars' laser missile shield named Iron Beam that will cover entire region." *Daily Mail*, 14 February 2014.

108 Ryan Whitwam, "Petawatt 'Death Star' laser prepares to investigate quantum mechanics, chemistry, and more." *ExtremeTech.com*, February 10, 2014.

109 Brad Reed, "Japanese scientists fire a 2 quadrillion watt laser, the most powerful ever." *Yahoo.com*, July 30, 2015.

110 Marciana Cooper-White, "Scientists Use Super-Powerful Laser To Simulate 'First Spark of Life.'" *Huffington Post*, December 10, 2014.

111 www.space.com/24897-air-force-space-surveillance.

magic shows to wow and frighten the masses—like the blue plasma spiral near Tromsø, Norway (December 9, 2009), the plasma spiral the following day over the Kapustin Yar missile range on the lower Volga of Russia, and a similar plasma spiral the year before over China[112]—a paranoid, aggressive Space Age is upon us, "incorporating data from multiple external sources, not just military radars and telescopes."[113]

Lasers play a role in producing the scalar or interfering waves (interferometry) discussed in Chapter 2. In 2014, CBS News made what may have been the first ever mainstream reference to this technology, describing it as "dual beams" from two transmitters, one beam stripping electrons from an area in the atmosphere, the other pulsing the plasma created by the stripped electrons to form CCN (cloud condensation nuclei) for the production of raindrops.[114] Two interfering beams of light requiring the presence of strontium barium niobate (SBN) in the atmosphere can also produce *holograms* or laser-induced plasma shapes in the air:

> . . .The resultant image is a hologram that can be viewed in real time over a wide perspective or field of view (FOV). The holographic image is free from system-induced aberrations and has a uniform, high quality over the entire FOV. The enhanced image quality results from using a phase conjugate read beam generated from a second photorefractive crystal acting as a double pumped phase conjugate mirror (DPPCM). Multiple three-dimensional images have been stored in the crystal via wavelength multiplexing.[115]

Haptic holography—holographs you can touch—is also now possible. With femtosecond laser technology,[116] Japanese researchers have developed "Fairy Lights, a system that can fire high frequency laser pulses that last one millionth of one billionth of a second" and "respond to human touch, so that—when interrupted—the hologram's pixels can be manipulated in mid-air."[117] An earlier 2008 account of "Fairy Lights" holography points out the *dangers* inherent in high-frequency ultrasound generated by high-intensity femtosecond laser (not to mention the atmospheric heat it produces):

112 See Chinese YouTube at www.youtube.com/watch?v=ixLE3iuszbU. Note how at 1:11 the direction of the spiral shifts from anticlockwise to clockwise.

113 See "Space Situational Awareness," SSS Working Group, Space Generation Advisory Council, 2012; www.agi.com/resources/educational-alliance-program/curriculum_exercises_labs/SGAC_Space%20Generation%20Advisory%20Council/space_situational_awareness.pdf.

114 Michael Janitch, "4/23/14 – CBS News – Scalar Weather Modification Proved – Dual Beams Cause Instant Storms," April 23, 2014, dutchsinse.com/4232014-cbs-news-scalar-weather-modification-proved-dual-beams-cause-instant-storms/.

115 Christy A. Heid et al. "3-D Holographic Display Using Strontium Barium Niobate." Army Research Lab, Adelphi, Maryland, February 1998.

116 See Yoichi Ochiai et al., "Fairy Lights in Femtoseconds." Digital Nature Group, June 22, 2015.

117 "Japanese scientists create touchable holograms." *Reuters.com*, November 30, 2015.

Researchers at the University of Tokyo have demonstrated a device that can create touchable, creepily invisible floating 'objects' using focused ultrasound waves...There's a major catch, though: the virtual objects won't provide much resistance or seem very 'hard' because at high enough levels the aurally imperceptible ultrasound will *destroy your eardrums*.[118]

Is Project Blue Beam coming to fruition at last? Satellites and laser-based projection systems mounted on aircraft,[119] thanks to computer software, project 3D optical holograms of multiple images onto the sodium layer screen sixty miles above, "deep perspective images that appear to emanate from the very depths of space."[120] Now picture a tractor beam[121] lifting masses of people into the sky as the ELF, VLF, or LF "voice of God" speaks to every cultural and religious persuasion as computer memory banks pour historical vistas into every mind, as well as what humanity must now do. Alien invasion, Rapture, and then the most fantastic of all:

ELF/ULF waves [and plasma] will allow "supernatural forces" to travel through optical fibers, coaxial cables (TV), electrical and phone lines in order to penetrate everyone at once through major appliances. Embedded [nano-] chips will already be in place. Global Satanic ghosts will be projected everywhere to push populations to the edge of hysteria and madness and drown them in a wave of suicide, murder, and permanent psychological disorder . . .[122]

On June 29, 2015 a cube poked through the plasma cloud cover over McGregor, Texas, site of Elon Musk's SpaceX testing facility. Its appearance was accompanied by wind and swirling jet-black plasma clouds, as if it were entering from another dimension, like the rectangular obelisk in Stanley Kubrick's seminal *2001: A Space Odyssey* (1968). After a brief tour through the cloud cover, it disappeared. *Kabba,* as in the Kabbalistic Tree, refers to the cube in the Tree, the cube being a symbol of Saturn a.k.a. Satan.[123] Could the cube's visitation be a *Lord of the Rings* holographic psyop referencing the metallic chemical rings being laid around the Earth?

118 "Ultrasound Haptic Devices Can Project Tactile Shapes Into Thin Air." *Gizmodo*, September 2, 2008.

119 "Ghost rider in the sky: Scientists use lasers to project movie onto clouds." *RT.com*, 6 July 2015.

120 Serge Monast, "NASA's Project Blue Beam," 1994. Monast was a French Canadian journalist whose two homeschooled children were taken away and made wards of the state in September 1996, followed by fifty-one-year-old Monast's psychotronic heart attack on December 5, 1996.

121 Alistair Munro, "Dundee scientists create real-life tractor beam." *The Scotsman*, 30 May 2014.

122 Ibid.

123 "Texas UFO 'Cube' – Multiple View Witnesses and Photos," in5d.com/texas-ufo-cube-multiple-view-witnesses-and-photos/.

RADAR

The high-frequency (HF) emission in near-Earth space from various powerful transmitters (radio communications, radars, broadcasting, universal time and navigation stations, etc.) form an integral part of the modern world that it cannot do without. In particular, special-purpose research facilities equipped with powerful HF transmitters are used successfully for plasma experiments and local modification of the ionosphere. In this work, we are using the results of a complex space-ground experiment to show that exposure of the subauroral region to HF emission can not only cause local changes in the ionosphere, but can also trigger processes in the magnetosphere-ionosphere system that result in intensive substorm activity (precipitations of high-energy particles, aurorae, significant variations in the ionospheric parameters and, as a consequence, in radio propagation conditions).

— V.D. Kuznetsov and Yu Ya Ruzhin,
"Anthropogenic trigger of substorms and energetic particles
precipitations," December 20, 2014[124]

Radar (Radio Detection And Ranging) is a primary player in obtaining and maintaining planetary full spectrum dominance for the *Under the Dome*[125] Space Fence lockdown of planet Earth.

In the 1930s, the major players in two world wars discovered just how far pulsed radio waves could take communications and surveillance. With antenna direction and timed pulses on an oscilloscope, the range and location of a target could be exactly determined. This discovery transformed warfare, and radar has come a long way since then. Beaming phased-array millimeter waves is now essential to priming the power density of our lower atmosphere for military operations, including weather engineering:

> Phased array Doppler radar facilities (ground / air / satellite based) can provide the focused microwave energy needed to create this cascade of atmospheric ions. Through this process, ions are multiplied and released into the lower atmosphere to increase power density in those layers. These modern tools of manipulation are essential to control of the weather. By controlling the electromagnetic potential in the Earth's atmospheric layers, man can affect natural processes of weather. Subtle perturbations of these natural phenomena

---

124 *Advances in Space Research*, Volume 54, Issue 12, pp. 2549–2558; December 20, 2014. adsabs. harvard.edu/abs/2014AdSpR..54.2549K.

125 A 2013 CBS science fiction television series.

can direct the distribution and discharge of the Earth's EM forces.[126]

Wireless energy transfers (WETs) are essential not just for billions of cell phones and computer transmissions on Earth but for "next generation aerospace systems":

> Our typical system for wireless energy transfer includes high power microwave generators, antennas or antenna arrays, side lobe suppression radomes (SLSRs), and a tracking/control system.[127]

Dependable WETs need a balanced *acoustic resonance frequency* between the ground and the upper mesosphere within range of the Schumann resonance.[128]

Thus one lifespan beyond World War II radar, the public is now bathed in the radiation of a wireless world, convinced that radiation's ubiquity is a necessary evil for national security while teaching their children that cooking with microwaves and holding transmitters against their heads are safe practices. Is it any wonder that people remain ignorant of EM manipulation of weather events (droughts, floods, earthquakes, tornadoes, hurricanes) and geoengineered artificial plasma cloud cover being delivered by jets? And don't forget the military ground forces now requiring a dense network of stationary and mobile ionospheric heaters, radar installations, NexRads, GWEN and cell towers, wind farms, ship tracks, and fiber optic cable—all to keep our atmosphere battery-ready and antenna-charged for wireless military operations.[129]

SUPER-DARN (SUPER DUAL AURORAL RADAR NETWORK).

Described as "an international collaboration involving scientists and engineers in more than a dozen countries," SuperDARN HF radar installations (8–22 MHz) in the Northern and Southern Hemispheres "provide global, instantaneous maps of plasma convection in the Earth's atmosphere."[130] Among the university collaborators operating SuperDARN radars is HAARP's overseer, the Geophysical Institute at the University of Alaska Fairbanks (UAF).[131]

---

126 Email from Christopher Fontenot, 2015.

127 "Microwave Power Beaming Milestone." Escape Dynamics press release, July 17, 2015.

128 Makoto Tahira, "Acoustic Resonance of the Atmosphere at 3.7Hz." *Journal of Atmospheric Science*, 52, 2670–2674.

129 See Jim Lee's *climateviewer.org* map that he has also allowed me to use in this book.

130 Virginia Tech Department of Electrical & Computer Engineering, "News/Events."

131 Other collaborating universities: Virginia Tech, Dartmouth College, the Institute of Space and Atmospheric Studies at the University of Saskatchewan, the University of Leicester, La Trobe University, and the Solar-Terrestrial Environment Laboratory at Nagoya University. The Geophysical Institute is affiliated with the U.S. Geological Survey, NASA, and NOAA, and has international agreements with

Covering northern, middle, and southern latitudes, SuperDARN is constantly adding more radar installations: two in Ft. Hays, Kansas (Virginia Tech); two in Christmas Valley, Oregon (Dartmouth College); two in the Aleutian Islands (U.S.); two in the Azores Islands (Portugal); etc.[132] The full scan of *each* of the thirty-five radar installations covers 52° in azimuth and over 3,000 km in range—that's one million square kilometers, much like ionospheric heater components.

Like other installations, the Ft. Hays, Kansas installation is now fully automated under NASA/SuperDARN out of Wallops Missile Launch Range in Virginia and the new supercomputer space systems at Virginia Tech. North of Ft. Hays is a wind farm constructed in a 102° arc with the SuperDARN site as its focal point. The Arecibo Observatory in Puerto Rico has also been calibrated to match the pulsed frequencies of wind farms, fracking wells, SBXs, NexRads, etc. This is how pathways are built up to maintain charges in the atmosphere *by weather alone*. In other words, the large sweeps of conductive metal nanoparticles and wind farm pulses work together to maintain an atmospheric power supply for all of the above systems subject to SuperDARN and GWEN tower high frequency. All use effective radiated power (ERP) for multiplication gain.

But the Space Fence radar network that includes SuperDARN is not just about the new (like the Kwajalein Atoll installation mentioned earlier) but about recalibrating ("upgrading") the old, like the Raytheon AN/FPS-115 "PAVE PAWS" phased-array missile-warning radars once in service to Strategic Air Command's Directorate of Space and Missile Warning Systems (SAC/SX) during the SDI buildup in the early 1980s. Mounted on a ziggurat model, the octagonal radar panels are ninety feet in diameter and can detect targets three thousand miles away. Originally, there were four PAVE PAWS in CONUS but now there are thirteen on or below 33°N, with one in Anchorage, Alaska and two at USAF Thule Air Base in Greenland (76°N) and RAF Fylingdales in the UK (54°N). While we are told PAVE PAWS has a narrow 5 MHz bandwidth and not to worry, the truth is the move from analog to digital has superseded the large/small bandwidth game.[133]

Recently, powerful radar units—again, like ionospheric heaters—have gone mobile. According to Raytheon propaganda, ballistic missiles of "rogue regimes" (excluding the U.S., NATO nations, Russia, and China) are now developing nuclear, chemical / biological warheads that only X-band radar (8–12 GHz)

Japan, Russia, China, Australia, Germany, the UK, and Norway.

132 Some SuperDARN installations are listed by coordinates under ionospheric heaters [transmitters] and observatories [receivers] in *Chemtrails, HAARP, and the Full Spectrum Dominance of Planet Earth,* Chapter 1, "A Thumbnail History."

133 For more on "Strategic Warning Radars," see what-when-how.com/military-weapons/strategic-warning-radars-military-weapons/.

will be able to detect in real time. In forward-based mode (FBX-T), Raytheon's extremely powerful long-range AN/TPY-2 X-Band radar on wheels searches the sky, detects, tracks, discriminates phase of flight, then informs Command and Control Battle Management. If the radar is deployed in terminal mode, AN/TPY-2 will *not* inform Command and Control but will on its own cognizance launch a THAAD (terminal high altitude area defense) interceptor missile.[134] Task Force Talon, the world's only deployed THAAD battery site, is on the American Territory of Guam while its three Delta 2 THAAD Batteries rotate out of Fort Bliss, Texas.

> As part of the military shift to the Pacific ordered by President Obama, the military should deploy a second THAAD battery from Fort Bliss to Guam, create a new Army Air Defense Brigade, and provide additional Patriot Batteries in the Pacific. These critical elements are needed for strategic deterrence and reassurance, and to enhance the limited resources of our Pacific allies and ourselves.[135]

"The military shift to the Pacific ordered by President Obama" is in part to provide protection for the Kwajalein Atoll installation, as well as to send a message to BRICS nations that U.S. presence is not going away.

Also mobile are electronic warfare (EW) programs, given that "the war of electrons may decide the outcome of the war of missiles."[136] What this means is that warfare is no longer necessarily *kinetic* (shooting a missile with a missile) but can be as simple as a communication between an EA-18G Growler jet and its on-the-ground Next Generation Jammers. Russia is pursuing a similar course of ground-based and offshore "automated real-time intelligence data exchange with airspace defense task force...jamming and influencing adversaries' command and control systems at long-range by emitting a complex, powerful digital signal."[137]

> Electronic warfare (EW) is any action involving the use of the electromagnetic spectrum or directed energy to control the spectrum, attack of an enemy, or impede enemy assaults via the spectrum. The purpose of electronic warfare is to deny the opponent the advantage of, and ensure friendly unimpeded access to, the EM spectrum. EW can be applied from air, sea, land, and space

134 "Army Navy / Transportable Radar Surveillance (AN/TPY-2)," Raytheon fact sheet, no date. According to Jim Lee at climateviewer.org, THAAD is modern SDI (Facebook message, March 15, 2015), and modern SDI is Space Fence.

135 "Guam — A Fiesta." Missile Defense Advocacy Alliance, November 30, 2015.

136 Sydney J. Freedberg, Jr. "Work Elevates Electronic Warfare, Eye On Missile Defense." *Defense Industry News*, March 17, 2015.

137 Jack Phillips, "Russia Starts Trial of New Electromagnetic Warfare System." *Epoch Times*, April 26, 2016.

by manned and unmanned systems, and can target humans, communications, radar, or other assets.[138]

DARPA's Mobile Hotspots program fits in nicely with Growlers and jammers. High-speed millimeter-wave backhaul networks (1 Gb/s) mounted on unmanned aerial vehicles (UAVs) in remote, forward-operating geography with little or no connectivity to tactical operation centers (or in urban areas with dense wireless cross-traffic on varying frequencies) can keep intelligence, surveillance, and reconnaissance (ISR) data going 24/7.[139]

The S-band radar essential to Space Fence lockdown will be covered at length in the next chapter. For now, let's examine another radar casting a wide and deep net over the urban / civilian battlespace: the Gotcha Spiral II, a work-in-progress since 2010 built to spy on "city-sized" areas of 10–20 km. When I asked Billy Hayes "The HAARP Man" why it was called "gotcha," he laughed and said, "It's a deep-penetrating spy satellite. 'Gotcha!'" Here is the 2010 U.S. Air Force description advertised through Federal Business Opportunities:

> . . .a dual-band (X/UHF) radar system capable of performing persistent, wide-area surveillance. The Gotcha radar concept is an airborne, wide angle, staring radar. . .The data is collected in a single radar mode, but is processed into several different data products such as video Synthetic Aperture Radar (SAR), Ground Moving Target Indication (GMTI) with Minimum Detectable Velocity (MDV), Coherent Change Detection (CCD), Super-resolution 2D imagery, and 3D SAR imagery.[140]

Gotcha sees through *everything*, given that the higher the gigahertz (Gotcha is 2.7 GHz), the greater the ground- and heat-penetration capability. 3D SAR imagery leads to 3D video detection and ranging (ViDAR), a Doppler radar imaging system:

> A moving sensor suite for imaging a scene has three Doppler radars, two moving and one fixed, a fixed video camera and a fixed GPS receiver. The Doppler radars measure the relative velocities between the radars and the scene, as well as the scene's electromagnetic reflectivity, while the video camera records the motion of the camera and the optical property of the scene. The correct registration of the Doppler radars and the camera is established by

---

138 Joint Publication 3-13.1: Electronic Warfare. Chairman of the Joint Chiefs of Staff (CJCS), Armed Forces of the United States of America, 25 January 2007.

139 "Remote Troops Closer to Having High-Speed Wireless Networks Mounted on UAVs." DARPA. mil, April 7, 2014.

140 "Gotcha Spiral II Radar System," Solicitation Number GOTCHA-09-01-PKS. Federal Business Opportunities, March 2, 2010.

finding the intersections of the moving Doppler radar motion vectors with the image plane of the video camera . . .[141]

Three intersecting radars mean 3D electro-optics. Crossing beams creates a "frosting" of IR heat or thermal sensing, but with ViDAR the picture will be *clear*. No more blobs of IR heat walking across the room. This is spying with a vengeance.[142]

Moving into space, we encounter proof that the missile era is being replaced by a directed energy (non-kinetic) era. *Particle beam weapons* shoot streams of quick-moving projectiles filled with super-accelerated atoms and molecules. Supposedly, particle (ion) accelerators are needed for particle beam "gunpowder," but with ionospheric heaters and engineered weather keeping our atmosphere and near-earth space primed. OTHR (over-the-horizon radar) EMP plasma wave systems are game-changers—like the Brilliant Pebbles particle-beam weapon system currently deployed in Australia at Laverton, Alice Springs (near the Joint Defence Space Research Facility at Pine Gap), and Longreach.

> Sold to the public as simply OTHR utilising HF Hertzian waves, these systems actually have other hidden effects. By utilising pulsed radar beams of high power, they can create and project a charged EM plasma via waveguide layers in the ionosphere. This can be triggered to create severe Earth dielectric induced currents, causing electronic damage, human electrocution, and other collateral damage upon targets located thousands of miles distant—a modern on-off switch equivalent of a nuclear bomb EMP blast. . .[T]he Exmouth [Australia] base is indeed a deployment site of *a major EM weapon system that can at the very least create and fire EM plasma energy pulses into space.*
>
> It is about charged-particle plasma beam weapons even more covert than Boeing's CHAMP (Counter-electronics High-powered Advanced Missile Project) "non-kinetic" missile firing bursts of high-powered microwaves (HPMs) for taking out electronics and data systems.[143] [Emphasis added.]

Boeing's CHAMP fires pulsed HPMs just like in *Star Trek*, and Raytheon's "lights out" CHAMP-derived EMP missile fires HPMs to disable electronics "with little collateral damage. . .[a] huge advancement forward in nonlethal warfare."[144] In the 2012 Utah test, "lights out" translated to fried electronics after

---

141 Lang Hong, Steven Hong, "3D video-Doppler-radar (VIDAR) imaging system," US8009081 B2, August 30, 2011. NexRads are Doppler radar systems.

142 See *The Listening* (2006), en.wikipedia.org/wiki/The_Listening_(film).

143 Nancy Owano, "Boeing CHAMP weapon passes test in sci-fi style (w/Video)." *Phys.org*, October 27, 2012. See video of CHAMP at Randy Jackson, "CHAMP — lights out," www.boeing.com/Features/2012/10/bds_champ_10_22_12.html.

144 "Boeing Successfully Tests Missile That Takes Out Electronic Targets." *Student News Daily*, November 30, 2012.

a silent ("acoustic retardation") initiating explosion.

The term "nonlethal" is truly Orwellian.

Still, interfering beams from two vehicle-mounted transmitters of EMP weapons like CHAMP can also be made to create a plasma arc as protection from EMPs:

> [Patent No. 8,981,261, March 17, 2015, "Method and system for shockwave attenuation via electromagnetic arc"] is for a shockwave attenuation system, which consists of a sensor capable of detecting a shockwave-generating explosion and an arc generator that receives the signal from the sensor to ionize a small region, producing a plasma field between the target and the explosion using lasers, electricity and microwaves.[145]

Block EMPs, create EMPs. . .Scalar tech means generating energy "out of thin air" (æther), lightning (sprites), and heat (plasma).[146]

SENSORS

Laser, radar, and now sensors. Take, for example, the infrared laser radar or LiDAR (light detection and ranging), a remote sensing instrument that meets a variety of military detection and targeting needs. LiDAR is such a versatile weapon that its use is directly overseen by the spooky NGA discussed in Chapter 5.

By bouncing light off of targets, LiDAR is far more accurate than radar when it comes to ranges and distances, plus LiDAR can carry digital information rather like fiber optics without the need for glass fibers. Stealth aircraft covered with radar-absorbent materials (RAM) that don't absorb laser signals use LiDAR, as do bomb damage assessment (BDA), reconnaissance, and chemical warfare agent detection. Unlike emission spectography, LiDAR can read target signature wavelengths without a high-temperature medium:

> Atmospheric pollutants [like chemical trails] are monitored by bouncing a laser beam off clouds which are overhead the measurement apparatus, or terrain behind the area of interest, or simply by analyzing the backscatter from the atmosphere. The backscattered light from the laser detected by the apparatus has traveled twice through the volume of atmosphere, once outbound and once inbound to the detection apparatus. The laser wavelengths absorbed by the

---

145 Michelle Starr, "Boeing patents 'Star Wars'-style force fields." *CNET*, March 22, 2015. Patent at patft.uspto.gov/netacgi/nph-Parser?Sect1=PTO1&Sect2=HITOFF&d=PALL&p=1&u=%2Fnetahtml%2FPTO%2Fsrchnum.htm&r=1&f=G&l=50&s1=8981261.PN.&OS=PN/8981261&RS=PN/8981261

146 See Dutchsinse YouTube, "3/23/2015 -- Force Fields are here! Boeing patents Microwave Forcefield - STOPS EXPLOSIVE BLASTS!"

passage through the air give an accurate indication of the presence of particular chemical species, as well as their concentration.[147]

Once again, we encounter the necessity for two interfering beams—the LiDAR beam and the designator laser beam—for "reading" the target.

Hyperspectral LiDAR imaging is the remote use of sensors to measure reflected light signatures that identify the composition of materials like minerals, snow, and vegetation—a lucrative tool for finding rich deposits of minerals in rugged topography like Afghanistan. Not only can it map decades in advance of what conventional ground mapping can do, but it multiplies the booty of war.[148] Hyperspectral LiDAR also excels in monitoring and collecting signature data on citizens. Though the U.S. Army Corps of Engineers claims that low-flying aircraft shining green lasers into Honolulu neighborhoods in the middle of the night was about high-resolution LiDAR mapping, citizens remained unconvinced.[149]

LiDAR is used in industry, as well—for example, the Irish corporation Treemetrics manages its forests by combining LiDAR information, the European Space Agency's Sentinel satellite aerial imagery, and drone photography.[150] The Global Ecosystem Dynamics Investigation (GEDI) managed by University of Maryland also uses it "to study forest canopy structure in a range of ecosystems, from the tropics to the high northern latitudes."

The question remains if these industry and research instruments are actually "dual use" for planetary surveillance. The Cloud-Aerosol Transport System (CATS) on the International Space Station (ISS) employs LiDAR to measure location, composition and distribution of aerosols, pollution, "dusty plasma," smoke, and other particulates in the atmosphere, while the ISS-RapidScat radar scatterometer gauges solar winds. Then there's the "fleet of 17 NASA Earth-observing missions currently [providing] data on the dynamic and complex Earth system" like the Global Precipitation Measurement Core Observatory, the Orbiting Carbon Observatory-2, the Stratospheric Aerosol and Gas Experiment III (SAGE III), the Lightning Image Sensor (LIS), the Global Ecosystem Dynamics Investigation (GEDI), the ECOsystem Spaceborne Thermal Radiometer Experiment on Space Station (ECOSTRESS), etc.[151] All of these programs employ multiple sensors to taste, touch, see, hear, and smell all that is going on around *and below* them.

---

147 Carlo Kopp, "Laser Remote Sensing — A New Tool for Air Warfare." *U.S. Air Force Air & Space Power Journal*, January 27, 2014.

148 "Afghanistan the First Country Mapped using Broad Scale Hyperspectral Data." U.S. Geological Survey, July 17, 2012. The Afghanistan war ran from 2001 to 2014.

149 "Low-flying aircraft with laser to scope Hawaii for months." *Hawaiinewsnow.com*, September 16, 2013.

150 Aaron Souppouris, "The environmentally friendly rainbow laser forest." *engadget*, February 8, 2016.

151 "NASA Launches New Era of Earth Science from Space Station," NASA/JPL/CIT, September 8, 2014.

Earlier, I discussed microscopic sensors narrower than a human hair, each with its own antenna and power source, released into the environment since at least 2002. Micron-thin Global Environmental MEMS (GEMS)—MEMS being microelectromechanical systems—collect "real-time atmospheric data essential for weather forecasting" (weather engineering) and meet military standards for radar reflection, according to staff scientist John Manobianco at the aerospace sciences and engineering division of Ensco, a systems integration and research corporation:

> As a cloud of such probes dispersed on wind currents inside a storm, they would measure atmospheric pressure, temperature, humidity and other factors with real-time, 3-D resolution not possible with radar or satellite sensors, Manobianco told United Press International. Computer models show the sensors would stay aloft for days if released from several miles up in Earth's atmosphere, he said, and other missions are possible.[152]

NASA's monitoring of Earth's "vital signs" and military surveillance are at least kissing cousins. CATS was built by NASA's Goddard Space Flight Center in Greenbelt, Maryland; ECOSTRESS, a high-resolution multiple-wavelength thermal imaging spectrometer, is managed by NASA's Jet Propulsion Lab (JPL) in Pasadena, California; LIS (lightning imaging sensor) was developed by NASA's Marshall Space Flight Center in Huntsville, Alabama; and SAGE III by NASA's Langley Research Center (Virginia) and Ball Aerospace (Boulder, Colorado). All are NASA and therefore military intelligence, not just ecological do-gooders. *Dual-use technologies go hand in glove with full spectrum dominance.* ECOSTRESS and CATS may actually be more about checking on Cloverleaf's impact having on the "vital signs" of the Earth and its inhabitants. We must learn to read military contexts carefully omitted from media releases in order to understand what actually lies behind NASA space weather programs.

> From space, streaks of white clouds can be seen moving across the Earth's surface. Other tiny solid and liquid particles called aerosols are also being transported around the atmosphere, but these are largely invisible to our eyes. Aerosols are both natural and man-made, and include windblown desert dust, sea salt, smoke from fires, sulfurous particles from volcanic eruptions, and particles from fossil fuel combustion.[153]

Surveillance by satellite depends upon sensors [visible light, ultraviolet (UV), infrared (IR), or radar technologies] tracking missiles above or populations and individuals below. I mentioned ASAT Brilliant Eyes with its long-wavelength

152 "Dust-sized sensors could monitor weather." *UPI.com*, October 30, 2002.

153 "NASA's CATS Eyes Clouds, Smoke and Dust from the Space Station." NASA, December 1, 2014.

infrared detector focal plane in the SDI surveillance satellite system. Later, it was renamed the Space and Missile Tracking System (SMTS), and now it is a low earth orbit (LEO) component of the U.S. Air Force Space Based Infrared System (SBIRS). William R. Burrows in *Deep Black*:

> The infrared imagery would pass through the scanner and register on the [charged-couple] array to form a moving infrared picture, which would then be amplified, digitalized, encrypted, and transmitted up to one of the spacecraft for downlink.[154]

Multispectral scanners, interferometers, IR spin scan radiometers, cryocoolers, hydride sorption beds. . .Sensing from miles above the Earth is a high-tech feat, and advances in *very-high-speed integrated circuits (VHSIC)* and parallel processing have led to criminal exploitation of space-based sensor data. Part of the "space garbage" being monitored and mapped by Space Fence outposts include billions of sensors dedicated to surveillance.

---

154 William R. Burrows, *Deep Black: Space Espionage and National Security.* Berkeley Books, 1988.

CHAPTER 8

Boots on the Ground

▼

[Igor Smirnov:] People's actions can in fact be controlled by unnoticed acoustic influence. Look, it's easy. All I have to do is record my voice, apply special coding which converts my voice to mere noise, and afterwards all we have to do is record some music on top of that. The words are indistinguishable to your conscious; however, your unconscious can hear them clearly. If we were to play this music over and over again on the radio, for instance, people will soon start developing paranoia. This is the simplest weapon. An image can also be coded...

— "Mind control: The Zombie Effect," Pravda.ru, November 10, 2004

Sometimes a weapon comes along that is so crazy in design that we have a hard time putting it into a category. We have decided to call them exotic weapons.

— Swords of Might[1]

Full spectrum dominance of planet Earth is shaping up nicely, now that our plasma-ized atmosphere is being constantly calibrated to keep a wireless Smart Grid up and running. And thanks to the conductive metals and nanobots we breathe, ingest, and absorb into bodies composed of 65 percent water, our nervous systems are plugged in and primed.[2]

Meanwhile, disaster capitalists multiply the towers and power lines, the Smart Meters and Internet of Things (IoT) around us as non-ionized (non-thermal) radiation wreaks havoc on human and animal health. Surges of current pulse along cables and pipelines into neighborhood power lines and 60 Hz homes, resonating in brains and bodies. HD televisions with far more detail in their scanning lines demand much more focus and data input of the viewer's brain, coupled with crystal-clear images far more hypnotic than what old analog television receivers offered. Hundreds of television towers have gone up to

---

1 www.swordsofmight.com/exoticweapons.aspx

2 Thanks to John St. John, "Smart Meters and Chemtrails." *PQBNews.com*, April 9, 2013.

accommodate the HDs at \$2 million/tower, some as tall as 2,049 feet.[3] Thus, zombifying HD screens and visual and audio subliminals are capturing whole generations addicted to ear buds and iPods and palm-sized electronic devices as nature and face-to-face human relationships take a backseat to an increasingly EM-mediated world.

None of this is "accidental." A global grid into which all of the biosphere and its inhabitants are in one way or another "plugged" has been the intent for decades, and now that the atmosphere is primed and the Space Fence ready, it's all systems go.

ECHELON / FIVE EYES

The erasure of the lines between telecom corporations and military intelligence began decades ago in secret with Echelon, now known as the "Five-Eyes" ("F-VEY"), the English-speaking (American NSA, British GCHQ, Australian DSD, New Zealand GCSB and Canadian CSE) intelligence agencies. Only recently has the fact that telecoms are defense contractors been allowed to go public—for example, with Amazon's agreement to provide the CIA with cloud-computing services:

> The sinister implications of Amazon's new CIA role have received scant public attention so far. As the largest Web retailer in the world, Amazon has built its business model on the secure accumulation and analysis of massive personal data. The firm's Amazon Web Services division gained the CIA contract amid fervent hopes that the collaboration will open up vast new vistas for the further melding of surveillance and warfare. . .Amazon is now integral to the U.S. government's foreign policy of threatening and killing.[4]

Echelon/Five Eyes began in 1948 with a secret treaty called UKUSA. Besides the U.S. (North and South Americas), there were First and Second Party UK (Europe, Africa, Russia west of the Urals) and Commonwealth partners Australia (Southeast Asia, Southwest Pacific and Eastern Indian Oceans), New Zealand (South Pacific island nations), and Canada (northern Russia, northern Europe, and American communications), plus Third Party Germany, Japan, Norway, South Korea, and Turkey. Only Echelon members had access to data captured by satellite,[5] microwave radio relay, undersea cable, and the Internet.

---

3 Robert Farago, "HDTV Mass Hypnosis." *Journal of Hypnotism,* September 1991.

4 "Amazon, the CIA and Assassinations." *Consortiumnews.com,* February 12, 2014. Current cloud-computing platforms are direct descendants of the Beowulf Linux clustered computers approach pioneered by the Universities Space Research Association and NASA.

5 Ferret satellites (1960s), Canyon, Rhyolite and Aquacade (1970s), Chalet, Vortex, Magnum,

Echelon's primary downlink facilities are still at Menwith Hill in England and Pine Gap in Australia, with various fiber optic trunk lines for international calls and emails. Besides the U.S. Navy-run receiving station at Sugar Grove, West Virginia in the Shenandoah Mountains collecting SIGINT from civilian satellites, there's Sabana Seca, Puerto Rico; Leitrim, Canada; Morwenstow and London, UK; Bad Aibling near Munich, Germany; Kojarena near Geraldton in Western Australia and Shoal Bay, Northern Territory; and Waihopai near Blenheim, South Island, New Zealand.

Menwith's SILKWORTH software with MAGISTRAND and PATHFINDER driving keyword search programs based on complex algorithms were light years ahead of what the public knew about VOICECST (voice recognition) and OCR (optical character recognition). When British MP Bob Cryer spoke out about the status and role of Menwith Hill in 1994, he ended up dead. Nineteen years before, in November 1975, the entire Australian Whitlam government was dismissed for questioning the CIA's three bases at Pine Gap, Nurrungar, and North-West Cape, where staff earmarked for duty underwent brainwashing and implantation.

By 1995, the NSA had expanded Echelon/Five Eyes capabilities to intercept ever more sophisticated and encrypted traffic, including microwave, cellular, and fiber optics. Under "upstream" Operation Tempora, huge volumes of data are drawn from fiber optic cables and stored for sifting and analysis.[6] Echelon/ Five Eyes supercomputers vacuum everything up, foreign and domestic, and military contractors like Lockheed Martin, TRW, Boeing, Hughes (Raytheon), and Aerojet make out like bandits.[7]

Once captured, phone calls, faxes, and emails are scanned by Dictionary computer programs employing keywords that constantly shift (like the COWBOY dictionary at the Yakima facility in Washington State and the FLINTLOCK dictionary at the Waihopai facility in New Zealand), then cherry-picked as to what to save for analysis. Around the clock, millions of messages an hour are processed, tagged, and filtered. Flagged messages get a four-digit code for the source or subject, date, time, and station code, and are then forwarded to agencies via PLATFORM, the global nervous system separate from the Internet. Once analyzed, the message and analysis are classified MORAY (secret), SPOKE (more secret), UMBRA (top secret), GAMMA (Russian intercepts), and DRUID (for non-UKUSA parties).

Five Eyes is the backbone of UKUSA Atlantic Alliance dominance. Given that the U.S. and UK view commercial espionage as a function of national security, Five Eyes radons eavesdrop on business as well as political conversations. In the

---

Orion, and Jumpseat (1980s), Mercury, Mentor, and Trumpet (1990s), etc.

6 "ECHELON Today: The Evolution of an NSA Black Program." *Antifascist Calling*, July 12, 2013.

7 James Bamford covers Echelon in *Body of Secrets: Anatomy of the Ultra-Secret National Security Agency* (Doubleday, 2001).

U.S., all SIGINT tending toward business is directed to the National Economic Council, which then feeds the intelligence to select corporations. Quibble over the presence of Five Eyes and you may be answering to the NSA, or dead.

TOWERS

Give me the money and three months, and I'll be able to affect the behavior of eighty percent of the people in this town without their knowing it.

— Dr. Elizabeth Rauscher, Director of Tecnic Research Laboratory, San Leandro, California, 1979–1988

The field was everywhere. Invisible, omnipresent, all-pervasive. A gigantic network of towers enmeshing the entire country emitted radiation around the clock. It purged tens of millions of souls of any doubts they might have about the All-Powerful Creators' words and deeds. The Creators controlled the minds and energy of millions. They inculcated in people an acceptance of the repugnant ideas of violence and aggression; they could compel these millions to kill one another in the name of anything they pleased; they could, should the whim strike them, stir up a mass epidemic of suicides. Nothing was beyond their control.

— Arkady and Boris Strugatsky, *Prisoners of Power*, 1969

kHz (kilohertz, 10^3 Hz)
MHz (megahertz, 10^6 Hz)
GHz (gigahertz, 10^9 Hz)
THz (terahertz, 10^{12} Hz)

When nifty consumer inventions like the cell phone are rolled out, the public rarely inquires if the device owes its existence to the military's electromagnetic learning curve. Mobile phones were used by military field officers and elite industrialists for half a century before they were made available to the public. Now, they are so ubiquitous that it's difficult to recall life without them. Approved for commercial use in 1983, they were big, cumbersome, and expensive. Over the decades they have gotten smaller, more compact, and faster in order to keep pace with the Internet.

But why the long wait on the part of the military-industrial complex, if profit is the sole motive? To assume that consumer convenience and profit drive agendas like total population surveillance and containment is shortsighted, if not ingenuous.

From the Russian Woodpecker to HAARP, from power lines to GWEN towers, decades of computation and experimentation have gone into mapping out and constructing an infrastructure that would consolidate a global Smart Grid under the aegis of a planetary space shield or fence. These objectives were envisioned in the 1980s under the "Star Wars" Strategic Defense Initiative (SDI) and are now coming to fruition, thanks to a 24/7 conductive, battery-ready atmosphere.

Sea-to-shining-sea transmitters / receivers / transceivers are now in place.

Since the 1950s, the military has spent a great deal of time, energy, and taxpayer money on suppressing the health effects that follow from the non-thermal (non-ionized) radiation technology that would assure planetary surveillance and containment. Chances were good that the public wouldn't inquire beyond official assurances because non-thermal radiation leaves no perceptual heat signature, and its effects take up to a decade or more to manifest. Whereas ionizing radiation damages cells and DNA immediately, nonionizing radiation is cumulative: the more received, the more cell and DNA damage over time. In fact, microwave poisoning is cumulative and can be passed on genetically, as can a weakened immune system and predisposition to cancer. As Will Thomas put it in *Scorched Earth* (1991), saying that microwaves are safe if they do not burn your skin is akin to saying smoking is safe if you don't singe your fingers.

The dual-use telecommunications industry provides cell phones and Internet for the public as well as whatever the military-industrial-intelligence complex needs under the authorization of the draconian 1996 Telecommunications Act. Three restrictions embedded in the Act make it clear that the Smart Grid is about far more than profits and public convenience:

(1) Telecommunications contractors (AT&T, Comcast, Verizon, Radian Corporation of ONEX, etc.) have jurisdiction over state and local zoning when it comes to construction, modification, and placement of microwave towers;

(2) Physical and mental health are barred from legal consideration;

(3) Independent scientific studies are to be ignored and marginalized.[8]

---

8 E.g., the Harvard Center for Risk Analysis (HCRA), an industry front for the world's largest corporations, and the Wireless Technology Research Foundation pay lawyers and researchers to bury and obscure research proving that cell phones cause brain tumors, EMR disrupts cell tissues, and the immune systems and brains of soft-tissue children are at the highest risk.

Now that the battlespace is everywhere, the public must be kept ignorant of the fact that they are funding their own surveillance, containment and dwindling health.

In January 1998, the European Parliament's Scientific and Technological Options Assessment (STOA) committee released "An Appraisal of Technologies of Political Control," which made a tremendous stir in European media but barely a ripple in America. For Americans, it was already too late—especially for Virginia Farver's twenty-nine-year-old son Rich, a San Diego State University (SDSU) graduate student and TA who died in 2008 of GBM (glioblastoma multiforme) brain cancer after having spent hours, day after day, meeting with students and grading papers on the bottom floor of Nasatir Hall.

Outside Nasatir Hall is the High Performance Wireless Research and Educational Network (HPWREN), the backbone node to University of California at San Diego (UCSD). Farver explains how HPWREN is a dual-use Smart Grid conduit:

> HPWREN is connected to the Lambda Rail (Grid), Tera Grid, PRAGAMA Grid. Combined, these are the "Smart Grid." It is the engineers and scientists behind the Grid who are developing Smart/ATI Meters, Smart Dust, Drones, RNM [remote neural monitoring] technologies, among many others. These Grids are located on many college campuses across the U.S. and other countries, forming a World Grid. There are 2 sites on the SDSU campus: one on top of the Communications Building (which towers above Nasatir Hall), and one on top of the KPBS news station on campus. Both are within 1/4 mile of each other...HPWREN includes Homeland Security, DOE, DoD, DARPA, Air Force, Navy, NIH (National Institute of Health), SPAWARS ["the Navy's Information Dominance systems command"], NASA Inmarsat satellite, Emergency Services, and many huge corporations. *The University of Alaska is connected through the Grid system, with research through the HAARP facility in Alaska.*[9] HPWREN is funded by the NSF (National Science Foundation), OUR tax dollars.[10] [Italics added.]

Since her son's death, Farver has been speaking up for the sake of other endangered students, including the six who died in a Nasatir Hall "cancer cluster." Each time another student dies, SDSU orders up more skewed epidemiological reports to bury the damning cause: the HPWREN tower outside Nasatir Hall. Farver has pushed for wide media exposure, but an anonymous *San Diego Union-Tribune* reporter told her that the story would never be allowed

---

9 "Thus HAARP fills a long-standing vacuum in controlled electromagnetic sources, with the potential to revolutionize low-frequency remote sensing and communications." www.inmarsat.com/cs/groups/inmarsat/documents/assets/

10 "Virginia Farver, Fort Collins CO, US," *EMR Action*, 21 April 2013.

to leave San Diego. When she asked why, he said *Money*.

Telephone poles, towers with multidirectional discs on them, legions of vans and trucks loaded with extension ladders and wire spools, fusion centers across the United States tracking and listening, invisible microwaves everywhere, transmitters on the roofs of schools, wireless Internet clogging up the ionized air. Teenagers go to bed talking to their friends on their iPhones and are incapable of entering the deep sleep that will repair the day's damage of their bodies. They experience mood and personality changes, lack of concentration, and eventually acoustic neuroma on their auditory nerve or glioma on their nerve sheaths, and still their doctors wonder *Why?*

From 1998 to 2003, Norwegian physician Gro Harlem Brundtland[11] was director general of the World Health Organization (WHO). After publicly revealing that she was electromagnetic hypersensitive (EHS), she disallowed all cell phones at WHO's Geneva headquarters. But calling public attention to the health impact of non-thermal radiation put out by cell phones, microwave towers and ovens, power lines, Wi-Fi, Smart Meters, etc., spelled trouble for Brundtland—trouble that may have included uterine cancer in 2002.[12]

In May 2011, the WHO finally reclassified non-thermal (nonionizing) radiation as a Class B carcinogen (like diesel exhaust, chloroform, jet fuel, DDT, and lead). At their Media Centre website in 2014, the reclassification had cooled to "The electromagnetic fields produced by mobile phones are classified by the International Agency for Research on Cancer as possibly carcinogenic to humans [Class 2B]."[13]

> WHO continues to ignore its own agency's recommendations and favors guidelines recommended by the International Commission on Nono-Ionizing Radiation Protection (ICNIRP). These guidelines, developed by a self-selected group of industry insiders, have long been criticized as non-protective. . .Martin Blank, Ph.D., of Columbia University, says, "International exposure guidelines for electromagnetic fields must be strengthened to reflect the reality of their impact on our bodies, especially on our DNA . . ."[14]

The Russians call electromagnetic hypersensitivity "microwave sickness," with symptoms of low blood pressure and a slow pulse followed by stress and high blood pressure, headaches, dizziness, eye pain, sleeplessness, irritability, anxiety, stomach pain, nervous tension, inability to concentrate, hair loss,

---

11 Brundtland served three terms as prime minister of Norway in 1981, 1986–89, and 1990–96.

12 "Repacholi Challenges Gro Brundtland, EHS," *Microwave News*, April 12, 2012.

13 "Electromagnetic fields and public health: mobile phones." WHO, October 2014, www.who.int/mediacentre/factsheets/fs193/en/.

14 "International Scientists Appeal to U.N. To Protect Humans and Wildlife from Electromagnetic Fields and Wireless Technology." Organic Consumers Association, May 11, 2015.

cataracts, reproductive problems, cancers of all kinds, appendicitis, depression, suicide ideation, etc.

The corporatized United States directs mainstream media to ignore the issue. Rare is the professional voice that courageously speaks up about the relationship between microwaves and carcinogenicity. James C. Lin, chairman of the Committee on Man and Radiation, Institute of Electrical and Electronic Engineers (IEEE), testified on July 25, 1990 before the Subcommittee on Natural Resources, Agriculture Research, and Environment Committee on Science, Space, and Technology in the U.S. House of Representatives about biological nonionizing radiation effects, including the microwave hearing effect.[15]

The Earth's "brainwaves" operate in the same spectrum as ours and all of life on Earth: the Schumann resonance of 7.83 Hz. Balancing the Earth's "brainwaves" are sixty-four elements in the ground and in our red blood corpuscles. The ancient resonance between our blood and the Earth's geomagnetic waves is now endangered by the tangle of ionized and non-ionized radiation waves we are caught in, all exacerbated by the Fukushima plume and artificial ELF ground waves driving our atmosphere and jet stream, heavy metal nanoparticles, "smart dust," polymers, fungi, and genetically engineered biologicals we are breathing and ingesting. Birds and bees and whole forests are dying, sea mammals are beaching themselves . . .

It is essential that we examine the power lines pulsing above our heads and underground, the microwave towers emanating on our highways, city streets, and school roofs. As with the television, computer, and cell phone, microwave towers too began with the military—first, the Soviet Tesla magnifying transmitter known as the "Russian signal" and then as the "Woodpecker," and then the U.S. Navy's Project Sanguine.

THE RUSSIAN WOODPECKER AND PROJECT SANGUINE

In 1953, the Soviets set up seven radio transmitters and began pulsing the American Embassy in Moscow with an ELF signal measuring 3.26–17.54 MHz. Embassy workers had no idea they were being pulsed while developing emotional and behavioral problems, leukemia (a 40 percent higher than average white blood cell count), cancer, and cataracts. Ambassadors Charles Bohlen and Llewellyn Thompson died; Ambassador Walter Stoessel, Jr. developed a rare blood disease and bled from the eyes. The Moscow Embassy "experiment" would run for thirty years, a full generation.

Meanwhile in the United States under DARPA's Project Pandora, scientists were studying how low-intensity microwaves might be used to induce heart attacks, blood/brain barrier leaks, auditory hallucinations, etc. This was the

---

15 See Lin's papers at www.ece.uic.edu/~lin/publications.htm.

MK-ULTRA mind control era whose overarching electromagnetic question was, *Can a microwave signal control the mind at a distance?* In 1962, the CIA dedicated Project Bizarre to studying the Moscow signal and made a crucial discovery: it was not the *strength* of the signal that was responsible for loss of biological health—a tiny fraction of the U.S. military (very high) "safe" exposure level— but the *pulsing*. This information was immediately classified and Pandora went black, along with other projects.

By July 1969, the U.S. had 71,524 microwave towers spaced according to the geometric harmonics of the world magnetic grid. SECOM II's[16] five towers broadcast in the 3–12 MHz range—within the Schumann range—in a round robin from Idaho and New Mexico to Missouri, South Carolina, and Maryland.

In 1968, the secretive JASON Group mounted Project Sanguine's 6,400-mile buried cable antenna for long radio wave transmissions (SLF 30–300Hz) out of upper Wisconsin. A transmitter on one side would pump ELF waves through the ground and out the other side so that, as bioelectromagnetics expert Robert O. Becker, MD, put it, "ELF waves issuing from it and resonating between the earth's surface and the ionosphere could be picked up anywhere on the globe."[17] Supposedly, Sanguine was for submarine communication (much like HAARP was sold to the public), but the antenna length produced minuscule frequencies (.3 kHz), which was odd, given that the best frequencies for defense purposes are much higher. The truth is that Sanguine was to be set in sync with the three Soviet Duga antennas at Chernobyl, eastern Siberia, and Ukraine/Croatia.[18]

Per the requirements of the Environmental Protection Act, U.S. Navy Captain Paul E. Tyler asked Dr. Becker to be on the scientific committee overseeing Sanguine. In 1984, Captain Tyler would author "The Electromagnetic Spectrum in Low-Intensity Conflict," a watershed paper which the International Committee on Offensive Microwave Weapons (ICOMW) described as "so important in the chain of evidence establishing the existence of an Electronic Concentration Camp System that if our Archive consisted of only two documents, the Tyler paper would surely be one of them!"[19]

In his 1985 book *The Body Electric: Electromagnetism and the Foundation of Life,* Dr. Becker detailed the committee's disturbing Project Sanguine findings: stress responses, desynchronized bio-cycles, cellular metabolism interference, increased cancer rates in hundreds of thousands of people living inside the

---

16 The SECOM II (SEcurity COMmunications) system is a High Frequency (HF) system that provides two-way digital communications between a central controller and vehicles carrying nuclear weapons and special nuclear material anywhere in CONUS.

17 Robert O. Becker, MD. *The Body Electric: Electromagnetism and the Foundation of Life,* Chapter 15, "Maxwell's Silver Hammer." Harper, 1985.

18 Wikipedia, "Duga-3."

19 www.icomw.org/archives/tyler.asp. Capt. Tyler: "[M]any studies . . . published in the last few years indicate that specific biological effects can be achieved by controlling the various parameters of the electromagnetic (EM) field."

antenna field, etc. What would happen when the long-wave signals resonated throughout the world? The committee recommended that Project Sanguine be shelved and that the 60 Hz power lines carrying far more power than the Sanguine antenna into homes across the nation be reexamined.

> As far as I know, our testimony was the first ever openly given by American scientists stating that electromagnetic energy had health effects in doses below those needed to heat tissue, and that power lines might therefore be hazardous to health. We criticized the White House Office of Telecommunications Policy for failing to follow up a tentative 1971 warning by advising the President that some harmful effects from electropollution were now proven. Moreover, although we didn't realize it at the time, we greatly embarrassed Captain Tyler and the Navy by publicly revealing the existent of the Sanguine report, *which had been secret until then.*[20] [Emphasis added.]

In 1973, Medford, Oregon became the suicide capital of the United States overnight, thanks to the ultra-low frequencies being beamed from a nearby military base to people's television antennas. The creation of a standing-wave resonance was connected to depression, *whether the television was on or not.* David Fraser, Ph.D., of the Department of Toxicology at the University of North Carolina, Chapel Hill, was paid by DARPA to run the Medford experiment.

The U.S. Navy buried the committee's report, moved the antenna to Michigan's upper peninsula, and renamed it Seafarer in 1975, then Austere ELF (extremely low frequency) in 1978. In 1981, an abbreviated Project ELF was constructed. When the New York Public Service Commission (PSC) had to decide about a network of 765-kilovolt power lines that would link nuclear reactors, they asked for a review copy of the Project Sanguine report and the Navy refused. (The nuclear industry and military stick together.)

Meanwhile, Dr. Becker's career was over:

Becker's involvement with high-voltage power lines and the U.S. Navy's submarine communications system (Project Sanguine, later Project Seafarer and still later Project ELF) proved to be his undoing. He was forced into retirement at the too-young age of fifty-six. As Becker wrote in the preface to *The Electric Wilderness,* a history of these struggles by Andy Marino and Joel Ray: "We faced a concerted and coordinated effort to suppress the truth which emanated from the military establishment and was simply aided and abetted by the greed of the utilities and the tarnished testimony of scientists for hire."[21]

In 1975, the Frank Church Committee briefly opened a window onto the "Moscow signal" (along with other Cold War sins, like MK-ULTRA), after which the window was nailed shut for decades. In 1976, while Americans were

---

20 Becker, *The Body Electric.*

21 *Microwave News,* May 28, 2008. (Dr. Becker died May 14, 2008.)

distracted by Bicentennial celebrations, the Soviets expanded the Duga-1 and Duga-2 Moscow signals to Duga-3, the over-the-horizon (OTH) broadcasts called the Russian Woodpecker due to the tapping (pulsing) sound it made. With full knowledge of the CIA, an electromagnetic interference grid was laid over the United States and—rather like HAARP twenty years later—a 10 Hz pulse (40 million watts per pulse) on 3–30 MHz bands was broadcast. The taps and TMT (Technology Media Telecommunications) transmitting them were powerful enough to disrupt radio and telecommunications and force unsuspecting brains into sympathetic resonance. When the signal hit U.S. power grids, it was picked up by power lines and pulsed into people's homes on light circuits—especially in targeted cities like Eugene, Oregon and Sausalito, California. By inducing nuclear magnetic resonance (NMR) in human tissues, the Woodpecker caused pressure and pain in the head, anxiety, fatigue, insomnia, lack of coordination, and numbness. A high-pitched ringing in the ears erupted everywhere the signal was picked up and resonating along power supply grids.

The following year, the U.S. government sold the Soviets a forty-ton magnet to help upgrade the Woodpecker, even sending a team of scientists to install it in Gomel, Belarus, due north of Chernobyl, Ukraine—a magnet capable of generating a magnetic field 250,000 times more powerful than the Earth's magnetic field, like the SQUID (superconducting quantum interference device).

THE TAOS HUM AND OTHER EARTH NOISES

Upgrades to the Woodpecker continued. Beginning in June 1991—six months before the official end of the "Soviet experiment"—a hum pulsing at 17 Hz with overtones of up to 70 Hz was heard in Taos, New Mexico. The Environmental Protection Agency (EPA) categorized these frequencies as "psychoactive," with harmful biological effects. Two percent of the 1,440 residents interviewed heard "the Hum." Some complained of dizziness, insomnia, pressure on the ears, severe headaches, nausea, nosebleeds, and broken marriages. Most could tell when whatever was producing the hum was switched on and off. Taos and Groom Lake are on the same 36° north latitude, so the Taos Hum could have been Groom Lake testing its own Woodpecker signal. Groom Lake might also explain why Catanya and Bob Saltzman received anonymous threatening phone calls after they brought in an acoustic engineer from Denver to check into the Hum. The *Taos News* received threats, too.

In Alaska, Nick Begich (author of the 1995 *Angels Don't Play This HAARP*) and Patrick Flanagan (inventor of the neurophone confiscated by the NSA twenty-five years earlier) wondered if the two percent who heard the hum were actually "hearing" ambient wireless devices operating at 60 Hz *through their skin*, given that electrodermal response is a favorite of psyops. For example, the

EC-130 Commando Solo flying out of the 193rd Special Operations Wing of Harrisburg, Pennsylvania broadcasts passwords that create capillary patterns on the skin.

Sounds are being heard around the world—growls, booms, hums, even "trumpets." Many are no doubt the varying acoustic signatures of magnetized plasma *Alfvén "whistler" waves* in *Birkeland currents* mentioned earlier in Chapters 1 and 3. (Low-frequency currents in magnetized plasma are *shear Alfvén waves*.) A strange sound resonating in the Schumann resonance well of the Earth's atmosphere may well be Alfvén waves or an *ion launcher* creating acoustic modes through the plasma densities of our conductive plasma cloud cover, now that the entire Earth has become a battlespace laboratory.[22]

Anecdotal data of people reporting the Hum seem to follow an earthquake curve (see the World Hum Map and Database Project, *thehum.info*), pointing to mantle movements and pressures resonating in crystals and ferrous components in the Earth that create a piezoelectric effect resonating through multiple harmonic frequencies under grids tuned to 50 Hz and 60 Hz. Mantle movements are part of the blowback of the ELV/VLF radio waves being broadcast deep into the Earth by ionospheric heater earth-penetrating tomography (EPT).[23]

As to why some hear the Hum and some don't, this is no more mysterious than individual differences:

> It is well established in the scientific literature that people can hear electromagnetic energy at certain frequencies and peak power levels. Previous studies have found that a subset of the population has an electromagnetic sensitivity that is significantly greater than the mean.[24]

GROUND WAVE EMERGENCY NETWORK (GWEN)

Under the Reagan-Bush-Cheney *troika* of the first SDI era, the U.S. Air Force built the pulse beam gyrotron system known as the Ground Wave Emergency Network (GWEN), which was pure Tesla technology. Tesla's Wardenclyffe Tower on Long Island was 187 feet high with a spherical terminal about 68 feet in diameter; GWEN tower broadcast hubs are 299 to 500 feet tall with spokes of copper wire 330 feet long radiating out just a few feet underground.[25] They have

22 See S. Dorfman and T. A. Carter, "Nonlinear excitation of acoustic modes by large amplitude Alfvén waves in a laboratory plasma." University of California Los Angeles, April 12, 2013. Thanks to Electric Universe proponent Christopher Fontenot for explaining Alfvén waves.

23 John Pike, "HAARP Detection and Imaging of Underground Structures Using ELF/VLF Radio Waves." *FAS Intelligence Resource Program*, August 9, 2002.

24 David Deming, "The Hum: An Anomalous Sound Heard Around the World." *Journal of Scientific Exploration*, Vol. 18, No. 4, pp. 571–595, 2004.

25 All new wind farm turbine supports are now 299 feet tall with a tuned grounding grid in the

six guy wires with insulator at 334 feet from the VHF translator antenna at the top to the ground, which means GWEN can emit waves between the upper VHF and lower UHF (225–400 MHz) as well as operate in the VLF range of 150–175 kHz traveling along the ground rather than through the air in twenty-minute or one-hour bursts to a distance of three hundred miles, sending messages with about two thousand watts of power while disrupting the Earth's magnetic field in a two-hundred-mile radius. Because VLF signals drop off over distance, GWENs have been built *every fifty miles* across the continental U.S. (CONUS), each costing $1 million.

Purportedly for military aviation (VHF) and continuity of government (COG) in the event of a national emergency (VLF), the VLF ground waves were initially harmonically tuned to the Russian Woodpecker, then to the HAARP (AK) - Arecibo (PR) - Millstone Hill (MA) geometric configuration of phased arrays, with specific frequencies tuned to the geomagnetic field strength of each geographic area so that magnetic fields across the nation could be individually altered, if necessary.

GWEN towers are military relays to *Milstar SCAMP* terminals. MILSTAR (Military Strategic and Tactical Relay) refers to USAF communications satellites in geostationary orbit, and SCAMP (Single Channel Anti-Jam Man Portable) Terminal (U) to the rapidly deployable component of the Army's MILSTAR Advanced Satellite Terminal program. This means that a GWEN electromagnetic web encloses CONUS in an artificial magnetic field five hundred feet high and girds CONUS with powerful radio frequency emissions that never sleep, replacing natural geomagnetic waves with artificial VLF waves and high levels of non-thermal radiation.

GWEN towers are used to fine-tune weather patterns and brain patterns. All that is needed are three satellites and interlocking towers spaced to allow for specific frequencies.

For the Great Flood of April-October 1993, the Russian Woodpecker altered the jet stream and set up weather blocks to target the American Midwest by relaying pulses along GWEN towers in Mechanicsville and Ladyard, Iowa; Chelsea, Wisconsin; Shephard, Minnesota; Curryville and Dudley, Missouri; Whitney, Nebraska; and on into Colorado and Montana—handy receivers for extending precipitation in the Upper Mississippi Valley and producing a flood with $15 billion in damages (1993 dollars). Woodpecker geoengineering was achieved by firing up the GWEN towers north of the two rivers in Mechanicsville and Ladyard and electromagnetically damming up and stalling one of the five atmospheric rivers moving 340 pounds of water per second 1.9 miles above our heads. Pump and dump, as in the days of Noah.

As for GWEN's role in fine-tuning brains, Dr. Becker wrote:

---

ground around them.

GWEN is a superb system, in combination with cyclotron resonance, for producing behavioral alterations in the civilian population. The average strength of the steady geomagnetic field varies from place to place across the United States. Therefore, if one wished to resonate a specific ion in living things in a specific locality, one would require a specific frequency for that location. The spacing of GWEN transmitters 200 miles apart across the United States would allow such specific frequencies to be "tailored" to the geomagnetic-field strength in each GWEN area.[26]

GWEN VLF waves provide on-the-ground power for microwave and radio frequency weapons (DEWs) to control and harass individuals or populations with mental and physical illness (nausea, diarrhea, vomiting, disorientation, organ damage, etc.) and death. VLF waves can act as carrier beams for pulsed ELF signals that can then pass through the skull, which otherwise without pulse resonance might be able to resist low electromagnetic frequencies. VLF does not lose power when traveling through the air and into fragile tissue, thanks to cyclotron resonance.

If microwave towers are pulse-modulated, all that is needed are the exact frequency signatures for inducing specific actions, emotions, or pathological states (anger, suicide, hysteria, trauma, serial killing, paranoia, lust, etc.). Dial in to the tower from a portable laptop-sized microwave transmitter, set the millimeter wave scanner to locate the victim behind the walls of his or her home or office via satellite and the Nationwide Differential Global Positioning System (NDGPS), point the device at the exact latitude/longitude of the target, and you're in the business of torture. Global positioning does not even need satellites if the signal can pick up three radio signals that can pick up the signature. The measure of the signal from point A to C may be four to five bars, and the measure to point B three to five bars. Calculate the distance between bars and the GPS will be displayed.

Portable "poppers" abroad, GWEN and cell towers at home. The neighborhoods and canyons of American cities are not yet patrolled by MRAPs or MRUVs, the mine-resistant ambush-protected utility vehicles loaded with CREW DUKE systems (Counter RCIED Electronics Warfare) like the AM-modulated microwave transmitters ("poppers") scanning and sweeping for frequencies, broadcasting "pacifying rays" from Iraqi rooftops in 2003–2011:[27]

The grunts call the plastic devices "poppers" or "domes." Once activated, each hidden transmitter emits a widening circle of invisible energy capable

---

26 Robert O. Becker, *Crosscurrents: The Perils of Electropollution / The Promise of Electromedicine*. Los Angeles: Jeremy P. Tarcher, 1990.

27 Will Thomas, "Microwaving Iraq: 'Pacifying rays' pose new hazards in Iraq." *Rense.com*, January 25, 2005.

of passing through metal, concrete and human skulls up to half a mile away. "They are saturating the area with ULF, VLF and UHF freqs," Hanks says, with equipment derived from US Navy undersea sonar and communications.[28]

Turn the dial and crank up cataracts, memory loss, numbing, tingling, buzzing, chronic fatigue, Alzheimer's, Parkinson's, and terminal cancer.

Thus the real mission of GWEN towers may be behavioral and cognitive population control coupled with weather engineering, all serving full spectrum dominance. With Tesla technology weaponized, Mother Nature and the sanctity of the human mind are clearly—and to a frightening degree—within the grasp of the military-industrial-intelligence complex.

CELL PHONE TOWERS / MOBILE MASTS

GWEN towers are guy towers, and the four other kinds of towers are lattice, monopole, concealed, and broadcast cell or microwave towers, at the base of which are T-lines of copper or fiber optic cables in the ground to carry masses of data; a box loaded with computerized switching control equipment, a GPS receiver, and power sources connecting the cables to the antennas on the mast. Somewhere nearby is a base transmitter station (BTS). On the tower are triangular antennas with directional rectangles or discs for receiving and transmitting, each triangular antenna belonging to a separate carrier like Verizon or AT&T, each panel covering 120 degrees with each point divided into 40-degree bits to pinpoint phone targets for better signal strength.

The cell tower itself accommodates antennas designed to work like a sprinkler radiating in all directions as well as other systems, like the microwave links or backhauls that look like drums or dishes, or the wideband code-division multiple-access (WCDMA) antenna that focuses signals in eight separate beams like a hose, thus doubling and tripling capacity.[29]

Long waves of tissues-penetrating radiation pulse from the top of a microwave tower, casting a long unseen shadow over the populations below. On the lower rungs are weather sensors, while the middle vertical feet are reserved for "national security" communications, surveillance, military / police radar tracking, etc. And there may be more: a brand new white utility truck with a hydraulic boom and basket may install a mysterious "black box" on a power pole or cell tower, as occurred in a Phoenix, Arizona neighborhood.[30] Mystery additions to cell towers do not bode well and may have to do with steering biological and mental frequencies.

---

28 William Thomas, "Scalar Weapons Used in Iraq, Britain." *Rense.com*, January 25, 2005.

29 Wade Roush, "A Cell-Phone Tower with Focus." *MIT Review*, May 1, 2006.

30 Ray Stern, "SRP [Salt River Project utility] Removes Mystery Box From Power Pole After Phoenix Homeowner's Complaint." *Phoenix New Times*, March 29, 2016.

According to Barrie Trower, a British physicist and microwave weapons expert who worked for the Royal Navy and British secret service, cell phone towers should be placed seven to nine kilometers from communities. Instead, corporations like Crown Castle and American Tower[31] are erecting towers fed by 400 KVA (kilovolt amperes) transformers every two miles.[32] Trower warns that beyond the proximity of the towers themselves, cell phones and routers are unsafe for children under sixteen years of age and lead to aggression, suicide ideation, ADD/ADHD symptoms, and leukemia.

In 2012, freelance writer Jim Stone blogged the burning question about the forest of cell towers around us: *Why the overpower?*

> In America, computer wifi is limited to 20 milliwatts transmit power and often even with that tiny amount you can connect from hundreds of feet away with no special antennas or hardware. Your cell phone can transmit 300 milliwatts. That's well over 10 times the power, and the range of your cell phone is that much better. All that is needed to make a great cellular node is a 10-watt transmitter (to make sure it gets real good penetration into the surrounding buildings) and a receiver that is more sensitive than the one in your cell phone because you are talking back with a lot less than 10 watts.
>
> 10 watts is 500 times as strong as your wireless N router that comes in perfect everywhere. A lucky neighbor might snag your unsecured router with only 20 milliwatts of output from over a block away with just a cheap netbook. Why then, since 10 watts will clearly do the job, are there these ENORMOUS out in the open giant goose-invading "cell towers" in America that are obviously capable of pumping many thousands of watts?. . .
>
> My guess is Japan, Hurricane Katrina, or Chiapas, Mexico. Whatever earthquake or disaster "they" need to accomplish a political objective. And there is something you need to know about radio wave propagation – it can be steered. So just because a cell tower appears to only be able to place all its RF output within a confined zone, it does not mean that signal cannot be diverted and sent elsewhere. This is because a neighboring cell tower can be synchronized in such a way that it steers the output from surrounding towers to a new remote area. And that is how HAARP works, BET ON IT.[33]

The network of microwave cell towers is designed to electromagnetically steer engineered weather systems into and out of geographic regions. Quicken local

---

31 See wirelessestimator.com/top-100-us-tower-companies-list/.

32 In 2002, American Tower collected $548,923,000 in tower rental and management fees, according to Keith Harmon Snow's "NetRad in the Neighborhood: Illuminating the Cell Tower Agenda." *Montague Reporter*, February 28, 2004.

33 Jim Stone, "Mind control via electronic manipulation," June 14, 2012. www.jimstonefreelance. com/cells.html.

towers and relays by means of *multilateration, trilateration,* or *triangulation*—basic geometry that any quantum computer can be programmed to do—and high-atmosphere weather systems can be moved over target areas.

> *Multilateration* — a navigation technique based on measuring the difference in distance to two or more stations at known locations that broadcast signals at known times. The LORAN-C (LOng-RAnge Navigation) uses low-frequency (LF) radio signals transmitted by fixed land-based radio beacons.
>
> *Trilateration* — the process of determining absolute or relative locations of points by measurement of distances, using the geometry of circles, spheres; trilateration is the basis of GPS.
>
> *Triangulation* — the process of determining the location of a point by measuring angles to it from known points at either end of a fixed baseline, rather than measuring distances to the point directly (trilateration).[34]

Microwave towers also act as autonomous switching stations. If city power goes out, two supercomputers cooled and powered by propane generators can handle traffic. If the generators fail, telecom corporations can set up temporary mobile cell phone towers. Conversely, since the PATRIOT Act (now quietly continued by Executive "autopen" as the PATRIOT Sunsets Extension Act of 2011), the FBI and Secret Service have standing authority to jam signals and shut down cell phone towers[35]—which is why it was more than suspicious that when Hurricane Katrina knocked out landlines, cell phones, and BlackBerries, no mobile towers arrived, nor was power reinstated, almost as if New Orleans was to be kept isolated. Even ham radio signals and frequencies were blocked.[36]

TOWERS MAKE ISOLATING CITIES POSSIBLE . . .

There are thousands of stories about the human fallout from cell towers. The 2010 documentary *Full Signal: The Hidden Cost of Cell Phones (fullsignalmovie. com)* exposes the dangers of living in proximity to these towers. One story the film tells is of the Kgosi Nkolo Kgafela II in Mochudi, Botswana who had to

---

34 All definitions extrapolated from Wikipedia entries.

35 "SOP 303," *Emergency Wireless Protocols.* "Chicago in a jam: Security services to block cell phone towers ahead of NATO summit?" *RT,* 17 May 2012.

36 Post by Robert Schoen, "Katrina: Natural Disaster or Sabotage and Institutional Terrorism?" September 7, 2005.

flee the palace in 2007 after his father died of a brain tumor due to microwave emissions from a Mascom tower. His people shut down the tower so the Kgosi Nkolo could return to rule from his palace, after which "unknown people" destroyed the tower.[37] Barrie Trower was subsequently invited to educate the people about microwave technology.

Since communications have gone digital, telecom defense contractors have been building and calibrating dual-use millimeter wave cellular systems. Video feeds to your wireless device depend upon satellites, base stations, and wireless providers balancing their 200 MHz piece of the bandwidth with other providers' pieces of the bandwidth for third generation (3G), fourth generation (4G), and now fifth generation (5G) Long Term Evolution (LTE-A). The 28–38 GHz frequencies and steerable direction antennas of millimeter waves overcome bandwidth shortages.

But wait a minute. The truth is there is no bandwidth shortage or need for wider bandwidth, thanks to the shift from analog to digital. In fact, masses of frequencies above 300 GHz remain unallocated by the Federal Communications Commission (FCC).[38] Now, it's about *high* digital bandwidths for data "stacking": the higher the frequency, the greater the bandwidth. Encrypted, digitized data are now stacked in digitized pulses. With our atmosphere ionized as an "antenna," all of this spells no limit to data transmission, and yet the military is grabbing all the bandwidth possible, even the garage door spectrum.[39]

Check *www.antennasearch.com* for the broadcast range in your neighborhood. (antennasearch doesn't report the wattage, so radiation exposure can't be computed.[40]) Is it in the 12–13 GHz and 18–23 GHz range (billions of cycles per second)—in other words, ultra-high microwaves transmitting WIMAX (WiFi), 5G LTE cell phones, and video? 4G doubled all cell and antenna strengths, meaning a higher penetration rate of buildings and bodies than higher frequency microwaves. (For more on cell phones, including 5G, go to Chapter 10.)

Terahertz waves (1 THz = 10^{12}) *are submillimeter* waves between the microwave and infrared, which means tremendously high frequency (100 GHz – 10 THz). Naturally, they're in solar and cosmic radiation; in Chapter 1, I mentioned that specific THz frequencies have contributed to spontaneous abortions and miscarriages among aircrew. Telecoms use THz for network backbones and broadcasting super-high definition 4K television signals with cameras in the

37 Monkagedi Gaotlhobogwe, "Mochudi Mascom tower Was Dangerous — British Scientist." *The Monitor* (Botswana), 19 April 2010. After Trower's presentation in France, the government began removing WiFi from their schools.

38 Michael J. Fitch and Robert Osiander, "Terahertz Waves for Communication and Sensing." *Johns Hopkins APL Technical Digest*, Vol. 25, No. 4, 2004.

39 Chris Williams, "US Marines stomp suburbia." *The Register*, 2 March 2007.

40 Thanks to Evelyn Savarin, M.U.P., Cellular Phone Task Force, March 22, 2012.

THz spectrum.[41] Cellular and Wi-Fi networks depend on microwaves in the gigahertz (your cell phone operates at 2.4 GHz), but THz radiation will kick what we're exposed to up to 1000 GHz.

Not surprisingly, THz waves like the millimeter waves in TSA airport scanners that "alter" DNA[42] are categorized as non-ionizing and therefore not harmful. While it is true that THz photons cannot break chemical bonds or ionize atoms or molecules like higher energy photons can (X-rays, UV rays), their nonlinear resonant effects in devices like the terahertz imaging detection (TID) camera[43] can "unzip" double-stranded DNA. Thus millions are daily subjected to terahertz lasers in TIDs at TSA checkpoints, supposedly to detect anything blocking the energy radiating from a body up to sixteen feet away and determine its chemical composition.[44]

Are TIDs probing like fMRIs do, given that fMRIs also utilize THz laser?

> The use of light to peer into the brain is almost certainly that of terahertz, which occurs in the wavelengths between 30mm and 1 mm of the electromagnetic spectrum. Terahertz has the ability to penetrate deep into organic materials without the damage associated with ionizing radiation such as x-rays. . .Terahertz can penetrate bricks and also human skulls. . .Medically, even if terahertz does not ionize, we do not yet know how the sustained application of intense light will affect the delicate workings of the brain and how cells might be damaged, dehydrated, stretched, obliterated.[45]

The terahertz wave extends the microwave and millimeter wavebands and therefore massively expands not just communications but *targeting* possibilities.

> If cellphones on a current "4G" network can download data at 10 to 15 megabits per second, terahertz technology can potentially send data back and forth at terabits per second (or millions of megabits per second). . .some researchers have already achieved lightning download speeds with wireless terahertz chips . . .[46]

The military-industrial-intelligence complex favors terahertz waves because they offer the most secure form of communications transmission. At-a-distance

41 "New filter could advance terahertz data transmission," *EurekAlert!* February 27, 2015.

42 Jason Prall, "Yes, Airport Millimeter Wave Scanners Alter DNA." *Jasonprall.com*, n.d.

43 TIDs are also available in portable size.

44 "How Terahertz Waves Tear Apart DNA." *Technology Review*, October 30, 2009. Also see B.S. Alexandrov et al. "DNA Breathing Dynamics in the Presence of a Terahertz Field." *Physics Letters A*, Vol. 374, Issue 10, 2010.

45 Smith, Carole. "Intrusive Brain Reading Surveillance Technology: Hacking the Mind." *Dissent* magazine, Australia, Summer 2007/2008.

46 Ibid.

interception of transmissions is not possible with THz; if a digitally encrypted communication is intercepted by those lacking the proper codes, the THz wave will simply drop off and shut down, which makes it perfect for "confined" digital upper-band (GHz) satellite phone systems.

> The beamlike properties of THz emission reduce the ability of distant adversaries to intercept these transmissions. The adversary may even lack the technological capability to detect, intercept, jam, or "spoof" a THz signal. In addition, atmospheric attenuation allows covert short-range communications, since these signals simply will not propagate to distant listening posts.[47]

Because terahertz lies between the electric and photonic (microwave and infrared), astrophysicists have been a driving force behind developing terahertz technologies that can detect plasma spectra (e.g., the far infrared wavelength is 15μm − 1 mm with a range of 20 THz to 300 GHz). THz frequencies are extensively employed in satellite-to-satellite communications where atmospheric absorption is not a problem, given that *THz waves propagate only in air.* Could THz wave propagation be one of the reasons for conductive desiccants like aluminum in chemical trails, given that both THz and millimeter waves experience less scattering loss from particulates than optical wavelengths?

Towers mean cell phone receptivity and profits, but more importantly they mean surveillance:

> . . .the laws relating to information sharing and wiretapping specifically regulate companies that provide services to the general public (such as AT&T and Verizon), but they do not cover the firms that provide services to the major carriers or connect communications companies to one another. . .[Spooks] can simply go to the companies that own and operate the wireless towers that the telecoms use for their networks and get accurate information on anyone using those towers—or go to other entities connecting the wireless network to the landline network.[48]

The transmission of image and video intelligence (IMINT) is particularly onerous for wireless, and there is no end to surveillance gewgaws that draw power from the wireless matrix. Cell phones tethered to microwave towers are perfect tracking devices for interception and eavesdropping,[49] with taxpayers once again

---

47 Fitch and Osiander.

48 Tom Burghardt, "Niche telecom providers assisting NSA Spy Operations." *Dissident Voice,* September 11, 2008.

49 Eric Lichtblau, "Police Are Using Phone Tracking as a Routine Tool." *New York Times,* March 31, 2012.

paying for their own surveillance.[50] Verizon Wireless, the largest cellular carrier in the U.S., monitors its customers and sells the information.[51] Then there are the "metasearch datamining" sensors. The Department of Homeland Security's Future Attribute Screening Technology (FAST) relies on "non-contact sensors." Palantir software lives up to its name as "one that sees from afar"[52] by vacuuming up financial records, DNA samples, voice, video, maps, floor plans, HUMINT, etc., until a "profile" emerges. Then there's the Defense Intelligence Agency's FICOR (foreign intelligence and counterintelligence operation records),[53] and on it goes from there.

Law enforcement can now use Cellebrite UFED (mobile forensic cell phone analyzers) during routine traffic stops and request that Apple or Google bypass a cell phone user's passcode.[54] Police officers can point the speech-jamming gun at you to listen with a directional microphone and play back your voice frequency with a directional speaker (delayed auditory feedback) to shut down your ability to speak. Or they can point the SMU100 laser at you to dazzle and temporarily blind you up to five hundred meters.

The FBI's CIPAV (computer and internet protocol address verifier) is illegal spyware on the order of DCSNet (digital collection system network), the renamed Carnivore (1997–2005) mentioned in Chapters 5 and 10. CIPAV is a point-and-click data gathering device able to capture IP and MAC addresses, open ports running programs, operating system, default web browser, last visited URL, and outbound emails with IP addresses. FBI wiretapping rooms have access to the entire civilian Smart Grid infrastructure. A high-speed DS-3 (T-3) digital line gives intelligence agencies and their contractors carte blanche when it comes to the carrier's wireless network.[55]

To the military mind, 24/7 access to cell towers and phones are filed under C4 cyberwarfare.

Whether used in the foreign or domestic battlespace, smartphones on military or FBI frequencies have become highly specialized and ubiquitous. They may be translators, field manual dictionaries, or apps like the Blue Force Tracker that tell where friendly forces are, or provide live video feed from drones

50 Andy Greenberg, "These Are The Prices AT&T, Verizon and Sprint Charge For Cellphone Wiretaps." *Forbes*, April 3, 2012.

51 Allison Reilly, "Protect Your Cell Phone from Prying Eyes in 2012." *Technorati*, December 30, 2011. Technorati recommends signing on for virtual private network (VPN).

52 The palantir is the seeing stone of the dark lords in *Lord of the Rings*. Palantir security teams wear black gloves and earpieces. Palantir Technologies resides in Facebook's former headquarters in Silicon Valley, and PayPal computer scientists created Palantir.

53 See www.defense.gov/releases/release.aspx?releaseid=13553. FICOR, Talon (threat and location observation notice, 2002-2007), and the e-Guardian all sound suspiciously alike.

54 Declan McCullagh, "Apple deluged by police demands to decrypt iPhones." *cnet.com*, May 10, 2013.

55 See "Verizon Law Enforcement Legal Compliance Guide," www.aclu.org/files/cellphonetracking/20120328/celltrackingpra_irvine7_irvineca.pdf

and robots like Lockheed Martin's Advanced Technology Laboratories robot fitted with a laser to covertly map in 3D the environment it perceives:

> The creation of robots that can hide from humans while spying on them brings autonomous spy machines one step closer. The spy approaches the target building under cover of darkness, taking a zigzag path to avoid well-lit areas and sentries. He selects a handy vantage point next to a dumpster, taking cover behind it when he hears the footsteps of an unseen guard. Once the coast is clear, he is on the move again—trundling along on four small wheels. . ."Lockheed Martin's approach does include a sort of basic theory of mind, in the sense that the robot makes assumptions about how to act covertly in the presence of humans," says Alan Wagner of the Georgia Institute of Technology in Atlanta, who works on artificial intelligence and robot deception.[56]

Femtocells are small mobile phone base stations, some with a range of a few hundred meters, others like Battlefield Connect with a 5G base station that can connect to big forward-operating networks accessed only by a special need-to-know SIM card or high-tech tattoo.[57] With a little altitude and a CSDA (Connecting Soldiers to Digital Applications), an agent can insert the SIM into a smartphone or laptop and with a microcontroller reminiscent of the *Star Trek* "collar of obedience" that kept the slaves of Triskelion on task[58] stimulate or reduce pain and vulnerability to stress in soldiers or riot police by remotely tweaking their brain with transcranial ultrasound pulses via their pain modulator-behavior helmet.

It's a brave new world.

C4 connectivity demands more and more towers and transceiver stations, some small and discrete, others going up on private land, like that of industrialist Klaus Groenke, whose Idaho property is conveniently located halfway between Lake Pend Oreille, home of the U.S. Navy's Cutthroat submarine, and a lightning generator encased in aluminum on Lunch Peak. A full section of the Berlin Wall encased in Plexiglas stands on Groenke's property, as do triangular satellite dishes, antenna arrays, a weather vane with three lightning-strikes similar to the *Schutzstaffel* insignia, and a red metal "sculpture" that looks like a Delta-T energy transmitter. As Idaho resident and well-known Big Pharma critic Len Horowitz says, "The Lunch Peak Navy station axis bisects a triangle made of three microwave 'Electronic Sites,' according to National Forest Service maps. These intriguing elements provide electromagnetic frequency generating capabilities

---

56 David Hambling, "Surveillance Robots Know When to Hide." *Telepresence Options*, March 25, 2011.

57 Victoria Woollaston, "Hi-tech tattoo that could replace ALL your passwords: Motorola reveals plans for ink and even pills to identify us." *MailOnline*, 30 May 2013.

58 See *MyBrainCloud.net*.

consistent with a long earthquake production and lightning generation."[59]

Our ionized atmosphere is inundated by ground-based radio signals bouncing off the ionosphere to satellites, cell or mobile phones. With 500,000 transistors and GPS, these digital devices are dependent upon semiotic codes. Who controls those codes? Barring knocking out the towers, you can no longer really turn off your little transceiver broadcasting at 884 MHz. Emission standards based on heat and not health are thousands of times higher than necessary for communication, and research into the powerful effects of pulsed low non-thermal levels is being suppressed.

This is our battlespace condition.

NEXRADS AND SBX "GOLF BALLS"

The Next Generation Weather Radar network (NexRad) with updated and recommissioned GWEN towers and cell towers complete the core ground-based Smart Grid relay system for communications and weather engineering. Radio station towers are dual-use—for example, the nine radio towers[60] around Moore, Oklahoma, leveled by an F5 tornado on May 22, 2011—and come equipped with S-band radar, Terminal Doppler Weather Radar (TDWR) and anemometers to gauge wind shear, NexRad "golf balls" for local weather engineering, and biological agent detectors.[61]

From 1,200 miles into space, S-band radar (2.2 – 2.3 GHz) can surveil objects less than two inches long. Masses of S-band data are fed to JSOC (Joint Space Operations Center) at Vandenberg Air Force Base for space situational awareness (SSA) not just of space debris but of trillions of sensors and 4" satellites like CubeSats (mentioned earlier) with attack mode capability.[62] And yet when Russia and China sought to release a free database listing of the thousands of near-earth orbit objects (including NATO and NORAD satellites) for the sake of more secure orbit operations (and to deter more militarization of space) the U.S. military objected.[63]

The Space Fence utilizes all ground-based S-band relays like NexRads but only as a "deterrent," public relations director Nicholas Mercurio of the Joint Functional Component Command for Space at Vandenberg AFB assures the public. Of course, Space Fence frequencies are classified and non-trackable, but there is

59 Elaine Zacky, "Len Horowitz — Wildfires Likely Linked to Energy Industry Plot." *Rense.com*, August 22, 2001.

60 KSGF, KEAX, KVNX, KICT, KTWX, KSRX, KTLX, KSGF, and KINX.

61 See www.wunderground.com/radar/map.asp.

62 Rob Verger, "How the Air Force's 'space fence' will keep American satellites safe." *FoxNews.com*, May 19, 2016.

63 "Russia to reveal location of US military satellites in free space database." *RT*, 22 June 2016.

no doubt that S-band recalibrations (2–4 GHz with wavelengths of 15–7.5cm) on multiple exchange frequencies are part of it. (Related bands are NATO's E / F.) In other words, the S-band Space Fence six-acre array at the Ronald Reagan Ballistic Missile Test Site on the Kwajalein Atoll 2,100 nautical miles southwest of Honolulu is connected to the four hundred evenly spaced NexRad transceivers across CONUS—all in the name of "space situational awareness."

S-band is about planetary lockdown and population control.

Each $5 million NexRad is armed with a *klystron*, a high-powered microwave (HPM) beam tube that can amplify high RF and convert a standard power of 50.8 kW of coherent energy to 750 kW, which can then be transmitted as pulsed rotating frequencies at varying angles (elevations). NexRads track and shepherd local or transiting weather systems by sending up radar/sonar signals and measuring the distance of clouds, moisture, particulates, etc.

The big NexRad "golf balls" perched on latticed towers were originally peddled by the Department of Homeland Security (DHS) as a national "weather and atmospheric surveillance" grid of miniature sensors and high-resolution video monitors, rapid signal processing and computing. NexRads and the look-alike SBX radomes offshore operate at similar frequencies, as both are *super* high-frequency (HF) Doppler radar systems. NexRads by land, SBXs by sea.

It is no mystery that NexRads are at the center of every tornado or storm vortex.[64] To create the wind shear for dozens of simultaneous tornadoes or super-cells, NexRad pulsed rotation frequencies need an artificial precipitation thick with metal particulate "chaff." Why? Because *frequency cannot be pulsed in pure air*. Industrial pollutants provide some of the needed particulates but not nearly enough for C4 operations. Chemtrail injections of conductive metals provide the necessary matter or matrix for building or moving storm systems. Once NexRad rotating frequency pulses strike the nanoparticles and H_2O, a collision occurs between RF- or microwave-heated artificial precipitation (HOT) and frequency-activated ice nucleation (COLD) to stir the brew into funnels that can then be steered while being fed.[65]

To engineer a supercell like a polar vortex, arctic hurricane, or "nor'easter," cold and warm air masses must be made to collide. With NexRads amplifying two clockwise winds at different altitudes and speeds, frequency-activated chemical or bacterial ice nucleation can be added to the mix to produce softball-sized hailstones and wet snowstorms while chemically altering the temperatures. NexRad frequencies can then make the precipitation "flash" into heavy wet "snow" in temperatures as high as 50°F. Once the supercell updraft tightens

---

64 "Massive Energy Pulses Knock Out Roanoke NEXRAD Station," January 7, 2013, endoftheage. blogspot.com/2013/01/massive-energy-pulses-knock-out-roanoke.html. The Roanoke NexRad Station KFCX was knocked off the air on January 6, 2013 by sixty-four electron beams launched by thirty-one different attacking radar stations.

65 See WeatherWar101's excellent NexRad YouTube series.

its spin and speeds up, a *wind shear* can occur—a horizontal rotating column of air (funnel cloud) that rain and hail cause to touch down as a tornado. NexRads operating in tandem can generate multiple tornadoes, like the sixty tornadoes that assaulted Granbury, Texas in May 2013.

NexRad data is collected and analyzed by signal processors and computers at seventy-two fusion centers and beyond. The Universities of Massachusetts, Oklahoma, Puerto Rico, and Colorado State are on NexRad teams, as are Raytheon, MACOM, Vaisala (Finnish), NOAA and National Severe Storms Laboratories. Like UMASS Microwave Infrared Remote Sensing Labs (MIRSL), all are military contractors laboring for military C4.

As was said earlier, the Sea-Based X-Band Radar (SBX) "Golf Ball" looks similar to a NexRad, operates at similar frequencies, and is also a super (high-frequency) radar system. As a floating, self-propelled mobile radar station (radio transmitter), the SBX is a mobile HAARP system more than capable of unconventional attacks such as triggering "atmospheric disturbances" (environmental warfare via geoengineering). For example, the SBX-1 was "conducting exercises" off the coast of Florida the day before the Haiti earthquake in 2010, and was off Japan just before the Fukushima disaster in 2011.

In 2015, the Pentagon attempted to convince the public that the SBX was a failed missile defense system.[66] The truth is otherwise: it too is part of the Space Fence infrastructure, its seaworthy mobility invaluable.

WIND FARMS AND FRACKING WELL STEMS

Visit a wind farm and you will quickly notice that life appears to have fled, which is strange, given how many think of wind farms as "green energy." While driving through western Ontario, Canadian activist Suzanne Maher noted that what was once beautiful landscape and farmland had been replaced by hundreds of massive, imposing wind turbines. People in homes close to the turbines do not seem to realize *they are living inside a power station* that operates by creating low-pressure systems, with the blades of each turbine at a 15° angle from the oncoming wind (7.5° / 7.5°) so as to produce a toroid straight-line wind tunnel. Public protests are usually about devaluation of real estate and noise, but the *pulsing* itself is dangerous to mental and physical health. The infrasound and low-frequency noise produce what is being called the *wind-turbine syndrome:* headaches, sleep problems, night terrors, learning disabilities, ringing in the ears (tinnitus), mood swings (irritability, anxiety), concentration and memory problems, and equilibrium issues like dizziness and nausea.

From the Mariana Islands to Hawaii (through the Kwajalein Atoll), a line of transmitters fires northeast in a repetitive pulse—an invisible wall of radio waves

---

66 David Willman, "The Pentagon's $10 billion bet gone bad." *Los Angeles Times*, April 5, 2015.

all the way to Alaska that creates a funneling effect. Fire radio frequency into a front that's been "seeded" with aluminum nanoparticles and a plasma-dense field arises, while the pressure wall off of Mexico acts like the bumper on a billiard table to roll the weather system north along the California-Oregon coastline to Vancouver Island and the jet stream ready to be pushed east and south through Wyoming and the Dakotas and into Kansas where NexRad and wind farm pulses will kick in to build high pressure for tornadoes, superstorms, and floods.[67]

Wind farms and hydraulic fracturing (fracking) wells are separated by exactly three hundred miles, like GWEN towers, then calibrated together. Fracking wells provide grounding points for wind farms while creating new fault lines for earthquake terraforming. According to Billy Hayes "The HAARP Man," the brine in the wellheads of all new wells is treated with absorbents like the aluminum oxide and barium oxide being laid in chemtrails. A directional explosive is then packed into the well to create perforations in the casing for about twenty feet out so the acid seeps down with the brine mix on top. This "re-dosing" must penetrate the casing because without water, it will float to the top of the well.

Wind turbines and oil well stems have the same resonant length of 1,282 feet. Each rotation of a wind turbine gives off a powerful static charge. In fact, every time the turbines pulse at 2.95 Hz in 1-nanosecond pulses, the liquid mixture at the wellhead jumps twenty pounds up and twenty pounds down 3X per second like a hammer. This is called "thumping." The discharge of the wind farms occurs in an arc at a certain length and a certain pulse that resonates with wells tuned to 2.95 Hz. Put your hand on the casing and feel it twitch, like it's alive.

Both wind farms and fracking well stems have a part to play in the Space Fence infrastructure, along ionospheric heaters, NexRads, cell and GWEN towers, etc. It is because of how they pulse together that nations like Scotland[68] and states like Oklahoma[69] won't be allowed to ban for long the unconventional practice of fracking.

FIBER-OPTIC CABLES

Today, there are 550,000 fiber-optic data channels cross the oceans. *Fortune* magazine graphics director Nicolas Rapp has a spectacular world map of the major seamless fiber-optic trunks:

---

67 See *waubrafoundation.org.au* and the YouTube "Are Wind Turbines Changing the Weather?" (HowStuffWorks, June 24, 2013).

68 Claire Bernish, "Scotland Just Banned Fracking Forever." *Elle*, August 6, 2016.

69 Emily Atkin, "Fracking Bans Are No Longer Allowed in Oklahoma." *Climate Progress*, June 1, 2015.

If the internet is a global phenomenon, it's because there are fiber-optic cables underneath the ocean. Light goes in on one shore and comes out the other, making these tubes the fundamental conduit of information throughout the global village. To make the light travel enormous distances, thousands of volts of electricity are sent through the cable's copper sleeve to power repeaters, each the size and roughly the shape of a six-hundred-pound bluefin tuna.

Once a cable reaches a coast, it enters a building known as a "landing station" that receives and transmits the flashes of light sent across the water. The fiber-optic lines then connect to key hubs, known as "Internet exchange points," which, for the most part, follow geography and population. The majority of transatlantic undersea cables land in downtown Manhattan where the result has been the creation of a parallel Wall Street geography, based not on the location of bustling trading floors but on proximity to the darkened buildings that house today's automated trading platforms. The surrounding space is at a premium, as companies strive to literally shorten the wire that connects them to the hubs.[70]

Fiber-optic signals are light passing in the form of a laser beam through thin strands of optical fiber or glass at very high speeds. The trunk line ("six-hundred-pound bluefin tuna") is made up of fiber optic cables bundled together to increase bandwidth and carry multiple "channels" for multiple networks (and agendas). Telecom contractors like AT&T Inc., MCI, Sprint, and CenturyLink own vast Internet backbone networks and sell their services to (and exert power over) Internet service providers (ISPs). On land, these same providers link hundreds of cell towers together with fiber-optic cable, thus pressuring local zoning boards to approve hundreds of cell towers in one fell swoop.

Fiber-optic cables, like wireless towers, spell power. (Whereas fiber-optic technology unlocks ultra-high speeds for wired connections, terahertz transmitters unlock fiber-optic speeds for wireless.[71]) First of all, whether on land or coming up out of the oceans, they are easy to tap with "intercept probes" and prisms:[72]

The tapping process apparently involves using so-called "intercept probes". . .[T]he intelligence agencies likely gain access to the landing stations. . .and use these small devices to capture the light being sent across the cable. The probe bounces the light through a prism, makes a copy of it, and turns it into binary data . . .[73]

---

70 Nicolas Rapp, "Mapping the Internet." *nicolasrapp.com*, July 9, 2012.

71 "Terahertz wireless could make spaceborne satellite links as fast as fiber-optic links." *Phys.org*, February 6, 2017.

72 The NSA's Internet surveillance program is called PRISM. See footnote in Chapter 5 and reference in Chapter 10.

73 Joe Wolverton II, "NSA Taps Directly Into Undersea Fiber-optic Data Cables." *USAHM News*, July 25, 2013.

Between 2003 and 2006, Room 641A at 611 Folsom Street in San Francisco was fed Internet backbone traffic (foreign and domestic) passing through the building along fiber-optic lines from beam splitters installed in fiber optic trunks. Room 641A was known as AT&T's SG3 [Study Group 3] Secure Room. J. Scott Marcus, a former chief technology officer for GTE (General Telephone & Electronics Corporation) and former adviser to the Federal Communications Commission (FCC), had access to all of it. Whistleblower William Binney, once director of the NSA's World Geopolitical and Military Analysis Reporting Group, estimates that ten to twenty such facilities have been installed throughout the United States.[74]

Because fiber-optic lines are strands of optically pure glass as thin as a human hair that can carry digital information, fiber optics are used in neuroengineering, along with molecular biology, optogenetic engineering, surgery, and lasers.[75]

Not entirely dissimilar to the "new science" of neuroengineering is the covert practice of mounting or implanting hair-like fiber-optic cameras into targets' bodies for remote surveillance. In fact, by studying how fiber-optic cable is utilized for brain-computer interface (BCI) in quadriplegics,[76] it's not difficult to imagine how fiber-optic cable in the home or neighborhood might be used to access minds: cortical neurons fire signals that are then picked up by an amplifier; the signals are converted to optical data and pulsed along fiber-optic cable to a computer that then converts the optical data into binary data for study.

We will go more deeply into these technologies when we study the WiFi neighborhood in Chapters 10 and 11.

---

74 "Room 641A," Wikipedia.

75 Quinn Norton, "Rewiring the Brain: Inside the New Science of Neuroengineering." *Wired*, March 2, 2009.

76 Richard Martin, "Mind Control." *Wired*, March 1, 2005.

CHAPTER NINE

The Temple of CERN

▼

The first gulp from the glass of natural sciences will turn you into an atheist, but at the bottom of the glass God is waiting for you.

— Werner Heisenberg

The Higgs potential has the worrisome feature that it might become mega-stable at energies above 100bn giga-electron-volts (GeV). This could mean that the universe could undergo catastrophic vacuum decay, with a bubble of the true vacuum expanding at the speed of light.

— Stephen Hawking, preface to *Starmus*

CERN is an acronym for *Conseil Européen pour la Recherche Nucléaire*, the European Organization for Nuclear Research. The site housing the massive particle accelerator known as the Large Hadron Collider (LHC)[1] was built (1998–2008) near Lake Geneva (Lac Léman). Two 1,000-ton superconducting magnets hang suspended down three-hundred-foot shafts into a cavern through which an underground river once ran until frozen with liquid nitrogen. The magnets are arranged like boxcars around the five-story, sixteen-mile (twenty-seven-kilometer) "ring" circling the cavern. Two detectors are built into the loop, the Compact Muon Solenoid (CMS) and A Toroidal LHC ApparatuS (ATLAS). Designed to detect tiny subatomic particles like Z bosons, pi mesons, strangelets (quarks bound by gluon), etc., the LHC "ring" is in search of "clues to the fabric of the universe":

> [Dan Green, project manager:] "We expect to see things which will change the way we view the universe"...No one really knows what the machine will give birth to. But the equations suggest that some weird stuff could be just around the corner— maybe "dark matter," the invisible stuff that seems to hang around galaxies."[2]

---

1 LHC is a synchrotron-type accelerator, a particular type of cyclic particle accelerator in which the accelerating particle beam travels around a fixed closed-loop path.

2 David Kestenbaum, "The World's Largest Particle Accelerator." *NPR*, May 18, 2008.

The phrase "No one really knows what the machine will give birth to" is reminiscent of what D-Wave's Geordie Rose says later in this chapter. *Haunting*.

CERN's massive superconducting electromagnets enhance and magnify the sustained reach of the Space Fence throughout our now-ionized atmosphere all the way to the magnetosphere. Certainly, CERN and other particle accelerators have been recalibrated to synchronize with the rest of the Space Fence infrastructure. CERN's two-year shutdown corresponded with HAARP's shutdown, and its restart on April 5, 2015 was just four months before HAARP ("UAF Gakona") resumed.[3] All systems connected with monitoring and controlling geospace, the ionosphere and magnetosphere, and harnessing cosmic processes to serve the Space Fence, have been upgraded and synchronized, including the International Space Station (ISS).

CERN studies how cosmic rays affect the plasma clouds now being produced in the troposphere and stratosphere by chemtrail jets, ships, and sounding rockets, as well as how to keep the electromagnetic circuit depositing ions to form the molecules that create plasma—molecules that CERN propagates via its CLOUD experiments.

The Cosmics Leaving Outdoor Droplets (CLOUD) experiment uses a special cloud chamber to study the possible link between galactic cosmic rays and cloud formation. Based at the Proton Synchrotron (PS) at CERN, this is the first time a high-energy physics accelerator has been used to study atmospheric and climate science. The results should contribute much to our fundamental understanding of aerosols and clouds, and their affect [sic] on climate.[4]

Remarkably similar to Tesla's Wardenclyffe Tower dome, CERN's Globe of Science and Innovations generates and feeds energy into the Global Atmospheric Electrical Grid. Similar to the Space Fence, CERN's mission includes creating and sustaining an atmospheric medium for wireless power (C4).

> CERN's mission of inducing voltage into the Global Atmospheric Electrical Grid is being accomplished by the ionic jet outflow perpendicular to the Globe of Science and Innovation. Ions are increased in energy and ejected into the Ionosphere by the phenomena patented by [David] La Point. Once that charge potential is induced into our environment, it can be used for wireless energy harvesting globally. The technology is being developed by DARPA: Fractal Rectenna [receiving antenna] Array technology.[5]

3 Like other "black" projects, confusion reigns over HAARP's resumption: "HAARP Facility to Reopen in 2017 under New Ownership," National Association for Amateur Radio, August 4, 2016.

4 "CLOUD," home.cern/about/experiments/cloud.

5 Christopher Fontenot, "CERN: Planetary Surface Power and Interstellar Propulsion," *A Microwaved Planet*, June 27, 2016; "Earth's Global Atmospheric Electric Grid Circuit," *A Microwaved Planet*, June 28, 2016. Also view "Bizarre Discovery at CERN," FreedomRebel, January 17, 2013.

The present count of particle accelerators affiliated with an international partnership of thirty-four institutions is as slippery as ionospheric heaters, but at least three in the United States are part of the more than 160 international support particle accelerator groups connected via ESnet5 (Energy Sciences Network)[6] fiber optics. Including CERN, particle accelerators generate 100X more magnetic energy than the entire planet naturally generates, especially when synchronized with mirrored satellite lasers creating electrical discharges (sprites).

Fermi National Accelerator Laboratory at the FermiLab outside Chicago in Batavia, Illinois, was recently upgraded with a 680-ton superconducting magnet for its Tevatron.

Los Alamos National Laboratories houses the High Current Proton Accelerator.

Brookhaven National Laboratory houses the Cosmotron. As I indicated in the Preface, BNL may have participated in the downing of the World Trade Towers in 2001. Here is David Masem, advanced technology engineer and senior Unix administrator who worked at BNL 2000–2003. As per his request, his story is quoted at length so readers can get a picture of what national labs are like:

> When I worked at BNL, director John Marburger III, Ph.D., had just been chosen to serve as science advisor to President George W. Bush. The head of the physics department then became the acting director. As he was a personal friend of my manager, I was given free rein at the RHIC particle accelerator (Relativistic Heavy Ion Collider) and collision detect for Project Aerosol's study of aerosol and nanoparticle impact on the environment under SAG (Stratospheric Aerosol Geoengineering) research. As a Unix Systems Administrator, I configured real-time Unix servers to perform samplings of collisions of positive and negative charged ions. I also configured Apache websites (just templates) for publishing the physicists' findings. During one of my Project Aerosol meetings, I found out that a mini-HAARP was under construction. There are a number of reactors there, several defunct and several more popular gradient & graphite reactors for which I had no clearance.
>
> Since I worked closely with the physicists, I was friended by several from Europe who were also working on the CERN. Over lunch one day at the BNL commissary, one physicist announced that he was going to publicly denounce BNL for its involvement in geoengineering research. But he never had the chance to do so as he had an "accident" in Pt. Jefferson while on his motorcycle and was pronounced dead at the scene. As a result of his "accident," I started to hack around a bit to find out more about Project Aerosol research, as all of this technology was new to me. Around that time, my manager was transferred to the physics department after the U.S. Navy brought in a lieutenant dressed in civilian clothes to manage the IT department. He made my life hell, moving me

___
6 See footnote in Chapter 6.

out of my capacity and taking me off all the projects I was involved with on the RHIC and the cluster server, then banishing me to do departmental backups. I knew it was because of my hacking around. Subsequently, I left BNL. Much of the IT department was outsourced after a new director arrived.

I can definitely testify that BNL has intranet access to CERN. I did an upgrade on BNL's cluster server and had to log in and test the link to make sure that our CERN link was back up and online. BNL also has a supercomputer that was built in-house with a proprietary link to both CERN and Japan's ILC (International Linear Collider) and the KEKB accelerator. I knew this for a fact as one weekend I worked on upgrading storage arrays on BNL's cluster while a physicist and a hardware technician I knew were doing a hardware upgrade on the BNL supercomputer. I remember they were quite concerned about the proprietary intranet links when they failed to come up after the upgrade.[7]

How many particle accelerator labs are still active and how many are proof of concept labs receiving data from LHC particle detectors for number-crunching and computer modeling is hard to say. *Wikipedia* (notoriously unreliable) lists many as having been shut down—or did they just "go black"? The site "Particle Accelerators Around the World" (www-elsa.physik.uni-bonn.de/accelerator_list. html) seems accurate, though again, who knows, given that most of the programs are geared to researching kinetic weapons and not physics per se?

Take the eight-hundred-foot "Desertron" Superconducting Super-Collider (SSC) in Waxahachie, Texas near Dallas, a fifty-four-mile-long tunnel 250 feet under the Earth's surface with a "ring" circumference of 87.1 kilometers and an energy of 20 TeV[8] per beam of protons—3X that of the LHC (7 TeV per beam). Protons hurtling at the speed of light (harmonic 16944), antimatter collisions shattering quantums of energy. The Desertron was to build and test superconducting magnets for other super-colliders (plus magnetic-levitation trains or maglevs, motors, energy storage, low-loss power distribution, etc.). Construction began in 1991 but by 1993 the story was that the Desertron was hopelessly over budget with just 22.5 km of tunnel and seventeen shafts completed, so it was shut down. In 2006, the Department of Energy purportedly sold it to Arkansas multimillionaire Johnnie Bryan Hunt whose dream was to build the Collider Data Center.

Hunt's unique selling point for Collider Data Center was its location and infrastructure. The collider sits on an independent power grid capable of delivering 10 megawatts of power (and up to 100 megawatts if needed), and it has its own dedicated fiber optic line. Its two warehouses can support floor loads of 500 pounds per square foot, perfect for the enormous servers that

7 Email, October 4, 2016.

8 Tera or trillion electron volts; 1 TeV = 1.6021773E-7 joule.

Hunt intended to buy. The entire complex is clear of flight paths and out of hurricane, tsunami, earthquake and flood zones.[9]

But then Hunt "slipped on ice" and died, so the half-dug underground facility again sat idle—or did it?

Earlier, I stressed that HAARP experiments have damaged the ionosphere and weakened the magnetosphere. CERN, in tandem with scalar interferometers like HAARP, is no doubt complicit, as well, but could also offset impact on the magnetosphere. "Ripples" in the magnetosphere can cause airplanes to literally drop from the sky. Rumors about CERN flew when on March 24, 2015, Germanwings Airbus A320 went from a cruising altitude of 38,000 feet to a rapid descent over the French Alps *127 miles from CERN*. Only 10 percent of airplane crashes occur after a plane reaches cruising altitude.

On April 23, 2016, the magnetosphere collapsed ("disappeared") for two hours. While it is tempting to blame the LHC superconducting magnets, CERN watcher Anthony Patch insists that at that time the LHC was running at low luminosity and producing corresponding low-duration energies and lower field intensities.[10] So no, it wasn't the LHC that caused the magnetosphere to "disappear" for two hours:

> We can, however, attribute these gaps in the electromagnetic lines of force to actions by the numerous HAARP antennas established around the planet. No longer are they simply heating the ionosphere and similarly targeting terrestrial areas to trigger earthquakes and volcanoes. Indeed, these are manmade events. Of recent, most HAARP installations have been revamped, resulting in higher output and range [due to working together at exact calibrations]. These are the cause and effects noted in our [magnetosphere] shields, not the reduced power levels of the superconducting magnets arrayed to the Main Ring of the Large Hadron Collider.[11]

Disruption of the magnetosphere means that more gamma rays and X-rays can access our atmosphere.

Echoing the Stephen Hawking quote at the beginning of this chapter, professor Otto Rossler at the University of Tubingen in late 2008 went so far as to file a lawsuit against CERN with the European Court of Human Rights on the grounds that the facility might trigger a mini-black hole that could annihilate the planet.[12] The lawsuit was thrown out.

---

9 "The Abandoned Remains of the Superconducting Super Collider." *Amusing Planet*, December 14, 2010.

10 If the LHC schedule can be believed: espace.cern.ch/be-dep/BEDepartmentalDocuments/BE/LHC_Schedule_2016.pdf.

11 Email from Anthony Patch to Kevin Baker of the Kevin Baker Show, April 25, 2016.

12 See YouTube "Higgs boson found, now what? (Inside Story – 5th July 2012) [HD]." TheNovadex,

However, if Rossler was referring to CERN's *antimatter* experiments, he was right: when matter comes in contact with antimatter, it is destroyed. Unstable and needing careful containment, if just one-millionth of a gram is loosed, it equals 37.8 kilograms (83 pounds) of TNT or 10 billion times a high explosive.

ANTIMATTER

The study of matter, what it is and isn't, has obsessed Western science since the eighteenth century, as has the determination to probe its origins. We now know that the reality our five senses perceive is actually a complex pattern of interlocking wave forms, much as was envisioned in the final scenes of the 2003 film *The Matrix Revolutions*. What we see as "matter" is the positive cycle of a wave pulsing into existence while in a hyperdimension its negative cycle manifests as antimatter. Quantum physics and harmonics mathematics may be preparing humanity for encounters beyond what occultists have for centuries termed the Threshold or Veil.[13] (At CERN, it's known as the "containment wall.")

Think of antimatter/matter as the yin/yang of reality, the building blocks of atoms being electrons (-) and protons (+), each with its own antimatter and opposite charge, as was mentioned in Chapter 7 while discussing lightning, a natural particle accelerator producing antimatter. Chapter 6 mentioned how lightning is the natural "particle accelerator" that produces antimatter, and how sprites are now being manufactured to do the same.

With terms like "Higgs boson" and "God particle," CERN claims to seek the "glue" binding all matter together. What it really seeks is the *antimatter* rooted beyond the Threshold of earthly existence.

In 2004, the U.S. Air Force publicly announced that it was pursuing antimatter weapons, "the eerie 'mirror' of ordinary matter."[14] With antimatter in hand, positron bombs would be "clean" bombs that wouldn't eject plumes of radioactive debris. But on Eugene Mirman's *Star Talk* radio program, astrophysicist Neil de Grasse Tyson blew holes in the military's naïve way of thinking: "Ask yourself: How much energy is keeping [the planet] together? Then you put more than that amount of energy into the object. It will explode."

Are some of the world's particle accelerators already in the antimatter weapons business?

July 7, 2012.

13 As Chapter 2 discussed, scalar potentials hearken from a hyperdimension. Matter, gravity, and light of this 3-space dimension are connected, as harmonics mathematics makes plain. See New Zealander Bruce Cathie's *The Harmonic Conquest of Space* (1995).

14 Keay Davidson, "Air Force pursuing antimatter weapons / Program was touted publicly, then came the official gag order." *SFGate.com*, October 4, 2004.

Antimatter being on the other side of the containment wall, enough strangelets must be produced so that an explosive potential can tear the Veil, expose the antimatter realm, and alter the very fabric of Space. Anthony Patch says that the right number of strangelets can change a planet into a neutron star. (Shades of Stephen Hawking's misgivings.) Blast the containment wall and *psycho-energetics*[15] will leak through, operating on the same scalar "channel" as visions, dreams, "spooky action at a distance" (quantum entanglement), synchronicities, and *déjà vu*—all of which share a frequency signature and resonance with another, unseen side of existence.

Having deciphered antimatter's signature and breached the Threshold, CERN opens the way to impact everyone's antimatter "tether." For example, if it is true that CERN has breached the Veil and opened an aperture ("portal") welcoming antimatter into our dimension, it seems obvious that if antimatter is the yin to matter's yang, antimatter entities now in 3-space will need to either produce a plasma body (plasma being the fourth state of matter) *or occupy* a body possibly already occupied.

In a quantum entanglement era, such a line of thinking is no longer so far-fetched. After all, the military-intelligence *apparatchik* has been studying paranormal events and parapsychology for decades—ever since antimatter was discovered in 1955, as a matter of fact. And if dark matter and the Higgs Field are wherever matter *isn't*—like antimatter and æther[16]—then proximity to the Threshold probably quickens paranormal and parapsychological events. Which brings up questions regarding the relationship between the paranormal and antimatter, beginning with *What, if anything, are scientists and staff at CERN experiencing?*

CERN PSYOPS

The fact that the LHC's huge, circular *dharma*-like geometry has been compared to ancient technologies like the Egyptian pyramids points to its *mythical* stature: a vast, complex technology in quest of the tiny building blocks of the universe. Is CERN the global elites' Mecca or Rome "hidden in plain sight"? Headlines in publications for the masses like the British *Sunday Express*—"Scientists at Large Hadron Collider hope to make contact with PARALLEL UNIVERSE in days" (August 15, 2015)—make CERN larger than life. Rumors of subterranean CIA headquarters beneath Lake Geneva add to the diversion from what is really going on there, and at the same time assure the scientifically minded that

---

15 The Russians classify scalar wave technology in three parts: *Energetics*, Bio-energetics, and Psycho-energetics.

16 "Seifer to Time magazine: 'Higgs field' is 'ether.'" *Changingpower.net*, March 18, 2012.

nothing but "science" is going on at CERN and that the choreographed shiny LHC-like revolving sigils and Saturn "rings" zipping around the Earth on news shows as tones deepen the state of TV trance are simply coincidental.[17]

The statue of the Hindu goddess Shiva Nataraja began dancing the *Anandatandava* or Cosmic Dance of creation and destruction outside CERN headquarters in 2004. The town in France that CERN's presence dominates is called Saint-Genis-Pouilly, Pouilly referring to the Latin *Appolliacum*, a Roman temple honoring Apollo and therefore a gateway to the underworld. But perhaps Shiva is dancing for the Indian *Apollyon*, primordial destructive force of the universe . . .

Meanwhile, ritual and "mirrored" events are being leaked to the Internet and press to further obfuscate public perception:

~*March 9, 2015*, the Shiva "Dance of Destruction" around the Large Hadron Collider a.k.a. Hindu *dharma* wheel in the cavern,[18] Shiva's dance being the dance of subatomic matter.

Wednesday, August 12, 2015, mirror event in New York City: a hologram of the Hindu goddess Kali cast upon the Empire State Building.

~*August 11, 2016*, a seemingly Satanic (or Saturnalian?) night ritual replete with a female sacrifice in front of Shiva in the main square.[19]

Wednesday, June 1, 2016, mirror event in Switzerland; a two-part ritual spectacle in Switzerland to christen the opening of the $12 billion Gotthard Base Tunnel (35.4 miles long) buried a mile and a half under the Gotthard mountain range of the Swiss Alps.[20] Though not on site with CERN, the latitude of Gotthard Base Tunnel is 46.600°N and of CERN 46.233°N. (Serious ritual always takes into account planetary and geodetic conditions.) Four heads of state and multiple global elites first watched the ritual spectacle of robotic workers in orange jumpsuits peeling down to their underwear *inside* the cavern, then witnessed the Mystery drama of Baphomet simulating anal rape as three scarabs, multiple mad souls and chaotic entities from other dimensions undermined toiling humanity, and saving angels *outside* the cavern. The

---

17 See YouTube "Illuminati Black Magick Television Tricks REVEALED!! Part 2," The Black Child, January 9, 2015.

18 Katherine Brooks, "Here's The Dance Opera That's Being Filmed Inside CERN's Large Hadron Collider." *Huffington Post*, March 9, 2015. Analysis of the dance-opera at thecreatorsproject.vice.com/blog/epic-dance-opera-filmed-inside-cerns-large-hadron-collider.

19 "Sacrifice to Shiva at CERN? Officials launch investigation after video of 'spoof' ritual emerges." *RT*, 18 August 2016.

20 *Inside:* "Uncut Occult ceremony at the Opening of World's longest tunnel near CERN in Gotthard massif.+ sound," Soul Rescue, June 2, 2016. *Outside:* "2/2 Uncut Occult ceremony at the Opening of World's longest tunnel near CERN in Gotthard," Soul Rescue, June 14, 2016.

spectacle might have been entitled "Ripping the Veil" or "Storming the Threshold."

The Gotthard Base Tunnel "outing" of an authentic global elite spectacle-cum-ritual-cum-psyop no doubt shocked those of the lower classes who were there as well as television audiences who eventually saw what had been filmed, but it was pagan business as usual for the high-ranking elites in attendance. No public debate over the artistic or religious merits of the "choreography" has been forthcoming. In fact, the entire event has dropped neatly into the Orwellian Memory Hole (the subconscious). Such an *Eyes Wide Shut* peep into the elite underworld was more than enough for the 99.9% conditioned to believe whatever they are told and not necessarily what they are *shown*.

These rituals are connected with the intent that CERN throw open the gates of the Abyss to antimatter. *On an energetic level,* "being human" straddles both matter and antimatter; thus it seems obvious that these rituals are intended to open us to our antimatter side, given that *our thoughts and feelings tend toward one side or the other.* Entities drawn to antimatter are *demonic* to human beings, so increased production and extraction of antimatter will increase demonic presence. By wielding antimatter's energy signature, the LHC will "call" the antimatter signatures in human beings to activate.

It seems that the temple of CERN is about altering reality as we have known it.[21] Theoretical physicist Feynman had it right: "If you aren't deeply disturbed by quantum mechanics, you clearly haven't understood it."

SATURN AND THE SPACE FENCE "RINGS"

On May 25, 2007—Bernard Eastlund died 12/12/2007—Bernard, Tom Bearden, Fred Bell, a filmmaker and I had a conversation in Texas about the possible development of Saturn rings around the Earth like a celestial space collider. It seemed like every time the LHC was activated, there were cyclotronic reverberations like shockwaves of earthquakes, volcanic and massive spontaneous gravity (scalar) wave activity.

— Billy Hayes, "The HAARP Man"

Even more startling than CERN's antimatter quest may be its mysterious connection to NASA's intent to create Saturn-like "rings" around planet Earth—*synchrotron* rings that will serve interplanetary communications, quantum

---

21 For more behind this interpretation, please listen to an alleged CERN insider: "Darkest Side of CERN — Destruction of Souls — Critical!!!!" Margaret Schaut, April 3, 2015.

shielding and surveillance, and the *real* SSS Space Fence lockdown. First, let's follow the psyop breadcrumbs.

Saturn has six letters and is the sixth planet from the Sun with a six-sided polygon as its north pole. Even the sixth day, Saturday—still a holy day for some—is named after Saturn. CERN's logo is a quantum entangled 666.

In Chapter 7, I mentioned NASA space scientist Robert Pfaff's justification for altering the near-earth space environment by referencing other *dynamos*[22] on Jupiter, Saturn, Uranus and Neptune.[23] In the same context, I discussed how "dusty plasma" is being used to construct a conductive Saturn-like "CD disc" around the equator bearing a chemical signature very like the conductive metal brew daily dumped by jets and zapped by radio frequency—following the ancient *As above, so below* alchemical dictate by weaving electro-chemical processes of the ionosphere and lower atmosphere into a plasma mesh or grid.

China is already working on a massive supercollider twice the size of any synchrotron or linear particle accelerator and 7X more powerful than CERN to produce millions of—what? Is China's supercollider to serve as our planetary dynamo? Or will it be the Compact Linear Collider (CLIC) or International Linear Collider (ILC) no longer firing electrons at each other but firing electrons and positrons (antimatter) at each other? China's quantum satellite is now experimenting with quantum entanglement.[24] Anthony Patch wonders if the quantum satellite will open portals to electrically link the Earth's rings with Saturn's rings via Birkeland currents (plasma conduits).[25]

Now, let's go a little deeper into what Saturn represents to various global elites.

The incursion of thousands of Nazis into the United States under Operation Paperclip lends credence to the possibility of a still-thriving secret Saturnalian Brotherhood in the United States, along with the normal fare of Freemasons, Knights of Malta, Opus Dei, Satanists, etc.

The 33° system of initiation known as the Fraternitas Saturni or Saturnalian Brotherhood is a German magical order founded in 1926. Originally, it was

---

22 The dynamo theory proposes a mechanism by which a celestial body such as Earth or a star generates a magnetic field. A dynamo is thought to be the source of the Earth's magnetic field, as well as the magnetic fields of other planets.

23 Tariq Malik, "NASA's Fourth of July Fireworks: 2 Rockets Launching Today." *Space.com*, July 4, 2013.

24 Elizabeth Gibney, "Chinese Satellite Is 1 Giant Step for the Quantum Internet." *Scientific American*, July 27, 2016.

25 Anthony Patch email, June 15, 2017: "China's satellites are for encrypted communications employing *their* version of the D-Wave quantum computer system. These will not open portals, nor will they produce or connect Birkeland currents to Saturn or other planets. Only the LHC's Main Ring conjoined with AWAKE, concurrent with the opening of the portal at CERN's North Area Impact Point, produce the Birkeland currents connecting first with Saturn, then with the other remaining planets. The Chinese have nothing to do with this. I do agree that Saturn-like rings are being produced around Earth as a result of spraying and Synchrotron energies produced by circulating particles within the LHC Main Ring. These energies are the same as with Saturn: gamma and X-rays, along with magnetic lines of force extending at right angles (orthogonal) to the Main Ring's superconducting magnets." (In the Bible, Birkeland currents are "Jacob's ladder." For a review, go back to Chapter 7.)

connected with Aleister Crowley and his Ordo Templi Orientis (OTO); now, "Saturn-Magic" lodges are primarily headquartered in Germany, Austria, and Switzerland, including offshoots like Ordo Saturni and Communitas Saturni.

I would not be surprised if NASA hosts a Saturnalian Lodge. *Wikipedia* defines NASA as "an independent agency of the executive branch of the United States federal government," but neglects to mention that it was founded under SS officer Wernher von Braun (1912–1977) and Paperclip Nazis who claimed to have derived technological insight from ancient "mythological" texts and entities beyond the Veil through occult rituals. Thus, NASA may be a very different kind of "independent agency of the executive branch."

The most high-profile NASA move toward Saturn is surely the *Cassini* space probe launched with its lander the *Huygens* on a Titan IVB/Centaur on October 15, 1997 and entering Saturn's orbit in 2004 with *Huygens* landing on Saturn's moon Titan on January 14, 2005. Data has been transmitting back to Earth ever since by using the orbiter as a relay, and the *Cassini* has been executing fly-bys of Saturn's other moons. However, once it enters Saturn's rings and atmosphere, it is slated for destruction on September 15, 2017, supposedly "to ensure protection and prevent biological contamination to any of the moons of Saturn thought to offer potential habitability."[26]

Questions regarding Saturn abound. Why and how does Saturn emit more energy than it absorbs? Why do Saturn's rings make those eerie resonance sounds? Why are the rings separate? What are the objects caught up in the rings? From ancient lore, we might ask, *Is it true that hundreds of thousands of years ago Saturn was Earth's Black or Midnight Sun? Were Saturn, Mars, and Venus once in polar alignment with Earth, and if so, what event ended that alignment, causing them to "float away" to their present orbits?* Of NASA, I would ask, *Is the purpose to go to Saturn or to make Earth into a Saturn?*

Shades of Immanuel Velikovsky (1895–1979).

Certainly, the LHC, AWAKE, and D-Wave are engaged in reconnecting the Earth to Saturn as NASA and other agencies and private deep pockets seed the Earth's atmosphere with nano-metals and build nano-metal Saturn rings around the Earth that surely will act as transceivers with Saturn's rings.

AWAKE AND D-WAVE

And the fifth angel sounded, and I saw a star fall from heaven unto the earth: and to him was given the key of the bottomless pit.

— Revelations 9:1

---

26 Phillippa Blaber and Angélique Verrecchia, "Cassini-Huygens: Preventing Biological Contamination." *Space Safety,* April 3, 2014.

The recent addition of the smaller (thirty-meter) Advanced Wakefield Experiment (AWAKE) plasma linear accelerator with a focusing beam and capability of accelerating particles to 20 PeV (peta electron volts, peta meaning one thousand million million) by "surfing" them on waves of electric charge (protons) created in a plasma (ionized gas) will generate 1,000X the particles that the LHC generates. "Because protons have greater mass than electrons, each proton pulse penetrates further into the plasma, setting up a longer series of charged regions, which in turn provides greater acceleration per pulse."[27]

Anthony Patch believes that the AWAKE will assure a connection to Saturn.

The AWAKE focusing beam means low luminosity and less fragmentation, and the "accelerator on a chip"[28] means moving away from the mechanical power of the ring-based synchrotron LHC smashing and crashing protons to the real source of greater energy power, which is plasma. Billy Hayes "The HAARP Man" well understands that size is not the issue anymore (other than smaller and smaller):

> Each [accelerator on a chip] can independently act as an optical laser amp or memory chip and redirect what it receives. Hence a new versatile super weapon one step from airborne nanoparticle accelerators launched into orbit by the trillions to nano-tag and digitally control life forms.[29]

As the LHC smashes and AWAKE drills holes via quantum tunneling into hyperdimensions, the new *VHEeP (very high energy electron-proton collider)*, superconducting magnets, plasma, and electron-proton instead of proton-proton collisions will generate the power needed to "pierce the Veil." 20 PeV with a focused beam at a quantum level (thanks to the D-Wave quantum computer) is powerful enough to break the nuclear bonds—the force binding quantum particles at a quantum level—that have heretofore prevented access to the dimensions beyond the Threshold.

In other words, the VHEeP is a kinetic particle energy weapon whose assault on the Threshold is an act of war, not an astronomical "experiment."

CERN's D-Wave adiabatic quantum computer system controls the portals—their rate of opening, their size, what comes through, etc. Whereas the nominal operational temperature for the LHC is 1.9 degrees above zero, the tiny niobium D-Wave chip in its ten-foot black box must be kept colder than deep space at just above absolute zero.[30] *It is the cold that transforms a mere chip into a qubit*

27 Elizabeth Gibney, "CERN prepares to test revolutionary mini-accelerator." *Nature*, 7 October 2015.

28 "Researchers Demonstrate 'Accelerator on a Chip'." SLAC National Accelerator Laboratory (Stanford University) press release, September 27, 2013.

29 In response to the SLAC press release.

30 By international agreement, absolute zero is −273.15° on the Celsius scale (−459.67° on the Fahrenheit scale).

superconductor. D-Wave also requires an extremely low magnetic environment, 50,000X lower than the Earth's ambient magnetic field—the exact opposite of CERN's massive superconducting magnets.

Co-founder and chief scientist Eric Ladizinsky equates D-Wave's qubit chip with the Manhattan Project *and magic,* as per Arthur C. Clarke's famous "Any sufficiently advanced technology is indistinguishable from magic." In a 2013 speech at Idea City, D-Wave's Geordie Rose revealed D-Wave's hyperdimensional capability by quoting physicist David Deutsch: "Quantum computation. . .will be the first technology that allows useful tasks to be performed in collaboration between parallel universes." Rose went on to mention the pulse tube dilation refrigerator unit "heartbeat" of the twelve-foot D-Wave, and confessed that standing beside the D-Wave was "awe-inspiring. . .like an altar to an alien god."[31] (See Chapter 6 for more on D-Wave.)

It is true: D-Wave appears to function interdimensionally. First came D-Wave Model 512 (qubits) linked with Josephson junctions, then the 1024, the 2048, and finally the 4096 that clears the way through the Veil (4,096 qubits in the coded "key"). D-Wave's 512 cracked the uncrackable Shor's algorithm of 2048; now, D-Waves are writing their own code that only other adiabatic quantum computers can crack. How many dimensions has D-Wave accessed so far with VHEeP?

In quantum worlds, (1) dimensions are frequencies, (2) everything locked in matter has a dual nature of waves (spiritual) and particles (physical), (3) quantum tunneling reaches into spiritual dimensions, and (4) antimatter frequencies entering our reality are subject to the same 3-space laws we are.

Is the Transhumanist agenda looming behind the Space Fence, CERN, and the D-Wave "alien god" to digitally augment our reality to the point that our DNA matches the digitized DNA of antimatter demons/aliens? Are we to one day be only virtual and exist entirely in the digital like cartoon characters?

The *Mandela Effect*—when someone has a clear memory of something that never happened in this reality—may be an early ripple foreshadowing such a future, or it could be yet another psyop via pulsing and Internet hacking. But it does seem that CERN and D-Wave are bent on opening and cross-fertilizing dimensions. Meanwhile, the rings of Saturn are going up around the Earth.[32]

---

31 YouTube "D-Wave lecture by Geordie Rose (IdeaCity 2013)." Paul Calhoun, July 9, 2013.

32 Listen to an interview with Jay Weidner and Troy McLachlan on "Windows on the World — The Cult of Saturn," RevelationMediaCo, June 20, 2014; an interview with Anthony Patch, "CERN The Next Generation: LHC's New Name & Energy Levels Revealed," The Kev Baker Show, July 28, 2016, and "Forget Mandela Effects, Think QUANTUM POLLUTION w/Anthony Patch," The Kev Baker Show, August 11, 2016.

"PEACE SUPPORT OPERATIONS" SPELL LOCKDOWN

THE YEAR WAS 2081, and everybody was finally equal. They weren't only equal before God and the law. They were equal every which way. Nobody was smarter than anybody else. Nobody was better looking than anybody else. Nobody was stronger or quicker than anybody else. All this equality was due to the 211th, 212th, and 213th Amendments to the Constitution, and to the unceasing vigilance of agents of the United States Handicapper General.

Some things about living still weren't quite right, though. April for instance, still drove people crazy by not being springtime. And it was in that clammy month that the H-G men took George and Hazel Bergeron's fourteen-year-old son, Harrison, away.

It was tragic, all right, but George and Hazel couldn't think about it very hard. Hazel had a perfectly average intelligence, which meant she couldn't think about anything except in short bursts. And George, while his intelligence was way above normal, had a little mental handicap radio in his ear. He was required by law to wear it at all times. It was tuned to a government transmitter. Every twenty seconds or so, the transmitter would send out some sharp noise to keep people like George from taking unfair advantage of their brains . . .

— Kurt Vonnegut, Jr., "Harrison Bergeron," 1961

And a stranger will they not follow, but will flee from him: for they know not the voice of strangers.

— John 10:5

CHAPTER TEN

The Covert Ascendance of Technocracy

▼

Technocracy is a totalitarian system of government where scientists, engineers and technicians monitor and control all facets of personal and civic life—economic, social and political. . .Smart Grid is born out of technocracy and not the other way around.
    — Patrick Wood, Editor, *The August Review*, June 23, 2011

Ever since Captain Kirk teleported to the surface of an alien planet and whipped out his portable communicator, everyone Earthside wanted one. And the telecos gave us flip-phones, along with compulsive texting, streaming video, online gaming and banking, the worldwide web and more.
    — Will Thomas, "Wireless, Chemtrails, and You," 2013

While a few millicuries (mCi) of cancer-causing ionized radiation can terrify people, not a word is said about the ubiquitous non-ionized microwave transmitters towering over our neighborhoods or up against our heads and in our pockets. Had the Space Preservation Act of 2001 (HR2977) not been torn to pieces in committee, we might have eventually realized that cell phones and microwave towers and the Internet of Things (IoT) are just more of the exotic weapons system described in HR2977:

- Electronic, psychotronic, or information weapons;

- Chemtrails;

- High-altitude ultra-low-frequency weapons systems;

- Plasma, electromagnetic, sonic, or ultrasonic weapons;

- Laser weapons systems;

- Strategic, theater, tactical, or extraterrestrial weapons; and

- Chemical, biological, environmental, climate, or tectonic weapons.

In its infinite wisdom, HR2977 forbade:

> . . .inflicting death or injury on, or damaging or destroying, a person (or the biological life, bodily health, mental health, or physical and economic well-being of a person) through the use of land-based, sea-based, or space-based systems using radiation, electromagnetic, psychotronic, sonic, laser, or other energies directed at individual persons or targeted populations for the purpose of information war, mood management, or mind control of such persons or populations; or by expelling chemical or biological agents in the vicinity of a person . . .

Of course, HR2977 was not allowed to see the light of day because exotic "nonlethal" weapons were exactly what were being "tested" for military peace support operations (PSO)—"peacemaking, peace enforcement and peace building" being Orwellian for the covert *technocracy* that would supplant human society in the name of progress.

Technocracy means rule by technology. The term was born in 1932 with the Technocracy, Inc. movement. Inspired by IBM (then collaborating with the Nazis), geoscientist M. King Hubbert (1903–1989) sought to measure and profile energy production, conversion, flow of goods and services, and consumption. Patrick Wood, author of *Technocracy Rising: The Trojan Horse of Global Transformation* (2015), puts it more succinctly: "Technocracy is a totalitarian system of government where scientists, engineers and technicians monitor and control all facets of personal and civic life—economic, social and political."[1]

In 1970, geostrategist Zbigniew Brzezinski—adviser to four administrations, including the Obama administration—prophesied the advent of a "technetronic age":

> Another threat, less overt but no less basic, confronts liberal democracy. More directly linked to the impact of technology, it involves the gradual appearance of a more controlled and directed society. Such a society would be dominated by an elite whose claim to political power would rest on allegedly superior scientific knowhow. Unhindered by the restraints of traditional liberal values, this elite would not hesitate to achieve its political ends by using the latest modern techniques for influencing public behavior and keeping society under close surveillance and control. Under such circumstances, the scientific and technological momentum of the country would not be reversed but would actually feed on the situation it exploits. . .The traditionally democratic American society could, because of its

---

1 Wood, Patrick, "Technocracy Endgame: Global Smart Grid." *The August Forecast & Review*, May 16, 2013. Also see Marion King Hubbert, Howard Scott, Technocracy Inc., *Technocracy Study Course*. New York, 1934.

fascination with technical efficiency, become an extremely controlled society, and its humane and individualistic qualities would thereby be lost . . .[2]

In 1983, Samuel Koslov—at one time involved in the Moscow Signal that eventually became Project Pandora (1965–70)—opened the classified Conference on Nonlinear Electrodynamics in Biological Systems at Johns Hopkins University by comparing the use of "external electric fields" to what "faced the physics community in 1939 when the long-time predicted fissionability of the nucleus was actually demonstrated"—meaning the secret Manhattan Project that produced the atomic bomb and inducted humanity into the Nuclear Age.[3]

Koslov was right. The rise of "nonlethal" weapons in the 1990s was pivotal to the ascendancy of the technocracy. "Penguin"[4] Col. John B. Alexander, author of the 1980 article "The New Mental Battlefield" (*Military Review*, December 1980), eventually became the Los Alamos National Laboratories kingpin of nonlethal weapons for incremental aggression, peace enforcement, weaponized electromagnetic fields, chemical and biological "anti-terrorism," high-powered microwave (HPM) technology, fracture and dynamic behavior, biotechnology, and acoustic technologies like synthetic brain-computer interface (BCI). Quiet, unseen weapons for a quiet, unseen war against human society.

In 1993, Igor Smirnov of the Moscow Institute of Psycho-Correction at Moscow Medical Academy gave closed session presentations for the National Academy of Sciences' 21st Century Army Technologies panel to FBI, CIA, DIA, DARPA, military contractor corporate executives, National Institutes of Health, and National Institute of Mental Health regarding the latest incarnations of the LIDA acoustic psycho-correction device able to remotely implant thoughts in minds, even over the telephone.[5] Smirnov taught attendees how to use the electroencephalograph (EEG) to remotely measure brainwaves, then demonstrated the computer software that could create an accurate brain map from those measurements, after which he sent a synthetic telepathy message to a target brain with the LIDA. This technology promised real-time alteration of decision-makers and key personnel, he said, and was happily less violent than the strong-arm techniques of yesteryear.[6]

---

2 Zbigniew Brzezinski, *Between Two Ages: America's Role in the Technetronic Era.* New York: Viking Press, 1970.

3 Peter Kirby, *Chemtrails Exposed: A History of the New Manhattan Project,* 2012.

4 The Aviary was a collection of military and intelligence scientists and officers in the 1980s studying artificial telepathy (remote viewing). See artificialtelepathy.blogspot.com/2006/06/aviary-key-players-of-1980s-and-1990s.html.

5 Smirnov is credited with inventing the LIDA in the 1970s, but it was actually invented by Dr. José Delgado, U.S. Patent # 3,773,049. For a fictionalized account of how it works, read Evelyn Waugh's 1957 novel *The Ordeal of Gilbert Pinfold.*

6 "Mind control: The Zombie Effect." *Pravda.ru,* November 10, 2004, www.pravdareport.com/science/tech/10-11-2004/7346-0/

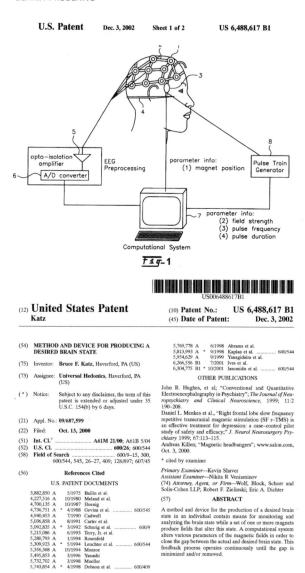

U.S. Patent   Dec. 3, 2002   Sheet 1 of 2   US 6,488,617 B1

Computational System

Fig. 1

US006488617B1

(12) **United States Patent**
Katz

(10) Patent No.: **US 6,488,617 B1**
(45) Date of Patent: **Dec. 3, 2002**

(54) **METHOD AND DEVICE FOR PRODUCING A DESIRED BRAIN STATE**

(75) Inventor: **Bruce F. Katz**, Haverford, PA (US)

(73) Assignee: **Universal Hedonics**, Haverford, PA (US)

(*) Notice: Subject to any disclaimer, the term of this patent is extended or adjusted under 35 U.S.C. 154(b) by 6 days.

(21) Appl. No.: 09/687,599

(22) Filed: **Oct. 13, 2000**

(51) Int. Cl.[7] A61M 21/00; A61B 5/04
(52) U.S. Cl. 600/26; 600/544
(58) Field of Search 600/9–15, 300, 600/544, 545, 26–27, 409; 128/897; 607/45

(56) **References Cited**

U.S. PATENT DOCUMENTS

3,882,850 A	5/1975	Bailin et al.
4,227,516 A	10/1980	Meland et al.
4,700,135 A	10/1987	Hoenig
4,736,751 A *	4/1988	Gevins et al. 600/544
4,940,453 A	7/1990	Cadwell
5,036,858 A	8/1991	Carter et al.
5,092,835 A *	3/1992	Schurig et al. 600/9
5,215,086 A	6/1993	Terry, Jr. et al.
5,280,793 A	1/1994	Rosenfeld
5,309,923 A *	5/1994	Leuchter et al. 600/544
5,356,368 A	10/1994	Monroe
5,495,853 A	3/1996	Yasushi
5,732,702 A	3/1998	Mueller
5,743,854 A *	4/1998	Dobson et al. 600/409

5,769,778 A	6/1998	Abrams et al.
5,813,993 A *	9/1998	Kaplan et al. 600/544
5,954,629 A	9/1999	Yanagidaira et al.
6,266,556 B1	7/2001	Ives et al.
6,304,775 B1 *	10/2001	Iasemidis et al. 600/544

OTHER PUBLICATIONS

John R. Hughes, et al; "Conventional and Quantitative Electroencephalography in Psychiatry"; *The Journal of Neuropsychiatry and Clinical Neuroscience*, 1999; 11:2 190–208.

Daniel L. Menkes et al., "Right frontal lobe slow frequency repetitive transcranial magnetic stimulation (SF r–TMS) is an effective treatment for depression: a case–control pilot study of safety and efficacy;" *J. Neurol Neurosurgery Psychiatry* 1999; 67:113–115.

Andreas Killen; "Magnetic headbangers"; www.salon.com, Oct. 3, 2000.

* cited by examiner

Primary Examiner—Kevin Shaver
Assistant Examiner—Nikita R Veniaminov
(74) *Attorney, Agent, or Firm*—Wolf, Block, Schorr and Solis-Cohen LLP; Robert F. Zielinski; Eric A. Dichter

(57) **ABSTRACT**

A method and device for the production of a desired brain state in an individual contain means for monitoring and analyzing the brain state while a set of one or more magnets produce fields that alter this state. A computational system alters various parameters of the magnetic fields in order to close the gap between the actual and desired brain state. This feedback process operates continuously until the gap is minimized and/or removed.

This is transcranial magnetic stimulation (TMS). Note the "EEG Preprocessing" (now mobile). Note the emphasis on amplification and pulsing. This was in 2002. Now, TMS can remotely stimulate the temporal lobe by manipulating *in loco* electromagnetic fields. The temporal lobe processes visual and auditory input, creates new memories, and is command central for emotional associations.

"In 2013, Arizona State University's Center for Strategic Communication ran a program called "Toward Narrative Disruptors and Inductors: Mapping the Narrative Comprehension Network and Its Persuasive Effects." "Strategic communication" refers to counterterrorist tactics against political dissidents, and "narrative disruptors and inductors" refer to jamming the brain's thoughts and implanting other "narratives" by means of *transcranial magnetic stimulation (TMS)* of the brain's temporal lobe. ("Secret DARPA Mind Control Project Revealed: Leaked Document." *Activist Post*, July 29, 2013.) The claim that implants can't be read without a scanner is not true; they can be read anywhere there are microwave/cell towers (wireless transmitters)."

— *Under An Ionized Sky: From Chemtrails to Space Fence Lockdown*

At yet another classified conference in November 1993, George Baker, Ph.D., from the Defense Nuclear Agency, and Clay Easterly, Ph.D., of the Oak Ridge National Laboratory praised nonlethals to the skies. In 1994, Steven Metz and James Kievit of the U.S. Army War College came out with *The Revolution in Military Affairs and Conflict Short of War*, an ominous techno-echo of George Orwell's *1984* and the film trilogy *The Matrix*. In 1996, the U.S. Air Force Scientific Advisory Board published a fourteen-volume study of future weapons development called *New World Vistas*. The dystopic military vision of *NBIC*— *nanotechnology, biotechnology, information technology, and cognitive technology*— had taken hold. The "soft kill option" and no-touch torture were no longer a conspiracy theory.

SMART GRID SURVEILLANCE

The technocratic city is a post-9/11 city. Federal troops can now enter cities with impunity, thanks to timed false-flag traumas. Tactically trained paramilitary SWAT teams roll down the street in $296,000 Lenco Bearcats paid for by the spoils of the so-called Drug War. "Urban training exercises" invite low-flying whisper-mode helicopters, explosions, flares, smoke bombs, and SOCOM military personnel. Shooting drills are common at high schools,[7] recalling Benjamin Franklin's warning that those who give up essential liberty for a little temporary safety deserve neither liberty nor safety.

Public protests are rare, but not because there is nothing to protest. There is *everything* to protest, but we have all watched the six o'clock news in horror as antipersonnel HSS (hypersonic sound) pain ray "risk management tools" like the Humvee-mounted LRADs (long-range acoustic devices) roll into Baghdad or Kabul for "area denial, perimeter security and crowd control." Like the HIDA (high-intensity directed acoustics "sonic bullet"), the LRAD can damage hearing while broadcasting speech up to three hundred meters ("voice of God" weapon[8]). The Thunder Generator—a cannon using a mixture of liquefied petroleum, cooking gas, and air—is similar, discharging a blast that can inflict permanent damage or even death if one is standing within ten meters of it; and the "heat ray" Active Denial System (ADS) is designed to behave like a particle beam weapon utilizing a narrow solid-state plasma antenna for individual pain beams.

Besides "risk management tools," a technocracy needs security police forces and border patrols. Multiple-tour infantry and Special Forces now return home to patrol shopping malls in cities that look more and more like Baghdad and Kabul. The millions of dollars that went into their counterinsurgency training

---

7 "Cary Grove Drill To Include Shooting Blanks In Hallways." *CBS Chicago*, January 29, 2013.

8 Sharon Weinberger, "The 'Voice of God' Weapon Returns." *Wired*, December 21, 2007.

need not go to waste as long as the public can be kept on edge with lone nuts, drug trafficking, terror networks, and standard violent Hollywood fare—

> Everywhere in films, in popular books on the grocery store shelves and in video games, there's an obsession with hit men, serial killers, sexual psychopaths and government agents with a license to kill; popular killers range from those in an underground, criminal world to those wearing badges and working under the lethal rights granted by [a] national sovereign.[9]

When the CIA's former director George H.W. Bush referenced "a thousand points of light" in his Presidential candidate acceptance speech in 1988,[10] he was probably referring to the advent of the Smart Grid and its satellite "eyes in the sky" above and glowing microwave towers below in an atmosphere bristling with a driftnet of conductive metal particulates, nanobot sensors and microprocessors outside and inside human bodies. As activist and targeted individual (TI) Carolyn Williams Palit put it in 2007:

> What they are ultimately trying to do up there is create charged-particle, plasma beam weapons. Chemtrails are the medium — GWEN pulse radars, the various HAARPs, and space-based lasers are the method, or more simply: Chemtrails are the medium — directed energy is the method. Spray and Zap.[11]

An efficient technocracy is absolutely dependent upon a consolidated digital wireless grid:

> a single, integrated communication-enabled electric delivery and monitoring system, collectively called Smart Grid. . .Thirty years ago, a researcher's mantra was "Follow the money, follow the power." This must now be restated: "Follow the energy, follow the power". . .Global companies like IBM, GE and Siemens are putting their full effort behind the 'build-out' that will consolidate all of America into a single, integrated, communication-enabled electric delivery and monitoring system, collectively called Smart Grid . . .[12]

The fact that cable companies like Time Warner, AT&T, and Comcast aren't subject to common carriage regulations points clearly to their "insider" military status. Telecom mergers have carte blanche. Monopolies control telephone and cable lines. After the AOL-Time Warner merger and AT&T-MediaOne merger,

9 John Grant, "Americans Love a Good Killer." *CounterPunch*, May 11–13, 2012.

10 1988 Presidential candidate acceptance speech at the Republican National Convention in New Orleans.

11 Carolyn Williams Palit, "What Chemtrails Really Are." *Rense.com*, November 9, 2007.

12 Patrick Wood, "Technocracy Endgame: Global Smart Grid." Also read Wood's "Smart Grid: The Implementation of Technocracy?" *The August Review,* March 2, 2010.

the FCC granted cable companies the right to manage the speed at which sites appear, block content, and deny ISP access. Comcast, the world's largest cable company, controls one-third of U.S. households.

AT&T and Verizon collaborate with Israel's "superintrusive" Verint and Narus to perform mass surveillance on American communications, sifting traffic 24/7 at key Internet gateways around the U.S. James Bamford, author of *The Shadow Factory: The Ultra-Secret NSA from 9/11 to the Eavesdropping on America,* writes that Israel is the eavesdropping capital of the world[13] and never need to be concerned about congressional oversight of its activities in the U.S., as the CIA and Mossad are Gog and Magog.

The global Smart Grid is up and running, jarringly symbolized by the NSA's megalithic Intelligence Community Comprehensive National Cybersecurity Initiative Data Center in Bluffdale, Utah, twenty-five miles south of Salt Lake City (mentioned in Chapter 6). Above ground, the Utah Data Center occupies 1.5 million square feet and quantifies in yottabytes (1 yottabyte = 1 trillion terabytes, or 1 quadrillion gigabytes) what is vacuumed up from satellite and underwater ocean cable[14] intercepts.

Southeast of the Utah Data Center in the Mimbres territory of Catron County, New Mexico, AT&T has an underground communications facility. In 1977, the Dia Art Foundation commissioned American sculptor Walter De Maria to build a piece of land art called *The Lightning Field* right over that underground facility: four hundred stainless steel poles precisely two inches in diameter and twenty feet 7.5 inches in height, spaced 220 feet apart, their pointed tips defining a horizontal plane in a grid array measuring one mile by one kilometer.

SMART METERS AND THE INTERNET OF THINGS (IOT)

The technocracy has had high hopes for smart meters, and not just as a 24/7 home energy accounting system counting kilowatts. In 2013, 35.5 percent of all U.S. electrical customers had smart meters, a 20 percent increase over 2012. By 2014, 58,545,938 smart meters had been installed, with 51,710,725 being residential.[15] In the developing world, there were 5.16 million units in Europe, the Middle East, and Africa in 2011. A Market Research reports for global smart meter sales listed the English-speaking spy network Echelon (Five Eyes) as a "top player."[16]

---

13 Ali Abunimah, "How Israel helps eavesdrop on US citizens." *The Electronic Intifada,* 3 November 2008. Israel's NSA is called Unit 8200.

14 See *www.submarinecablemap.com/.*

15 Ibid.

16 Market Research Reports, www.marketresearchreports.com/qyresearch/global-and-china-smart-meters-sales-market-report-2020.

In fact, read between the lines of Title XIII of the Clean Energy / Energy Independence and Security Act, signed into law on December 19, 2007, and you will see that smart meters are really about surveillance, period. Note the repetition of the (military) term "deployment":

(1)   Increased use of digital information and controls technology to improve reliability, security, and efficiency of the electric grid;

(2)   Dynamic optimization of grid operations and resources, with full cyber- security;

(3)   Deployment and integration of distributed resources and generation, including renewable resources;

(4)   Development and incorporation of demand response, demand-side resources, and energy-efficiency resources;

(5)   Deployment of 'smart' technologies (real-time, automated, interactive technologies that optimize the physical operation of appliances and consumer devices) for metering, communications concerning grid operations and status, and distribution automation;

(6)   Integration of 'smart' appliances and consumer devices;

(7)   Deployment and integration of advanced electricity storage and peak-shaving technologies, including plug-in electric and hybrid electric vehicles, and thermal-storage air conditioning;

(8)   Provision to consumers of timely information and control options;

(9)   Development of standards for communication and interoperability of appliances and equipment connected to the electric grid, including the infrastructure serving the grid;

(10) Identification and lowering of unreasonable or unnecessary barriers to adoption of smart grid technologies, practices, and services.

Unlike regular electric AMR (automated meter reading) meters, smart meters are *two-way* AMI (advanced meter infrastructure), supposedly to "allow utilities and customers to interact to support smart consumption applications."[17]

Smart meters are joined at the hip with the Internet of Things (IoT). By 2020, every American home is to have a *two-way* gas and electricity AMI logging energy use of smart appliances at two-second intervals, as per power signatures. If Li-Fi[18]

17 U.S. Energy Information Administration.

18 "Shanghai Fudan University develops new Li-Fi technology." *WantChinaTimes.com*, October 18, 2013.

complements your WiFi, its 1-watt LED [light-emitting diode] has a microchip in it that can simultaneously connect multiple computers to the Internet. Every plug-in and wireless appliance, every film you watch, whether you are home or not, your state of wakefulness or sleep, is monitored, thanks to the ZigBee microchip in each smart meter that wirelessly communicates from smart appliances to utility poles, central utilities offices, police stations, and fusion centers.

And ZigBee chips have a kill switch.

Home energy accounting and surveillance are two smart meter agendas under the technocracy. The third agenda is to serve as a node in the computing architecture, given that the computer cards in the communications modules of smart meters have the same computing power as a cell phone. With Linux software, Hive Computing, smart meter vendor Itron and partner Cisco are building a *mesh network* of nodes from millions of smart meters for a distributed intelligence platform of greater Smart Grid control—for "socially useful computing tasks," of course.[19]

> . . .the latent capacity of the world's smart meter network approaches that of the world's better known supercomputers. For example, 3,000 smart meters have nearly the same amount of processing power and memory capability as Deep Blue, the IBM supercomputer that beat Garry Kasparov in a game of virtual chess in 1997, and 150,000 meters add up to about half the computing power of IBM's Watson supercomputer . . .[20]

That's one teraflop (1 trillion floating operations per second) of processing power. One million smart meters is the equivalent of the world's twentieth fastest supercomputer.

Lovely: a supercomputer module active 24/7 on the outer wall of your home or business in a neighborhood filled with other modules, all receiving and sending transmissions, paid for by consumers who must deal with the health issues and cybersecurity issues of third-party software nosing around through smart meters.[21]

The IoT overseen by smart meters makes every neighborhood a veritable antenna farm that affects the health of all life forms within the target zone. Daniel Hirsch, a lecturer on nuclear policy at University of California Santa Cruz, stresses that one smart meter exposes people to 100X the microwave radiation of a cell phone.[22] Symptoms following from exposure range from

19 Jeff St. John, "Could Millions of Smart Meters Be Used to Create a Powerful Supercomputer?" *Greentech Media*, August 7, 2015.

20 Ibid.

21 K.T. Weaver, SkyVision Solutions, "Insanity: Turning Smart Meters into a Supercomputer Platform." *Smartgridawareness.org*, July 20, 2015.

22 J. Hart, "Nuclear expert: Smart Meters 100X Radiation Exposure of a Cell Phone." *Open Publishing*

an inability to sleep deeply to headaches, heart palpitations, chronic fatigue, difficulty concentrating, and cancer.

> Opponents nationwide have testified that radiofrequency emissions from Smart Meters cause headaches, nausea, and insomnia. A Hurst woman told the *Dallas Morning News* in June that her autistic daughter began having seizures when their Smart Meter was installed. . .[O]pponents of Smart Meters nationwide have complained about inaccurate readings, radiofrequency interference with medical and other devices in their homes, and privacy and security concerns.[23]

The electromagnetic impact of smart meters extends far beyond the 900 MHz resonating from home wiring. Add the 2.4 GHz frequencies of the machine-to-machine Home Area Network (HAN) overseeing the smart meters in your area and you have RFR (radio frequency radiation) in the harmful microwave range of 1–300 GHz. Now add the collector meter antennas affixed to telephone poles to serve as transmission hubs for 500–1,000 homes, with frequencies from 850 MHz to 1.9 GHz for cell phone connectivity.

The cumulative effect of wireless 900 MHz, 1.9 GHz, and 2.4 GHz,[24] coupled with conductive nanoparticle "clouds," points directly to the neighborhood "cancer clusters" popping up everywhere. This from Canada:

> When the suburb is all transmitting, for example, in parallel signals to NAN M2M [Neighbourhood Area Network machine-to-machine] contact, not only are there clouds inside the residences but also an *overcloud* at about 2.5 meters and above in streets and above boundary layers of rooftops where the noise [of the Alasdair Philips "electrosmog" detector] can increase from reasonable background levels of below 0.05 microWatt/cm² to 400 microWatt/cm² in ambient air.
>
> Clearly the impact of SMART meter networks is phenomenal, especially when clustered together, inducing Doppler effects for long distances, *sometimes hundreds of meters in series of clusters, reflections, generation of "noise" and probably thousands of new signals/cubic meters.*[25] [Emphasis added.]

Is it any wonder that Lloyd's of London, the world's foremost insurer of risk, refuses to cover health effects from wireless devices, including cell towers? Here is Sharon Noble, director of Coalition to Stop Smart Meters:

Newswire, April 21, 2011.

23 Garry Reed, "Texas committee hears Smart Meter health hazards." *Examiner.com,* October 12, 2012.

24 Thanks to "Smart Meters Are An Even Larger Threat Than I Had Thought." *The Non-toxic nurse,* March 25, 2012.

25 Dr. Andrew Michrowski, "Emerging phenomena with smart meters imposed in Quebec." Dr. Michrowski is president of the Ottawa-based Planetary Association for Clean Energy.

Lloyd's of London is one of the largest insurers in the world and often leads the way in protection, taking on risks that no one else will. Attached is a recent renewal policy which, as of Feb. 7, 2015, excludes any coverage associated with exposure to non-ionizing radiation. In response to clarification, this response was received on Feb. 18, 2015 from CFC Underwriting LTD, London, UK agent for Lloyd's: "The Electromagnetic Fields Exclusion (Exclusion 32) is a General Insurance Exclusion and is applied across the market as standard. The purpose of the exclusion is to exclude cover for illnesses caused by continuous long-term non-ionising radiation exposure i.e. through mobile phone usage."[26]

Smart Meter arcs are burning out computers and shorting out surge protectors and appliances. Add the complicating factor of the conductive metal nanoparticulates falling from chemical jet trails and power poles bursting into flame make more sense than "a build-up of dust on transformers that shorted out due to low-lying fog."[27] Ionized "precipitation-inducing materials" in electrosmog[28] are more to the point.

Not all Americans are lying down for the Smart Grid rollout of smart meters. For example, a fifty-five-year-old Houston woman ordered a CenterPoint Energy worker attempting to install a smart meter off her property at gunpoint,[29] having astutely perceived that smart meter installation represents a clear and present danger of explosions, hot sockets, flashovers, and deadly fires, not to mention the loud, obnoxious hum[30] foretelling endless health problems.

SMARTPHONES

Life evolved with negligible levels of microwave radiation. An increasing number of scientists speculate that our own cells, in fact, use the microwave spectrum to communicate with one another, like children whispering in the dark, and that cell phones, like jackhammers, interfere with their signaling.

— Arthur Firstenberg,
"The Largest Biological Experiment Ever," 2006

---

26 www.rfsafe.com/lloyds-of-london-insurance-wont-cover-smartphones-wifi-smart-meters-cell-phone-towers-by-excluding-all-wireless-radiation-hazards/.

27 "Power Pole Fires Leave Thousands Without Power Thursday." KWTX News 10, December 6, 2012.

28 Electrosmog is accompanied by a 7–9% drop in humidity, is highly charged and highly alkaline.

29 Vicente Arenas, "Power struggle: Local woman uses gun to stop worker from installing smart meter." *KHOU 11 News*, July 18, 2012.

30 "Smart Meter Fires and Explosions," *EMF Safety Network*.

What is emanating from the forest of cell phone (microwave) towers going up invisibly interweaves with the chemical trail grids overhead. In 2013, it was finally made public that the CIA's division of *Behavioral* and Social Sciences and Education was funding a geoengineering study in concert with the National Academy of Sciences (NAS), NOAA, and NASA:

> The National Academy has held two previous workshops on geoengineering, but neither was funded by the intelligence community, says Edward Dunlea, the study director of the latest project. *The CIA would not say why it had decided to fund the project at this time*. . .The last time the government tried to do cutting-edge research related to the atmosphere—with the High-frequency Active Auroral Research Project (HAARP), which aimed to protect satellites from nuclear blasts—people speculated that it might be a death ray, a mind control weapon, or, worst of all. . .a way to control the weather.[31] [Emphasis added.]

The presence of the word "behavioral" should have sounded the alarm.

Hacking masses of cell phones, computers, and brains is easy in a conductive metal-rich atmosphere. WikiLeaks revealed a package of spy tools called the FinIntrusion Kit for hacking into WiFi ("IT intrusion" = hacking), while FinFisher's FinSpy takes over smartphones like iPhones or Windows Mobile, then with FinUSBSuite transfers information to fusion centers, police stations, corporations, etc. The Fin malware is produced by Gamma Group, a shadowy UK corporation.[32]

The little transceivers we call cell or mobile phones are perfect for spying. More important are the mental and physical health effects of what amounts to cell phone addiction with cell phone radiation dialing into our cells. We've been bioelectrical beings for millions of years, but now our cells are forced to interact with artificial EM signals as our tower-like bodies process the pulsed frequencies of AM/FM (amplitude/frequency modulation) radio, metal appliances, cars, ionized rain and snow, glass, towers, antennas, and the constantly zapped metal nanoparticulates raining down and into our lungs. Headaches and migraines, foggy thinking, forgetfulness, weak concentration, impaired learning, and accelerated dementia reflect the rising ricochet of EM signals and the ionized nanoparticles breaching our blood-brain barrier. The first mile above the Earth is buzzing with *two million times* the amount of electromagnetic radiation (EMR) of 1900.

Neither telecom giants nor Big Pharma are keen on the public learning that two minutes on a cell phone or sitting near a cordless phone base or wireless

---

31 Dana Liebelson and Chris Mooney, "CIA Backs $630,000 Scientific Study on Controlling Global Climate." *Mother Jones*, July 17, 2013.

32 Daniel G.J. "The Malware That Can Spy on You through Your Own Smartphone." *IntelliHub*, July 15, 2013.

computer router will short-circuit the body-brain electrical ion flow and end in clumps of fibril proteins growing in electromagnetically altered brains. It seems that the World Health Organization (WHO) reclassification of cell phone and WiFi emissions as possible carcinogens was not enough; few Americans know or care that a federal appeals court has blocked the City of San Francisco from requiring retailers to issue warnings to cell phone buyers.[33] Western Europe and Israel have had wireless technologies a decade longer than North America, which may be why 250,000 Swedish citizens are on permanent disability for microwave sickness, early Alzheimer's, Parkinson's, fibromyalgia, multiple sclerosis, and cancer.[34] As the price of our love for convenience and comfort climbs, it is just a matter of time before landlines are quietly discontinued.[35]

Despite all the cell phone radiation lawsuits,[36] smartphones continue to be the favored Trojan horse of the Smart Grid. 7.3 billion cell phone users don't realize they are now the proud owners of two-way microwave radios pulsing low levels of EMR and producing brain cancers like ipsilateral glioma.[37] WiFi has gone from radiating a one-hundred-yard microwave cloud ("hotspot") for people to tap into, to wireless MiFi routers that can be carried wherever one goes. Meanwhile, the iPhone 6, Moto X, Nexus 6, etc., with their built-in hotspots are demanding ever more Towers of Babel and satellites for voice, Internet, music, video games, TV reception, and video streaming. WiMAX upgrades to LTE now grace towers and broadcast in the 5.8 GHz spectrum to fill in those frustrating receptivity gaps between towers.

Add satellites with steerable spot beams and a barium-enriched atmosphere and you have a weaponized Smart Grid with sizzling microwave "clouds" ready and waiting everywhere. *Cloud-based systems are highly hackable corporate-owned and corporate-managed shared-space server farms not subject to public oversight.*

Are specific absorption rates (SARs) of radiation being field-tested on large populations? The maximum legal SAR is 1.6 watts per kilogram of tissue, an adult male head standard, not an adult female or child head, and concerned only with overheating body tissue, not with subthermal effects.[38] AT&T and T-Mobile

---

33 Bob Egelko, "Court blocks S.F. warning on cell phones." *San Francisco Chronicle*, June 26, 2013.

34 Will Thomas, "Wireless, Chemtrails, and You." *WakeUpWorld*, 27 May 2013. Also see his "Dialing Our Cells."

35 "Verizon Seeks Purely Wireless Phone Network on Fire Island, Abandons Landlines." *LongIsland. com*, 5 July 2013.

36 "Cell Phone Radiation Lawsuits: Bernstein Liebhard LLP Notes New Study Finding Cell Phone Radiation May Cause Brain Tissue Damage." *Digital Journal*, March 27, 2013.

37 "Swedish review strengthens grounds for concluding that radiation from cellular and cordless phones is a probable human carcinogen." *Pathophysiology*, April 2013: 123–9; and Joshua Pramis, "Number of mobile phones to exceed world population by 2014." *Digital Trends*, February 28, 2013.

38 Joel M. Moskowitz, Ph.D., School of Public Health, UC Berkeley, "Your Cell Phone Company May Affect Your Risk of Brain Cancer." PRLog press release (March 27, 2013) for new study by Sven Kuehn et al., "Analysis of mobile phone design features affecting radiofrequency power absorbed in a human head phantom." *Bioelectromagnetics*, March 26, 2013.

employ the Global System for Mobile Communication (GSM) with a SAR 100X that of the Code Division Multiple Access (CDMA) system employed by Verizon and Sprint. And if 2G phones (GSM/UMTS and CDMA/W-CDMA) are "bioactive"—doublespeak for potentially harmful to living bodies—what does that make 5G and 6G smartphones?[39]

Frequencies 1–5 GHz penetrate organ systems. *Fifth-generation (5G) cell phones* operate in the 28–100 GHz range, the same frequency as the extremely high frequency (30–300 GHz) millimeter wavelength spectrum (10–1mm) used in Raytheon's active denial system (ADS) directed energy weapon for "perimeter security and crowd control."[40] The supposed justification for the millimeter / terahertz spectrum is that the 700 MHz to 2.6 GHz WiFi bandwidth is saturated and consumers are demanding yet more speed and bandwidth. The truth is closer to how 5G cell phones and the tower antennas they require will serve population control. 5G will mean cell antennas on residential streets, so the insides of homes will be irradiated, blasting through walls with millimeter wave technology. Cellular mm-wave propagation *every quarter of a mile* in densely populated areas is extremely unwise if safety is the primary consideration, which it is not.

Ronald Powell, Ph.D. (Harvard) in applied physics (*emfanalysis.com, smartgridawareness.org*), and Yael Stein, M.D. of Hadassah Medical Center, Jerusalem, are *begging* the FCC Commissioners to withdraw 5G:

> Computer simulations have demonstrated that sweat glands concentrate sub-terahertz waves in human skin. Humans sense these waves as heat. The use of sub-terahertz (millimeter wave) communications technology (cell phones, WiFi, antennas) could cause humans to perceive physical pain via nociceptors.[41]

The smaller the wave, the more damage to human and animal metabolism. Millimeter waves in 5G have 25X the WiFi microwave photon energy of 2.4 GHz. Translated into impact on living cells, millimeter wave photons disturb 25X as many molecules as current WiFi and 4G disturb. TETRA (Terrestrial Trunked Radio) phones used by police and military are the most dangerous,[42] with TETRA carrier frequencies typically at 400 MHz and modulation frequencies at 17 Hz, the brain's frequency range being 0.5–30 Hz.

Synergy is a term we must become familiar with as we weigh the risks of increasing SARs. Synergy goes beyond mere mathematical increase ("100X

---

39 "Smartphone Technology: What is 5G and 6G Wireless Technology: Difference between 5G and 6G Mobile Technology." technologyoffice.blogspot.com/2013/06/what-is-5g-and-6g-wireless-technology.html.

40 "Active Denial System," Wikipedia.

41 "Letter to the FCC from Dr. Yael Stein, M.D. of the Hadassah Medical Center, Jerusalem, Israel in Opposition to 5G Spectrum Frontiers Millimeter Wave Technology." ehtrust.org, July 9, 2016.

42 For more on TETRA, see Barrie Trower's "Secret Report On Cell Phone Dangers and Tetra,"

that of the CDMA") to an *exponential* impact of competing and overlapping non-ionized and ionized radiations. In a January 1, 2008 email, public health scientist, epidemiologist, and leading expert on electromagnetic radiation Dr. George Carlo hit the microwaved nail on the head:

> EMR [electromagnetic radiation] is most likely a synergen in these cases, including the case with the disappearing birds. The exposures such as pesticides, chemtrails, harvesting of forests, and urban sprawl are not causes competing with EMR — they are additive and synergistic causes. They are insults to the system that require strong biological compensation to overcome. EMR depletes that ability to compensate, and thus the person — or species — succumbs to the effects of the environmental insult more severely and more rapidly.[43]

PHASED ARRAYS, HOLOGRAMS, AND ELECTRONIC WARFARE (EW)

Thus far, we've determined that the average American neighborhood is now an electromagnetic interchange of crisscrossing signals from computers, television, smart meters, cell phones, microwave ovens, IoT, utilities, military frequencies, transportation guidance systems—all translating to more cancer clusters, compromised immune systems, miscarriages, premature aging, and DNA damage as signals invisibly jet from one emitter to another, pulsing over high-voltage lines, along pipelines and cable, railroad tracks, assaulting the natural frequencies of forests, bodies, and brains.

In suburbs, big city neighborhoods, and sparsely populated rural districts, omnidirectional and highly directional WiFi router antennas produce ambient aerial signals emissions *multiplied* by the presence of highly conductive aluminum and barium nanoparticulates. Above, television satellites in geosynchronous orbit are delivering programming to homes while spy satellite sensors probe the signatures of tens of millions of host computer systems as wireless and optical networking technologies pulse Internet over broadband landlines (coaxial cable, fiber optic, copper wires), WiFi, speed-of-light supercomputers, satellite and 5G/6G cell phones.

All of our comforting, convenient, entertaining electronics are digital and dual-use. The previous analog signal was unstable and made it difficult to manipulate moods via frequencies. Digital technologies, like Internet, reroute neural pathways and alter the brain.

Conventional arrays with analog beam formers are no more. Now, digital phased-array antennas with individual transmitters and receivers (T/R

---

43 Paul Raymond Doyon, "Are Microwaves Killing the Insects, Frogs, and Birds? And Are We Next?" *Rense.com*, March 20, 2008.

This photo is what our sidewalks, yes, in front of our homes and businesses, are going to have every 50 feet, including our bedrooms . . . SB 649 [the "small cells" like 5G bill] . . . would strip local governments of their property rights and force them to ignore safety, aesthetic, health and other issues, and approve all wireless company applications along public streets and on public properties. The California State Assn. of Counties has written a letter strongly opposing SB 649 . . . Lloyd's of London and Swiss Re, two major insurance companies of the world, will not cover medical expenses incurred due to exposure of electromagnetic radio frequency radiation (EMR) (i.e. cellphones, WiFi, cell towers, antennas, DAS, IoT devices, smart meters, etc.).
— Nevada City Councilwoman Reinette Senum, April 26, 2017

See Councilwoman Senum's April 3, 2017 article "The 5G Network: What You Don't Know May Kill You,"
www.thefoghornexpress.com/single-post/The-5G-Network-What-You-Dont-Know-Can-Kill-You

modules) are being reconfigured and their wideband performance upgraded for dual-use operations. Throughout neighborhoods, micro-level phased arrays connect to microwave transmission lines, and power dividers perch on telephone and electric poles, cell towers and less obvious platforms, invisibly capturing and transferring electromagnetic signals. As the 2009 paper "Distributed Phased Arrays and Wireless Beamforming Networks" put it, "Multiple radar, communications, *and electronic warfare functions* can be served by a single antenna having such an architecture."[44] (Emphasis added.)

44 David Jen, et al., "Distributed Phased Arrays and Wireless Beamforming Networks." *International Journal of Distributed Sensor Networks*, 5:283–302, 2009.

Basically, phased arrays are geometrically arranged collections of individually controlled transceiver antennas. Coupled with the ducting and amplifying of the "chemical antennas" being delivered by chemtrails, the phased array's geometric arrangement guarantees that the output field will be greater than the sum of the waves (exponential) emitted by individual antennas. Control the phases of the individual signals and the output field can be given any desired "shape" or direction, the only limitations being the length of the wavelength, the number of individual antennas, and the size of the array. The "shape" can be a narrow beam directed at one person in one building or hundreds of independent beams directed at hundreds of targets, depending on the "phasing." Phased-array radar can track many people simultaneously while the antennas remain absolutely stationary.

In its receiving mode, the phased array picks up much more information than an ordinary antenna because it operates as a hologram, measuring entire EM field geometries, not just one-dimensional EM signals. Its holographic capability can remotely create 3D EM field distributions around targets. While correcting for reflections and interferences, the field can then be transformed and moved to another target somewhere else in a fraction of a second.

A constant dissemination of barium nanoparticles enhances the holographic capability of phased arrays. Strontium barium niobate ($Sr_xBa_{1-x}Nb_2O_6$) is used in outdoor holographic 3D displays as a "screen" (Wikipedia). Barium stearate ("soap" bonded to metal particulates) also aids high-tech 3D radar imaging. As Will Thomas wrote in 2013, "Even if commercial wireless frequencies are not conducted as coherent signals by chemtrails specifically 'tuned' to HAARP frequencies, all commercial RF/microwave energy is going to be randomly confabulated and conducted by the barium-strontium chemtrail mix."[45] Thomas' use of the word "confabulated" points to the magician-like holographic "special effects" already being unveiled in our barium- and strontium-enriched plasma sky theater.

The U.S. Navy's Radio Frequency Mission Planner (RFMP) computer program known as the Variable Terrain Radio Parabolic Equation (VTRPE) utilized 3D battlefield imaging in the chemical skies over Iraq (2003–2011) and Afghanistan (2001–2014). Ionized barium and phased arrays work in concert with the VTRPE code, "a full-wave propagation model that solves EM wave equations for complex electric and magnetic radiation fields."[46] With satellite imagery, VTRPE can propagate radio waves in such a way as to generate a 3D version of terrain modeled on RFMP computer monitors: battlefields, fog, clouds, apartment complexes, houses, workplaces, government offices, etc. VTRPE can also cloak an aircraft behind signal inversion layers, similar to how submarines are cloaked by underwater inversion layers.

45 Will Thomas, "Wireless, Chemtrails, and You," 2013.

46 Frank J. Ryan, "User's Guide for the VTRPE Computer Model." Naval Ocean Systems Center, October 1991; funded by the Joint Electronic Warfare Center and the Office of Naval Technology.

Simply put, the phased array is an excellent dual-use weapon capable of accurately locking onto EM targets. Telemetric sensors measure the necessary geometry of attack pulses, then VTRPE and a sophisticated fifth-generation laptop or satellite computer analyze the geometrics for target acquisition and beam control, and the invisible torture begins, particularly in areas of maximum EM radiation absorption (brain, heart, genitals, etc.) dependent upon the geometry of the wave. Hitting a target with a frequency that excites the metal particulates breathed in from chemtrails and ingested from GMOs can ignite an implosion as organs hemorrhage.[47]

Geometric is the name of the game for electron or plasma beam weapons— not so much in the sense of theorems as in the sense of building the output signal by "tuning" the frequency, amplitude, and spacetime "shape" to the target's body. In 2008, TI Carolyn Williams Palit struck the heart of the matter: "The plasma beam is used much like an electric line to the targeted person."[48] Picture a microwave or radio wave piggybacking a plasma beam.

Phased-array radar may not have a beam weapon's firepower, but a suitcase system (piggybacking) can shape the pulse in spacetime with direct coupling[49] to a high-speed computer for non-linear biological effects. Former NSA analyst Robert Duncan explains in his 2010 Internet book *The Matrix Deciphered*:

> These beams can be steered by phase shifting other antenna signals. . .A phased array of thousands of elements can arbitrarily create one or more beams and the energy can be narrowly focused in constructive or destructive interference mode. This is the basis of directed energy weapons.
>
> Using high-powered steered phased arrays and focused directed energy from two sources. . .a nearly undetectable scalar wave—destructive interference at the point of interest—can be created. With just a minor energy interaction, the interfering beams bounce back with a strong signal-to-noise ratio to be resolved at the sources again. *This allows for any imaging technique to be done over extremely large distances,* in effect making distance irrelevant to the detection feature, be it RADAR, MRI, or ESR [electron spin resonance] imaging.[50] [Emphasis added.]

47 MrCati in the YouTube "The Chemtrails Vaporific Effect" calls this a *vaporific effect weapon* (December 27, 2011).

48 Carolyn Williams Palit, "The Air Force Wants Your LOV: Aerial Seeding of Biological Implants into Food, Water, & Air," February 28, 2008. www.whale.to/b/palit.html. Palit's website exoticwarfare.com is no longer.

49 See more on coupling in Chapter 11.

50 Robert Duncan, Ph.D., *The Matrix Deciphered* (self-published, 2006). Duncan has also written *Project: Soul Catcher: Secrets of Cyber and Cybernetic Warfare Revealed* (Amazon, 2010) and *How To Tame A Demon: A short practical guide to organized intimidation stalking, electronic torture, and mind control* (January 2014). He claims to have been tortured and programmed by the very technologies he worked on for the Department of Defense, U.S. Navy, and NATO. Documentation of his background is elusive.

A 2011 doctoral dissertation clarifies how phased-array antennas and satellites work in tandem as dual-use weapons:

> [T]he simulation of beamforming using phased array antenna is explored to obtain *multiple and also steerable beams for tracking the satellite smoothly.* . .Moreover, the adaptive beamforming allows mobile objects such as planes and vehicles to have access to satellite programs."[51] [Emphasis added.]

On September 23, 2010, the FCC opened up a new spectrum for dual-use no-touch torture deployments: the White Space TV band spectrum for Super WiFi "digital data communications." Super WiFi requires no license, so White Space frequencies are free, unregulated, and widely scattered throughout the radio spectrum. Super WiFi fixed base stations are 4 watts, well over the 100-milliwatt limit for home, office, or school WiFi base stations, and White Space devices use pulsed digital waveforms highly dangerous to human bodies. White Space antennas are on towers and buildings, and when they're not, Super WiFi can leapfrog from one tower to another on roofs or via Smart Meters and telephone poles.

The day after the FCC's move to open White Space frequencies, Magda Havas, associate professor ofEnvironmental & Resource Studies at Trent University in Peterborough, Ontario, wrote the following about how the dual-use Super WiFi weapon can go straight through your wall, your skull, and your body:

> WiFi currently uses PULSED DIGITAL on higher 2.4 gigahertz frequencies[52] that *cannot* penetrate thick concrete at long distances and thus you get a slow speed connection resembling dial up. By using the lower TV frequency spectrum (54–698 MHz) as a carrier wave, the PULSED DIGITAL WiFi signal will now be able to penetrate obstructions such as buildings over great distances and deliver high-speed broadband data.[53]

Neighborhoods have become wireless Homeland Security gulags teeming with invisible energies and agendas as ionospheric heaters move energy thousands of miles by ducting and zapping our conductive atmosphere. Smart meters, cell phones, towers, phased arrays, and satellites all play their part in an exotic dual-use weapon system exploiting citizens in the name of C4 full spectrum dominance, disaster capitalism profits, and a cyborg Transhumanist future.

---

51 Daniel Ehyaie, "Novel Approaches to the Design of Phased Arrays." Doctoral dissertation in Electrical Engineering, University of Michigan, 2011.

52 In 1990, the 2.4GHz frequency was released for unlicensed use in short-range wireless (three hundred feet).

53 Magda Havas, "Is White Space Super WiFi dangerous?" September 24, 2010. www.magdahavas.com/

WIFI AND SPYING IN THE 'HOOD

The technology is based on isolation, and the technical process isolates in turn. From the automobile to television, all the goods selected by the spectacular system are also its weapons for a constant reinforcement of the conditions of isolation of "lonely crowds." The spectacle constantly rediscovers its own assumptions more concretely. . .In the spectacle, which is the image of the ruling economy, the goal is nothing, development everything. The spectacle aims at nothing other than itself.

— Guy Debord, *Society of the Spectacle*, 1967

Imagine a UHF dome arching over your city as ELF signals hop cell towers and phones, penetrate bodies and brains. Digitized voice carried by computers, data arriving wirelessly from across the world, WiFi spilling over into one of the hundreds of WiFi transceivers in the footprint of one cell phone tower. This is the contiguous, barium-enhanced Smart Grid enveloping us all.

In the beginning, WiFi seemed to be about connecting your computer to the global Internet for data transfer, and cellular was about connecting your telephone to a network to transfer voice. WiFi had a radio transceiver (access point) with a range of two hundred feet or so, cellular range covered miles, and the WiFi access point and cellular cell site separately connected to a telephone central office via coaxial cable, fiber optics, or copper.

Now, we're enmeshed in WiFi and social media surveillance. All business, government, and personal communiqués are hoovered up and stored for possible "actionable intelligence." Even as Private First Class Bradley "Chelsea" Manning was being court-martialed at Fort Meade, Maryland for allegedly providing more than 700,000 national security files to WikiLeaks,[54] NSA whistleblower Edward Snowden was exposing the PRISM surveillance program, an expanded version of Stellar Wind run by the NSA and nine "friendly" social networking providers—Microsoft, Yahoo, Google, Facebook, PalTalk, AOL, Skype, YouTube, and Apple[55]—with their own "diplomats.[56])

At an In-Q-Tel[57] summit, former CIA director David Petraeus admitted that the Internet of Things (IoT) interfaces with CIA "tradecraft":

---

54 "WikiLeaks founder solicited U.S. secrets: court-martial witness." *Reuters*, June 11, 2013.

55 Dan Roberts and Spencer Ackerman, "Edward Snowden's explosive NSA leaks have US in damage control mode." *The Guardian*, 10 June 2013. Note: Apple was a holdout until Steve Jobs' death.

56 Swift, Mike. "Facebook to assemble global team of 'diplomats'." *Mercurynews.com*, May 22, 2011.

57 In-Q-Tel is a venture capital firm of the now-corporate CIA.

With the rise of the "smart home," you'd be sending tagged, geolocated data that a spy agency can intercept in real time when you use the lighting app on your phone to adjust your living room's ambiance. "Items of interest will be located, identified, monitored, and remotely controlled through technologies such as radio-frequency identification, sensor networks, tiny embedded servers, and energy harvesters—all connected to the next-generation internet using abundant, low-cost, and high-power computing," Petraeus said, "the latter now going to cloud computing, in many areas greater and greater supercomputing, and, ultimately, heading to quantum computing." Petraeus allowed that these household spy devices "change our notions of secrecy" and prompt a rethink of our "notions of identity and secrecy."[58]

An exotic technology has become the American dual-use heart of darkness. *Mister Rogers' Neighborhood*[59] is no more; instead, a WiFi 'hood pulsing with EM frequencies robs us of our Schumann resonance birthright. Walk out the door and chemical trails are waiting for you overhead with their conductive metal nanoparticles and polymers to breathe, eat, and drink. Crystalline fibers glitter from trees, cars, animal fur, and human skin as they facilitate signals tracking (SIGINT). Nano-barium enhances the "Wi-Vi" thermal imaging[60] fluorescing our bodies as nano-sensors and microprocessors slide silently through our bloodstream and into our brains, alert for remote communications.

The sheer number of EM signals per cubic meter from commercial and military sources is staggering—airport and navigational aid emissions communications, pipelines, M2M (machine to machine) infrastructure, transformers, power lines, fiber optic cable, phased-array antennas, smart meter Neighborhood Area Networks (NANs), etc. Our bodies grapple with signals from ground-based gyrotrons (EM field oscillators, GWEN pulse RADAR),[61] LIDAR/LADARs (light detection and ranging / laser detection and ranging or "laser radar"), OP-FTIR (remote sensing Open-Path Fourier Transform Infrared spectrometry), FLIR (forward-looking infrared) for "own the night" infrared imaging with photoconductive detectors,[62] ionosondes ("chirpsounder" radars), actinic rays (EM radiation producing photochemical reactions used in imaging technology)—and all the while ionospheric heaters are heating, multiplying, siphoning, and steering ionized electrons.

---

58 Spencer Ackerman, "CIA Chief: We'll Spy on You Through Your Dishwasher." *Wired*, March 15, 2012.

59 A children's TV show (1966–2001) in which Fred Rogers played the gentle father in the studio home that "latchkey" American children longed for.

60 Helen Knight, "New system uses low-power WiFi signal to track moving humans—even behind walls." *MIT News*, June 28, 2013.

61 Electron guns capable of generating, mirroring, and guiding EM waves to targets.

62 Patented in 1963 by Texas Instruments Defense Systems and Electronics Division, now owned by Raytheon, previous holder of HAARP patents.

If only we could see beyond our narrow five senses into the vast dimensions of the electromagnetic spectrum! Giant EM clouds several meters in diameter creep through our living zones—an interferometry of signals from thousands of emitters in all directions, thousands of different frequencies—and only our immune system senses it. With an electrostatic meter or magnetometer, you can at last perceive some of the new dimension we are living in:

When the suburb is all transmitting, for example, in parallel signals to NAN M2M contact, not only are there clouds inside the residences but also an *overcloud* at about 2.5 meters and above in the streets and above boundary layers of rooftops where the noise can increase from reasonable background levels of below 0.05 microWatt/cm2 to 400 microWatt/cm2 in ambient air.[63]

Daily distribution of RF-zapped metal particulates is all about keeping aloft a plasma "mirror" dense enough not just to detect and resist incoming cyber-attacks but to remotely utilize exotic weapons on populations. From "psy warrior" microcomputers to cellular phones and "critters" (hacker programs), our neighborhoods, homes, offices, and bodies are being hacked. Telephone microphones, listening posts and fusion centers, traffic cameras, surveillance aircraft like the U-2 Dragon Lady, satellite linkups with state-of-the-art software like British subsidiary BAE Systems' GXP 3D mapping and GOSHAWK big picture scrutiny—all are on constant EM real-time alert, whether it is for a dissident or license plate, an impending engineered weather event or forest fire. Everywhere is now the technocratic battlespace and everyone a twenty-first-century land warrior "either with us or against us."[64]

Slide the Bill of Rights into the Memory Hole.

If we finally understand what dual-use technology means, we may realize that the user-friendly home computer is little more than a base station for transmitting and receiving signals, its cables unshielded, its telephone lines antennas, its CRT video display emitting electromagnetic radiation at 2–20 MHz, and that cell phone handsets are little more than minicomputers with software that can modify the usual interface.

Freelance journalist Jim Stone says *dithering* is doing for PC monitors what digital does for television monitors,[65] and television signals are 10,000 times stronger than GPS signals (which may be why the CIA's investment arm In-Q-Tel initially tracked people anywhere in the world with just a $40 Rosum radio chip and TV signals).

It is technologies like these that are behind the demise of thinking and the deepening of cognitive dissonance.

---

63 Andrew Michrowski, Ph.D. "Emerging phenomena with smart meters imposed in Quebec." See his article "Electromagnetic fields: High-level microwave technology concerns," The Planetary Association for Clean Energy, Inc., media.withtank.com/42391c31ef.pdf.

64 Then-President George W. Bush to a joint session of Congress, September 20, 2001.

65 Jim Stone, "Mind control via electronic manipulation," June 14, 2012. www.jimstonefreelance.com/cells.html.

Computer monitors, like televisions, contain an electron gun (cathode ray) at the back of the tube transmitting beams of electrons at the screen to make pixels fluoresce into a picture. In New York, the NYPD might be driving their Z Backscatter Vans down your street and X-raying your vehicle, home, and body[66] while "war driving" Google Street View cars[67] hoover up payloads of data by means of the highly classified point-and-click TEMPEST (transient electromagnetic pulse standard), also known as Van Eck monitoring ("phreaking"), that receives, intercepts, views, and reconstitutes your incoherent signals—Google MAC (media access control) addresses, SSIDs (service set identifiers), signal strength measurements—before transmitting them to FBI techs, low-life contract agents, and NSA analysts.

C4 cyberwarfare is alive and well in the 'hood.

DCSNet of Communications Assistance for Law Enforcement Act (CALEA)—"The Gold Standard in Public Safety"—collects, sifts, and stores phone numbers, calls, and text messages. DCSNet endpoints connect to more than 350 switches. Besides monitoring servers and select emails of surveillance targets for capture, analysis, and storage, the FBI's Carnivore accommodates a sophisticated point-and-click surveillance device to cover signals, telephones, and Internet: DCS-3000 Red Hook (SIGINT) does pin-register (outgoing numbers) and trap-and-trace (incoming); DCS-5000 Red Wolf is responsible for counterintelligence; and DCS-6000 Digital Storm, wiretap, captures content. All were initially processed by Sprint on a private, encrypted network. FBI wiretap outposts are called Regional Information Sharing Systems (RISS). An FBI agent in New York can wiretap a phone in California, get the location, and receive pass codes and conversation in New York. Analysts interpret phone call patterns, then send them on to the Bureau's Telephone Application Database for link analysis.

Since CALEA, all cell phone manufacturers include Nextel software, which means that whether the phone is on or off, the microphone is a remote roving bug that transmits. When a call is in progress, the roving bug calls the FBI (or a corporate competitor) and activates the microphone.

Someone in your neighborhood may have a wideband radio receiver with spectrum display, horizontal/vertical sync generator, multi-scan video monitor and a phased-array antenna with shielded cables (like a radio telescope) and video taping equipment;[68] or may have leased the Trailblazer successor to

---

66 "The NYPD is using X-ray vans that can see into cars and homes." *Business Insider*, October 14, 2015.

67 David Kravets, "An Intentional Mistake: The Anatomy of Google's WiFi Sniffing Debacle." *Wired*, May 2, 2012. When confronted by the FCC about its war driving, Google maintained that it had broken no laws because "the data was flowing, unencrypted, over open air waves." And where have those specially rigged-out Street View cars gone?

68 Frank Jones, "The vulnerability of CRTs CPUs, and peripherals to TEMPEST monitoring in the real world," 1996, www.parallaxresearch.com/ dataclips/pub/infosec/emsec_tempest/info/ TEMPESTMonitor.txt.

ThinThread, a high-tech surveillance tool designed by NSA whistleblower William Binney to intercept, anonymize, and graph metadata according to relationships and patterns.

> It was around 2003 when they started putting optical fibers coming into the U.S. through Y-connector Narus devices. Basically these would duplicate the data coming across the Internet—one set of packets would go the normal route, the other set would go to NSA facilities. There, they collect all the data coming in through fiber optics, reassemble all the data packets into useable information—emails, file transfers, etc.—and then pass it along for storage. That means they are taking all that data off the fiber optic lines at 20 main convergence points in the U.S., collecting almost all of the Internet traffic passing through the U.S. This gets them pretty much control over the digital world.[69]

Time cloaking programs make whole data systems disappear into "time holes":

> This cloak works using a wave phenomenon called the Talbot effect. Manipulating the timing of light pulses so that the crest of one light wave interacts with the trough of another creates a zero light intensity—where the two signals cancel each other out—in which data can be hidden.[70]

Over Homeland neighborhoods, armed fifty-pound ShadowHawk helicopter drones (General Atomics and Lockheed Martin, $300,000 each) are executing low fly-bys while in wireless communication with whatever agency or corporation has been granted frequency command by measures and signatures intelligence (MASINT). DHS "public safety" surveillance drones and high-altitude killer drones—some remotely supervised, some programmed to be fully autonomous—are making their rounds, as well.

> . . .kill committee members from the National Security Agency, the Central Intelligence Agency, and the Department of Defense act as the prosecution, judge and jury for "low-level" targets. The president, consulting his kill list, makes the decision on "high-value" targets, including American citizens.[71]

In every urban or rural neighborhood of the Homeland, MASINT is being collected and collated as infrared peers through walls and ceilings. Thanks

---

69 Simon Black, "Digital Privacy Black Paper." www.commieblaster.com/dl/NSA-Black-Paper.pdf

70 Shaunacy Ferro, "Laser Time Cloak Hides The Fact That An Event Ever Happened." *Popular Science,* June 6, 2013.

71 Col. Ann Wright, "America's Drones Are Homeward Bound." *Truthdig.com,* July 17, 2012.

to the 1968 Wiretap Act[72] amended in 1986, wireless eavesdropping is not legally considered to be wiretapping. *War is peace, freedom is slavery, ignorance is strength. . .*

Beyond Carnivore (the Digital Collection System) and FinSpy hacks of iPhones,[73] there are dirty tricks programs like:

- Financial Crimes Enforcement Network (FinCen);

- Trapwire surveillance network with facial recognition run by Abraxas (CIA) and Mail Isolation Control and Tracking that photographs and records letters moving through the federal post office;

- Kapsch TrafficCom in-car transponders armed with in- and out-facing mini-cameras;[74]

- FBI FANTOM tracking data in 3D patterns;

- Encoded Apple signals commanding wireless devices to disable recording functions;[75]

- Skype messaging sent to Microsoft headquarters;[76]

- Stingray (Dirtbox on airplanes[77]) fake cellphone towers trick and track phones and their users.[78]

Do the three million university undergraduates glued to Facebook know that it was begun with venture capital provided by Stanford graduate Peter Thiel (PayPal, *The Diversity Myth*, on the board of VanguardPAC) and Accel Partners, whose manager chaired the National Venture Capital Association (NVAC) and was on the board of BBN Technologies that spearheaded ARPANET and thus the Internet? Or that the NVAC board was graced by the CEO of In-Q-Tel, whose mandate is to "nurture data mining technologies"?

---

72 Title III of the Omnibus Crime Control and Safe Streets Act of 1968.

73 G.J. Daniel, "The Malware That Can Spy on You Through Your Own Smartphone." *IntelliHub*, July 15, 2013. FinSpy's creator Gamma Group is "a British company owned by a shell corporation in the British Virgin Islands."

74 "Government Cameras In Your Car?" *MSNBC*, October 14, 2011.

75 Nick Farrell, "Apple patents tech to let cops switch off iPhone video, camera and wi-fi." *Techeye.net*, 7 August 2013.

76 "Skype with care — Microsoft is reading everything you write." *heise Security*, 14 May 2013.

77 Kim Zetter, "California Police Used Stingrays in Planes to Spy on Phones." *Wired*, January 27, 2016.

78 Bob Sullivan, "Pricy 'stingray' gadget lets cops track cell phones without Telco help." *NBC News*, April 3, 2012. Arizona police paid for their Stingray ($244,195) with a $150,000 grant from the State Homeland Security Program, plus RICO monies. For a fee, wireless carriers will track cell phone signals back to cell towers.

Facebook is actually a user-friendly human network analysis and behavior model-building engine. In fact, says computer engineer and founder of the Free Software Foundation Richard Stallman, Facebook is a surveillance engine[79] mapping your personality and selling your information, whose algorithms in Google Search are deciding what news and political opinions you should see while sending the rest elsewhere (or nowhere). In fact, *web personalization* is why real activism will never happen on Facebook or the Internet.[80] Eli Pariser explains in his book *The Filter Bubble: How the New Personalized Web is Changing What We Read and How We Think*:

> ...the Internet has unleashed the coordinated energy of a whole new generation of activists—it's easier than ever to find people who share your political passions. But while it's easier than ever to bring a group of people together, as personalization advances it'll become harder for any given group to reach a broad audience. In some ways, personalization poses a threat to public life itself.[81]

79 Sriram Srinivasan and Sangeetha Kandavel, "Facebook is a surveillance engine, not friend: Richard Stallman, Free Software Foundation." *The Economic Times*, February 7, 2012.

80 Jonathan Matthew Smucker, "What Facebook Is Hiding From You." *AlterNet*, May 29, 2011.

81 British version listed under *The Filter Bubble: What the Internet Is Hiding From You.* Penguin Books, 2012.

CHAPTER ELEVEN

The Unlawful "Silent War" Against Citizens

▼

Let the tuning commence.

— *Dark City*, 1998

"You took my sonar concept and applied it to every phone in the city. With half the city feeding you sonar, you can image all of Gotham. This is wrong. . .This is too much power for one person."

— Lucius Fox to Bruce Wayne, *The Dark Knight* (2008)

One can envision the development of electromagnetic energy sources, the output of which can be pulsed, shaped, and focused, that can couple with the human body in a fashion that will allow one to prevent voluntary muscular movements, control emotions (and thus actions), produce sleep, transmit suggestions, interfere with both short-term and long-term memory, produce an experience set, and delete an experience set.

It would also appear possible to create high fidelity speech in the human body, raising the possibility of covert suggestion and psychological direction. When a high power microwave pulse in the gigahertz range strikes the human body, a very small temperature perturbation occurs. This is associated with a sudden expansion of the slightly heated tissue, This expansion is fast enough to produce an acoustic wave. If a pulse stream is used, it should be possible to create an internal acoustic field in the 5–15 kilohertz range, which is audible. Thus, it may be possible to 'talk' to selected adversaries in a fashion that would be most disturbing to them.

— USAF Scientific Advisory Board, *New World Vistas,*
Air and Space Power For the 21st Century, Ancillary Volume, 1996

Forty years ago, geophysicist Dr. Gordon J.F. MacDonald, science advisor to former President Lyndon Johnson, spelled out the advent of the devil hiding in the woodshed of electromagnetic details to the House Subcommittee on Oceans and International Environment, but few understood references like "global," much less the remote possibility of "geoengineering" brain waves:

> The basic notion [in low-frequency research] was to create within the cavity between the electrically charged ionosphere in the higher part of the atmosphere and conducting layers of the surface of the earth, this neutral cavity to create waves, electrical waves, that would be tuned to the brainwaves. The natural electrical rhythm of most mammalian brains, including man, is about 10 cycles (hertz) per second, and there are indications that if you tune in at this frequency—that is, these low frequencies of about 10 cycles per second—you can produce changes in behavioral patterns or in responses.[1]

As the Bill of Rights is supplanted by technocracy, electromagnetics barrage Homeland neighborhoods as *Arbeit macht frei*[2] invisibly shimmers amidst the conducting aluminum, barium, and lithium nanoparticles drifting down with trillions of nanosensors. Meanwhile, people sleep in their chains of "progress" and convenience, many allowing dentists to implant sensors and tiny 2.45 GHz Bluetooth radios that wirelessly transmit data,[3] others ingesting Proteus Digital Health 1mm "digital pills":

> Digital Medicines are the same pharmaceuticals you take today, with one small change: each pill also contains a tiny sensor that can communicate, via our digital health feedback system, vital information about your medication-taking behaviors and how your body is responding.[4]

As charged antennas, we now interlock with the Earth's magnetic field,[5] ready targets for remote BioAPI's (biometric program interfaces), our bodies resonating like amplification capacitors. Wherever on Earth we go, we pick up and resonate with whatever is pulsing along the Earth's magnetic field lines.[6]

1 Gordon J.F. MacDonald, "Geophysical Warfare: How to Wreck the Environment." *Unless Peace Comes: A Scientific Forecast of New Weapons*. Ed. Nigel Calder. A. Lane, 1968. 181–205. A "ghost in the machine" arises with a 19Hz frequency.

2 "Labor makes you free," sign over the entrance to the Auschwitz concentration camp.

3 Paul Lilly, "Wi-Fi Enabled Tooth Sensor Tells Doctors If You've Been Overeating or Smoking." *HotHardware.com*, July 28, 2013.

4 Sayer Ji, "Microchip-Laden Drugs Given FDA Approval." *GreenMedInfo.com*, August 3, 2012. Also see Mordechai Harel, "Microparticles for Oral Delivery," US20080044481 A1, February 21, 2008.

5 Thomas H. Maugh II, "Caltech Scientists Find Magnetic Particles in Human Brains." *Los Angeles Times*, May 12, 1992.

6 Clifford E. Carnicom, "The Earth Is The Antenna," March 18, 2003.

> It has now been discovered the magnetic fields have a profound influence not
> only on human consciousness, but also on the human nervous system and
> immune system.[7]

The magnetic fields around power lines and transformers now act as open-air fMRIs. Heat the magnetic six-nanometer particles we inhale (created at SUNY's University of Buffalo Nature Nanotechnology labs) to 34°C (93.2°F)—the avoidance response threshold—and our cells are activated.

> "By developing a method that allows us to use magnetic fields to stimulate cells
> both *in vitro* [in glass] and *in vivo* [in a living organism], this research will help
> us unravel *the signaling networks that control animals behavior*," says Arnd Pralle,
> Ph.D., assistant professor of physics in the UB College of Arts and Sciences
> and senior/corresponding author on the paper.[8] [Emphasis added.]

Can you see now how the sensors, metal nanoparticles, and nano-fibers loaded with microprocessors ("nanobots") can be made to serve the programmed DNA microchips that will morph the masses into obedient cyborgs?

> The basic idea consists of a set of nano-wires [fibers] tethered to electronics
> [microprocessors] in the main catheter such that they will spread out in
> a "bouquet" arrangement into a particular portion of the brain's vascular
> system. Such arrangement could support a very large number of probes (in the
> millions). Each n-wire would be used to record, very securely, electrical activity
> of a single or small group of neurons without invading the brain parenchyma.[9]

Altering weak and vulnerable humanness lurks behind the military alteration of regions of the ionosphere and therefore the Earth. In 1968, geophysicist MacDonald talked about tuning brain waves via the ionosphere; in 1987, Bernard Eastlund's HAARP patent spelled out how to make cyclotronic resonance tune our lower atmosphere to the ionosphere.

Given that long-wave ELF propagation now traverses the globe 24/7, all of life depends upon studying and understanding exactly how frequencies and biological systems interact. In 2003, independent scientist Clifford Carnicom made a careful study of the ELFs pulsing via conductive metal particulates in the air over northern New Mexico. He determined that the pulse was arriving in

---

7 "Flaring hot summer," *Sunlive* (New Zealand), 11 January 2013.

8 Ellen Goldbaum, "With Magnetic Nanoparticles, Scientists Remotely Control Neurons and Animal Behavior." Press release, University of Buffalo SUNY, July 6, 2010.

9 www.dataasylum.com/mindcontrol-chemtrails-summary.html. This site deserves thorough exploration.

multiples of 4 Hz, and because the frequency range of the human brain is 4–52 Hz, the pulse was surely influencing biologically based systems:

(1) Changing human mental functioning, influence, and control;

(2) Disrupting cellular metabolism;

(3) Suppressing the immune system;

(4) Affecting free radical reactions;

(5) Changing DNA; and

(6) Changing cyclotronic resonance itself.[10]

Using his body as the sole antenna/receiver for the ELF circuit, Carnicom was able to pinpoint the locale of minimum ohmic resistance at the base of the skull behind the two ear lobes. By attaching an electrode and adjusting the gain of the circuit to the ELF range governing the human brain, he discovered that the frequencies received were subsequently amplified by the circuit.

[Note: Carnicom stresses caution in using the body as a receiver, as he suffered neural interruption, disorientation, and headaches for twelve to thirty-six hours after the experiment.[11] Given that electrical frequencies can be used to heal as well as destroy,[12] with a self-designed ELF detector (operational amplifiers, high inductance coil) and a frequency generator (30 kHz), Carnicom was able to alter the electromagnetic field around his body and effectively interrupt the fatigue being generated by ambient geometric ELF frequencies.]

Loading our brains with beacon-like metals and nano-sensors sounds distinctly like the sharks being weaponized by forcing them to follow chemical trails and sense delicate electrical gradients:

To transmit signals to the sharks, the team [of project engineer Walter Gomes of the Naval Undersea Warfare Center in Newport, Rhode Island] needs nothing less than a network of signaling towers in the area.[13]

Do we at last see that our transceiver towers with phased-array antennas are not just about profits and cell phone convenience? Coupled with the daily spraying of conductive metals, magnetic nanoparticles digging foxholes in our bodies and brains, and digital pills, it is now clear that we are being plugged into our Earth's magnetic field lines for a very different human future.

---

10 Clifford E. Carnicom, "ELF & The Human Antenna," January 19, 2003.

11 Clifford E. Carnicom, "A Direct Connection: The Human Antenna II," March 21, 2003.

12 See Hulda Clark, *The Cure For All Diseases: With Many Case Histories* (New Century Press, 1995); Rife Machine of Royal Raymond Rife.

13 Susan Brown, "Stealth sharks to patrol the high seas." *New Scientist*, March 1, 2006; referred to in Tom Engelhardt's "Shark and Awe, Weaponizing the Shark and Other Pentagon Dreams," *TomDispatch. com*, March 7, 2006.

EMK-ULTRA[14]

*Two-way communication with the depth of the brain makes it possible
to send and receive information to and from the brain, circumventing
physiological sensory receptors and motor effects. . .We can start, stop,
and modify a variety of autonomic, somatic, behavioral, and mental
manifestations. In these explorations of intracerebral physiology in behaving
subjects, we are reaching not only for the soma but for the psyche itself. . .
As no batteries are used, the life of the transmitters is indefinite.
Power and information are supplied by radio frequencies.*

— *Jose M.R. Delgado, "Man's Intervention in Intracerebral Functions,"
Department of Psychiatry, Yale University, 1967*

After 9/11, the term "terrorist" underwent a confusing revisionism to potentially include anyone critical of the (Deep) State. At the same time, domestic spying expanded exponentially and made all electronics fair game, from phones, televisions, and computers to geospatial intelligence (GEOINT) and MASINT for electronic harassment and no-touch torture in your home, at work, or in your car. *Full spectrum dominance.* While Americans were being conditioned to condone the torture going on at Guantánamo Bay and elsewhere, a C4 lockdown was being set up at home, thanks to the plasma-thickened, quartz-resonant ionized atmosphere overhead being ceaselessly zapped.

For readers unfamiliar with the Cold War game-changer known as MK-ULTRA, the following serves only as a thumbnail sketch. MK-ULTRA is the granddaddy of American nonconsensual brain and behavioral experimentation. Formally run by the CIA from 1953 to 1973 to the tune of $25 million in Cold War dollars,[15] MK-ULTRA actually inherited a jumpstart from the thousands of Paperclip Nazis ferreted into the United States and Canada with their intact wartime files of concentration camp experiments. Drugs, hypnosis, and pain induction were then the primary tools of the mind control trade, but remote electronic mind control was always the objective.

In the days preceding MK-ULTRA, electrodes were being inserted into babies' skulls.[16] In the 1950s, it was Yale University's *José Delgado, M.D.,* who initiated at-a-distance no-touch torture by pulsing frequencies and wirelessly triggering changes in consciousness without electrodes. In 1962, *Allen Frey, M.D.,* discovered *microwave hearing*, the transmission of spoken words into the auditory cortex via a pulsed microwave analog of the speaker's sound vibration.

---

14 The "E" stands for electronic and alludes to at-a-distance, no-touch torture.

15 Today, at 688.9% inflation, that would be over $197 million.

16 Rauni-Leena Luukanen-Kilde, M.D. "Microchip Implants, Mind Control, and Cybernetics."

Microwave hearing devices are now everywhere. In 2008, *Lev Sadovnik, Ph.D.*, of the Sierra Nevada Corporation created the Mob Excess Deterrent Using Silent Audio (MEDUSA), and the Disney Research microphone converts a voice message into an inaudible signal transmitted to the body of the person holding the microphone who can then transmit the message to another person by touching them.[17]

Patrick Flanagan, Ph.D., invented the neurophone in 1967 when he was fourteen, after which it was immediately classified by the NSA. The neurophone programmed suggestions through skin contact, due to the skin's many sensors for heat, touch, pain, vibration, and electrical fields. The neurophone is a low-voltage, high-frequency amplitude-modulated radio oscillator that converts modulated radio waves into a neural modulated signal bypassing the eighth cranial hearing nerve and transmitting words directly into the brain.

The Soviet vacuum tube LIDA machine was made by the Soviets in the 1950s, after which the North Koreans used it for brainwashing American POWs by placing the vertical plates along each side of the head to make the POW feel like he was in a dream. The LIDA is basically a 10 Hz pulsed radio signal transmitter that functions as a hypnotist's "swinging watch" to put people to sleep. (Quickening its pulse produces wakefulness.) It is listed with the IBM Intellectual Property Network (but not the U.S. Patent Office). By combining pulsed light, sound waves, and electromagnetic radiation, brain waves can be entrained and emotional states manipulated. *Both the neurophone and the LIDA machine are transmitters only, not transceivers.*

In the 1970s, *Andrija Puharich, M.D.*, studied the Soviet Woodpecker, then wired subjects, sealed them in a metal room, and beamed ELF waves at them. Waves below six cycles per second (6 Hz) emotionally upset subjects and disrupted their bodily functions; waves at 8.2 cps elevated their mood; waves at 11–11.3 cps depressed, agitated, and produced violent reactions. When Puharich later sought to warn the public about machine telepathy, he mysteriously disappeared from his home in Glen Cove, Maine,[18] ninety-six miles from Rangeley, where Wilhelm Reich, M.D. had lived.

Ross Adey, M.D., also used electromagnetic fields to remotely influence emotional states and behavior. Adey discovered that a $0.75mW/cm^2$ pulse-modulated microwave signal at a frequency of 450 MHz will control all aspects of human behavior. By directing a carrier frequency to stimulate the brain, then using amplitude modulation to shape the wave to mimic EEG frequencies, Adey could force a 4.5 cps *theta* rhythm (sleep) upon his subjects.

In 1973, *Joseph Sharp, M.D.*, successfully demonstrated artificial microwave voice-to-skull (V2K) or synthetic telepathy. In 1978, *James C. Lin, Ph.D.*, published *Microwave Auditory Effects and Applications* in which he related his

---

17 "Watch Disney Transmit Sound Through Human Touch." *Fastcodesign.com*, December 26, 2013.

18 Alex Constantine, *Psychic Dictatorship in the USA*. Feral House, 1995.

own demonstrations of microwave hearing with a pulsed microwave transmitter. Lin has published a great deal on biological nonionizing radiation effects,[19] and on July 25, 1990 testified before the Subcommittee on Natural Resources, Agriculture Research, and Environment Committee on Science, Space, and Technology in the U.S. House of Representatives in his capacity as chairman of the Committee on Man and Radiation, Institute of Electrical and Electronic Engineers (IEEE).

A library of exact frequencies and pulse rates of every emotion, passion, and state of mind in what the Soviets used to call *acoustic psycho-correction* has been compiled—for example, *Elizabeth Rauscher*, nuclear physicist at Technic Research Laboratory in San Leandro, California, identified the happiness and aggression frequencies, and *Michael Persinger, M.D.*, chief neurologist at Laurentian University Environmental Physiology Laboratory in Ontario, used time-varying fields of low-intensity 1–10 Hz ELFs to induce nausea.[20]

In June 1975, while Russian Woodpecker ELFs were zapping Eugene, Oregon brains, Leonid Brezhnev, General Secretary of the Central Committee of the [Soviet] Communist Party (1964–1982), called for a ban on EM weapons "more terrifying than nuclear arms." But it was too little, too late. As nuclear engineer Thomas E. Bearden told it in 1990

> [Brezhnev] stated the need for an "insurmountable barrier" to the development of such weapons. In July, he repeated his strange proposal to a group of visiting U.S. Senators. Ponomarev, a Soviet national party secretary, again raised the same issue to a delegation of visiting U.S. congressmen in August. At the United Nations' thirtieth Session of the General Assembly on Sept. 23,1975, Foreign Minister Andrei A. Gromyko strongly raised the same issue, warning that science can produce "ominous" new weapons of mass destruction. He urged that all countries, led first by the major powers, should sign an agreement to ban the development of these unspecified new weapons. He even offered a draft, entitled "Prohibition of the Development and Manufacture of New Types of Weapons of Mass Annihilation and of New Systems of Such Weapons"...*By its fixation on nuclear weapons and its ignorance of scalar EM, the West may have lost its only opportunity to prevent the spread of scalar EM weapons "more frightful than the mind of man has ever imagined," to use Brezhnev's characterization.*[21]
> [Emphasis added.]

---

19 www.ece.uic.edu/~lin/publications.htm.

20 Persinger's experiments with solenoids and the temporal lobes are intriguing. When he focused solenoids on the right hemisphere of the temporal lobe, the subject experienced a negative presence on the left side of the body, like an alien or demon; when focused on the left hemisphere, an angel or god would appear on the right side of the body. Focusing on the hippocampus produced the opiate effects of ecstasy; on the amygdala, sexual arousal.

21 Lt. Col. T.E. Bearden (ret.), "Historical Background of Scalar EM Weapons," 1990. www.cheniere. org/books/analysis/history.htm. See Appendix K.

Between 1975 and 1977, the U.S. Senate Select Committee on Intelligence and Subcommittee on Health and Scientific Research of the Committee on Human Resources (the Church Committee) attempted too little too late to get to the bottom of CIA mind control experiments, even unearthing ten large boxes of documents labeled "MK-ULTRA, 1952–62." John Marks, author of *The Search for the "Manchurian Candidate": The CIA and Mind Control: The Secret History of the Behavioral Sciences* (1979), requested that the CIA's Office of Research and Development (ORD) provide files "on behavioral research, including. . .activities related to bio-electrics, electric or radio stimulation of the brain, electronic destruction of memory, stereotaxic surgery, psychosurgery, hypnotism, parapsychology, radiation, microwaves and ultrasonics." He received 130 cubic feet of classified documents on "behavioral experiments" and learned that from 1950 to 1970, forty-four colleges and universities had hosted MK-ULTRA research with funds quietly flowing through CIA fronts like the Human Ecology Fund (Brown University).

Once the Church Committee furor died down, MK-ULTRA was recast under a variety of names while Wall Street family foundations continued to fund it through the ultraconservative American Family Foundation (AFF) formed by former CIA MK-ULTRA personnel operating out of the New York City law offices of Morris and McVeigh.[22]

In 1994, neurophysiologist Donald York and speech pathologist Thomas Jensen of the University of Missouri correlated brainwave patterns with spoken words and thought ("silent speech") even as the U.S. Army War College's infamous *Revolution in Military Affairs and Conflict Short of War* announced: "Behavior modification is a key component of peace enforcement" and "The advantage of directed energy systems is deniability." Jason Jeffrey in *New Dawn* magazine warned readers that the Revolution in Military Affairs (RMA) was all about EM technological "solutions" à la *New World Vistas*, and he was right:

> Individuals unwilling to go along with the revolutionary changes are "identified using comprehensive inter-agency integrated databases." They will then be "categorized" and "sophisticated computerized personality simulations" will be used "to develop, tailor and focus psychological campaigns for each."[23]

Meanwhile, HAARP was going up in Alaska and Project Cloverleaf chemtrails were uploading the sky.

The last round of public inquiry into MK-ULTRA occurred on the Ides of March 1995. Dutifully, then-President Clinton's Advisory Committee on

22 Alan Morrison, "Controlling the Opposition . . ." 16 May 2005. Morrison has been scrubbed from the web; his article can still be found at www.getbig.com/boards/index.php?topic=261151.0;imode. In 2004, the AFF was re-christened the International Cultic Studies Association (ICSA).

23 Jason Jeffrey, "Electronic Mind Control: Brain Zapping, Part One." *New Dawn*, March-April 2000.

Human Radiation Experiments met with victims who had been subjected to the military's desire to weaponize the brain. Like the Church Committee twenty years earlier, once testimony was heard and recorded, MK-ULTRA was made to appear as if it had ended years before. The truth was that it was not only alive and well and buried in deeper black, but it had gone electromagnetic.

FROM RFIDS TO REMOTE NEURAL MONITORING (RNM) AND BRAIN-COMPUTER INTERFACE (BCI)

Cybernetic mind control began with microchip-implanted brains linked to satellites controlled by ground-based, 20 million bits per second supercomputers. The FBI and CIA were partial to behavior-mod chips with their 250,000 parts and 400,000 bytes of information, their own little oscillator and lithium battery recharging from body heat, each chip the size of a grain of rice and—once Navstar satellites picked up its signal—capable of being remotely reprogrammed via pulsed frequencies on low-frequency radio waves. The next generation was the Destron-Fearing transponder sheathed in biomedical-grade glass imprinted with 34 billion bytes of information. At first, the implants were made of silicon, then of gallium arsenide. In Vietnam, soldiers were injected with the Rambo chip to increase adrenaline flow; in the 1991 Gulf War, the IMI biotic (intelligence-manned interface) "engaged" soldiers.

In 1999, two interrelated events occurred. The first was the suspension of psychology professor Kathryn Kelley of SUNY-Albany[24] after she had presented a paper in Orlando, Florida to a summer conference sponsored by the World Multiconference on Systemics, Cybernetics and Informatics, and the International Conference on Information Systems Analysis and Synthesis (based in Venezuela). In her presentation, she had mentioned that implant research was being funded by the NSA and DoD ($2 billion annually) and that uninformed, nonconsensual human guinea pigs were being used for "open field" experiments. The Institutional Review Board had approved of Kelley's research into "advances in technology that affect interpersonal communication" but not into monitoring and control technologies like RAATs (radio wave auditory assaultive transmitting implants). Kelley revealed that when [short-wave] operators transmitted to or scanned RAAT implants in victims, they could remotely and anonymously talk to the victims as well as hear their speech and thoughts.

Four months after Kelley's dismissal, Applied Digital Solutions in Palm Beach, Florida announced it had acquired the patent rights (U.S. Patent No. 5,629,678) to a miniature digital transceiver powered electromechanically by the host's muscle movement and body heat and had christened it the Digital Angel

---

24 Andrew Brownstein, "Human Brain Implant Research Suspended at Major University." *The Albany Times Union*, August 25, 1999.

"personal tracking and recovery system." Back in 1971, a similar microprocessor could only hold 2,300 tiny transistors; by 1996, it was 5.5 million.

From one centimeter to the size of a BB or a grain of rice to today's nano-sized microprocessors being inhaled and consumed, implants are still being tracked and brain functions altered by pulsing frequencies under the remote monitoring system that translates to surveillance, isolation, and torture. As thoughts, reactions, hearing, and visual observations cause a certain neurological potential, spike, or pattern in the brain, its electromagnetic field can now be decoded onto a monitor somewhere.[25] Once the unique bioelectrical resonance frequency is remotely coded,[26] EM signals can be pulsed to just that brain. In other words, by decoding the evoked potentials (3.50Hz, 5 milliwatt) the individual brain emits, billions can now be remotely monitored by NSA Signals Intelligence (SIGINT) via satellite electro-optical capabilities, supercomputers, AI, and ground-based infrastructure.

Between 2000 and 2009, four hundred neurotechnology patents were filed; in 2010, the number doubled, and in 2014 it doubled again to 1,600 patents.[27]

Just before he died, Australian journalist Joe Vialls wrote that the *tachistoscope* was the first step in manipulating the American mind by timing the insertion and duration of subliminal frames in movies and television shows at 1/25 per second. Next, low-light subliminal images were blended into show and film content to bypass the conscious mind and target the subconscious mind. The third step was hexidecimal color-coding, brazenly depicted by director Steven Spielberg in his 1977 film *Close Encounters of the Third Kind* and later perfected by digitalized television.[28]

Consider "free" HuLu, the digital high definition (HD) interface between home computers and television networks. U.S. Patent #6,488,617, "Nervous System Manipulation by EM Fields from Monitors" (December 2, 2002), describes using computer monitors and HD TV screens as broadcast media for digital (think *pulsed*) "electromagnetic fields capable of exciting sensory resonances in nearby human subjects" while "displayed images are pulsed with subliminal intensity." A. True Ott, Ph.D., asks, "Is this why over 20% of the new DIGITAL bandwidth was given to the Dept. of Homeland Security in the name of 'Public Safety'?"[29] Good question.

---

25 Ibid.

26 This process is called *prima freaking*, the mapping of primary frequency allocations, the distribution of biotelemetrically responsive frequencies. (Email, Alex Constantine.)

27 Andrew Griffin, "Patents for technology to read people's minds hugely increasing." *The Independent*, 8 May 2015.

28 Joe Vialls, "Danger: Mind Controllers At Work!" 15 May 2005. Vialls died at sixty-one of a "heart attack" just two months after his article went viral on the Internet and before Part Two explaining the mind control behind "government-selected 'patsies'" could follow.

29 A. True Ott, "HuLu, A Quantum Leap in Electronic Mind Control & Manipulation," February 16, 2009.

"FIELD EXPERIMENTS"

Five years after 9/11, military apologist Jonathan D. Moreno felt he was on moral high ground in the war on terrorism when he defended human experimentation with "nonlethals" (less-than-lethals / electronic weapons) in his book *Mind Wars: Brain Research and National Defense*:

> Unlike more familiar ballistic or explosive devices, which can often be tested on inert objects or animals to get the desired information, many of the measures classified as NLWs [nonlethal weapons] involve human perception and behavior, so they must be tested on humans. . .Now, what would be the specific justification for human testing of the pulsed energy projectile [PEP]? One reason would be the fact that pain is a somewhat subjective experience, and the behaviors of animals, even higher primates, might not be useful in setting dose limits. Also, pain experience is associated with brain receptor sites that could have lasting effects, perhaps triggering mental illness in some people . . .[30]

Moreno goes on to explain how "people in uniform" and prisoners are no longer the plentiful "volunteer" pools they once were. He then admits to the very tactic the military uses to get around the Nuremberg Code consent requirement:

> Cynics might point out that there is an easy way around the problem of experimentation ethics: just call the projects field trials instead of human experiments. That way, the medical ethics rules might be evaded. During the cold war, sailors and soldiers were exposed to the atomic bomb and chemical nerve agents. These incidents were categorized as field tests of environmental contaminants or training exercises rather than as human experiments. As a result, the rules then in place that required voluntary consent weren't technically relevant, even though in some cases soldiers wore radiation badges and had their bodily fluids checked . . .[31]

Whatever the delivery system, connect the Smart Grid dots and you eventually arrive at a new class of "silent warfare" citizens numbering in the *millions* around the world: *targeted individuals (TIs)*, those being stalked and invisibly tortured daily with directed energy weapons (DEWs) in their neighborhoods and workplaces of big cities, towns, and rural areas.[32] TIs give a whole new meaning to "home incarceration."

---

30 Jonathan D. Moreno, *Mind Wars: Brain Research and National Defense*. Dana Press, 2006.

31 Ibid.

32 To block OS/EH attacks through the power lines, Stetzer power line filters come highly recommended and are available at The Power Mall for $35 each.

The brains of gamers playing Emotiv are being hacked into,[33] memories are being remotely triggered and read, dreams engineered, illness frequencies remotely transmitted, heart palpitations and extreme pressure made to mimic panic attacks and pulmonary conditions. Microwave beams can do all of this and more, from creating an electromagnetic "flu" to ramping up an illness by frying the immune system. This list of symptoms that most physicians attribute to "psychological" maladies can just as well be symptoms of electromagnetic sensitive (EMS) and/or DEWs:

Neurological: headaches, dizziness, nausea, difficulty concentrating, memory loss, irritability, depression, anxiety, insomnia, fatigue, weakness, tremors, muscle spasms, numbness, tingling, altered reflexes, muscle and joint paint, leg/foot pain, "Flu-like" symptoms, fever. More severe reactions can include seizures, paralysis, psychosis and stroke.

Cardiac: palpitations, arrhythmias, pain or pressure in the chest, low or high blood pressure, slow or fast heart rate, shortness of breath.

Respiratory: sinusitis, bronchitis, pneumonia, asthma.

Dermatological: skin rash, itching, burning, facial flushing.

Ophthalmologic: pain or burning in the eyes, pressure in/behind the eyes, deteriorating vision, floaters, cataracts.

Others: digestive problems; abdominal pain; enlarged thyroid, testicular/ovarian pain; dryness of lips, tongue, mouth, eyes; great thirst; dehydration; nosebleeds; internal bleeding; altered sugar metabolism; immune abnormalities; redistribution of metals within the body; hair loss; pain in the teeth; deteriorating fillings; impaired sense of smell; ringing in the ears.[34]

At thehum.info, you will find an interactive World Hum Map and database. There, Todd West clarifies what cellular radar (*celldar*) can do:

...Wherever a cell signal can reach, passive radar can take and via computerized networks track your electromagnetic signature, including seeing and hearing inside many types of building structures. In the U.S., Homeland Security agents via State Police liaisons introduce and train local law enforcement as well as, and this is VERY important, classified contractors on the use of celldar units.

33 Ivan Martinovic et al. "On the Feasibility of Side-Channel Attacks with Brain-Computer Interfaces." 21st USENIX Security Symposium, August 8–10, 2012.

34 From *No Place To Hide*, April 2001.

Celldar can cause untold torture to electromagnetic sensitive people, and over time extensive use of the technology causes diabetes, cardiovascular distress, cancer and loss of cognitive functions, TO START WITH. By fine-tuning the units, provocateurs can use the technology to disable, harm and kill those it is used on, making it the ultimate covert directed enemy [*sic:* energy] weapon used under the guise of "homeland security." It has been determined "the hum" being reported worldwide is the cacophony of electromagnetic eavesdropping technology running on top of cellular-style networks. It is microwave madness to a degree of deceptive overkill.[35]

Quietly and all around us, the new Homeland battlespace bristles with contract agents and criminals logging EM experiments on nonconsensual citizens in the name of "field research" while law enforcement looks the other way. Agents armed with laptops and classified DEWs/NLWs move into neighborhoods and apartment complexes to conduct see-and-hear-through-walls surveillance. First, they build a profile of the target's habits, friends, family, love affairs, medical and mental history, problems, etc., then begin EM attacks to provoke:

- Hearing voices or strange noises no one else hears
- Jolts and burning sensations to the limbs
- Pain or numbness (head)
- Involuntary muscle movements (legs)
- Invisible "strikes"
- Snaps, clicks, tinnitus, etc.
- Strange dreams/nightmares and sleep deprivation

A senior Australian Federal Police executive told Paul Baird, author of the pirated and unpublished *In the Year 2252* by "P. Barber,"[36] that such torture might go on for years. Baird writes:

Human thought operates at 5,000 bits/sec, but satellites and various forms of biotelemetry can deliver those thoughts to supercomputers in Maryland, USA, Israel, etc. which have a speed of 20 BILLION bits/sec each. These, even today, monitor millions of people simultaneously.[37]

---

35 At the Hum site, see "The Physics of the Hum: A Primer on Electromagnetic Radiation and Wave Interference for the Lay Person."

36 www.surveillanceissues.com/case.htm.

37 Paul Baird, "Categories of Surveillance/Harassment Technologies," U.S. Congress Office of Technology Assessment. www.surveillanceissues.com.

Regional fusion centers screen candidates for targeting, record the "field experiments" and share them with local law enforcement and federal agencies (DOJ, DHS, CIA, NSA, etc.) as necessary. Satellite targeting entails electromagnetic "leashes" of high-powered microwaves (HPMs), subatomic particle beams, ELF/ULF, plasma orb "necklaces," etc. Land-based and sea-based systems transmit mood and mind control frequencies. Unmarked and bogus corporate logo vans parked outside your home, laptops in Colorado Springs, briefcase-sized active denial systems (ADS) in neighbors' homes triangulating with phased arrays on cell towers. . .GSM (Global System for Mobile Communications) digital cellular transmissions can piggyback behavioral patterns on ELF signals (3–3000Hz) and pulse them into the microwave network being fed into cell phones.[38] Even when no signature is being broadcast, microwaves cause neurons to release calcium ions that make targets feel tired, irritable, and hopeless.

According to UK scientist Turan "Tim" Rifat, the dual-use GM900 microwave network (Vodaphone, British Telecom) and GM1800 systems (Orange) use weapons-grade higher-frequency microwaves:

> It is a rule of the intelligence community that you hide things in plain view. Getting the public to accept microwave mind control weapons which affect their behavior under the guise of mobile phones was a stroke of genius. Getting the public to pay for these microwave mind control devices, so their brains and behavior can be damaged, to make them more docile and easy to control, was pure diabolical genius.[39]

Rifat describes how ELF microwaves can be made to mimic brainwaves, after which they are pulsed by transmitters into British brains to make "submissive zombies who cannot think clearly, become depressed, apathetic, and want to lounge around all day doing nothing: the inner city malaise found on Britain's streets. . .[and] microwave phones use pulse modulated microwaves of the correct intensity to pass through the skull into the brain and control behaviour."[40] Rifat says UK police use 450 MHz[41] for behavioral control and clearing areas of the homeless, and use 75–1000 MHz to induce nervous and physical collapse.

IRIS (interface region imaging spectrograph) infrared satellites track mobile phones, body heat, and brainwaves. An NSA SIGINT laser monitors a target's brainwaves so a computer can decode its evoked potentials (3.50 Hz 5

---

38 Tim Rifat, B.Sc., B.Ed., "Electromagnetic Murders & Suicides & How They're Done." *Psychotronics & Psychological Warfare* "Category Archives: Group Stalking," 1999.

39 Tim Rifat, "Microwave Mind Control." www.whale.to/b/rifat/html.

40 Ibid. Once James Clerk Maxwell (review Chapter 2) had been erased from twentieth-century physics, Tavistock Institute applied his theories to mass mind control.

41 Rifat recommended testing earpieces of mobile phones to see if they give off 0.75mW/cm2, discovered by Adey (discussed earlier) to control human behavior if the pulse-modulated microwave signal utilizes a frequency of 450MHz.

milliwatts). The telltale EEG is then fed into a supercomputer and an audible neurophone input/output or other NLW device pointed at the victim's brain so an AI algorithm, a human NSA operator, a bored Hollywood star or corporate player can eavesdrop on the target's thoughts and input their godlike response.

Pulse modulation is the name of the game for brain transmissions—

> Human behaviour and reactions can be entirely controlled by using pulse modulated microwave EM radiation. Pulse modulated microwaves are useful as the carrier for the mind control signals as they are able to pass through the skull, which is rather resistant to low level EM. . .These signals are extremely low frequency recordings of brain electrical [excitation] potentials which have been recorded by neuro-medical researchers. . .It is now possible to broadcast mind control commands directly into the brain by use of microwave beams. All that is needed is a catalogue of every specific brain frequency for each mood, action, and thought. . .[H]aving the excitation potential for suicide beamed into your brain day and night by microwave mind control weapons soon resets the brain into a cycle of depression that spirals out of control . . .[42]

—in tandem with the millimeter wave scanners exposing the target behind the walls of their home or workplace:

> When firing microwave beams through walls at one specific target, every material in the way of the microwave beam attenuates or modifies the intensity and frequency of the beam. Since precise frequencies and intensities are needed for mind control, very sophisticated microwave arrays and computer programmes had to be developed so that the microwave beam could be changed in response to the materials which lay between the target and the weapon, as the victim moved around the house. To do this, the reflectivity and refractivity of the materials between victim and weapon had to be analysed in real-time and fed to a computer which could change the microwave array in concert with the changing environment between victim and weapon, as the target moved around his/her home. Secondly, there had to be an automatic interrupter if another person walked in front of the beam. The victim needed to be driven mad or disabled without anyone else being aware that he or she was being targeted . . .[43]

42 Rifat, "Microwave Mind Control." www.whale.to/b/rifat/html.

43 Ibid. Marconi worked on such "smart weapons" in the 1980s. Between 1982 and 1990 twenty-five British-based GEC-Marconi scientists and engineers working on U.S. SDI-related projects ("Star Wars") died under mysterious circumstances. See www.whale.to/b/marconi.html.

COUPLING / HETERODYNING

Earlier, I discussed how the specific focus output power of electromagnetic pulse (EMP) high-power microwaves (HPMs) *couples* with (transfers energy to) a wide target area. In Chapter 10, I also mentioned how suitcase systems (piggybacking) can shape a pulse in spacetime with direct coupling to a high-speed computer for non-linear biological effects.

Coupling human beings with "nonlethal" machines is specifically addressed in the fifteenth ancillary volume of the 1995 USAF Scientific Advisory Board's fourteen-volume *New World Vistas: Air and Space Power for the 21st Century: Ancillary Volume*, whose "Biological Process Control" is framed in the usual misleading future tense:

> One can envision the development of electromagnetic energy sources, the output of which can be pulsed, shaped, and focused, that can couple with the human body in a fashion that will allow one to prevent voluntary muscular movements, control emotions (and thus actions), produce sleep, transmit suggestions, interfere with both short-term and long-term memory, produce an experience set, and delete an experience set . . .[44]

In *Psychotronic War and the Security of Russia* (Moscow, 1999), Russian ambassador V.N. Lopatin and engineer V.D. Tsygankov explain DEW-brain coupling:

> Electromagnetic pulse (EP) anti-personnel weapons have many scientific and technical features in common with the laser weapons under development in the American and Soviet anti-missile defense programs. Both use electromagnetic radiation, propagating at 300,000 kilometers per second, to achieve their destructive effect. Both require compact power sources, generators of electromagnetic radiation (e.g. lasers, magnetrons, gyrotrons, etc.), beam radiator and focusing apparatus (e.g., optics for lasers, wave guides and phased-array antennas for microwave weapons), and computerized control systems. In both cases also, the maximum effect of these weapons is obtained by "tuning" or "tailoring" the output to the characteristics of the target.
>
> The chief peculiarity of EP anti-personnel weapons lies in their exploitation of highly non-linear effects of electromagnetic radiation upon living organisms. Typically, these weapons employ complicated pulse shapes and pulse trains, involving several frequencies and modulations which can range over a wide spectrum from extremely low frequencies (ELF) into the hundred gigahertz range. Thus, although state-of-the-art technology permits construction of mobile systems of extremely high output power (up to 10 megawatts average power, peak

---

44 *New World Vistas: Air and Space Power for the 21st Century: Ancillary Volume*. Washington, D.C.: USAF Scientific Advisory Board, 1995.

pulsed powers of many gigawatts), *it is not the high power per se which determines the lethality of the system, but rather its ability to "couple" the output effectively into the target and to exploit non-linear biological action.* While high output power may be used to obtain range and breadth of effects and penetration into enclosures and defenses, the minimum lethal "dose" on the target will typically be orders of magnitude less than that which would be required to kill by mere heating, in the manner of a microwave oven. . .[Italics added.][45]

To electromagnetically read brainwave patterns so that the brain can be coupled to a machine, scanning and tracking systems first couple with the magnetic field surrounding the head, then wirelessly transmit patterns to sophisticated computers for interpretation and recording. Once the aural feedback loop is established, remote psychotronic torture can begin (hauntingly depicted in the 1998 film *Dark City*). For example, a laser can direct signal-to-noise live talk, noise, or computer-generated music at the head's magnetic field, and because the ears are bypassed ("microwave hearing"), the sound will first register as electrical impulses in the nervous system and then be picked up by the brain.

The software utilized for Smart Grid targeting must be capable of coupling brains with computers. In Chapter 10, I mentioned NSA analyst Robert Duncan and his 2010 Internet book *The Matrix Deciphered*. Duncan calls this coupling process *EEG cloning* or *heterodyning*,[46] a radio processing technique that combines two frequencies to create new frequencies. Often funded under *cognotherapy*, EEG heterodyning software programs are an NBIC convergence (nanotechnology, biotechnology, information technology, and cognitive technology), otherwise known as *biotelemetry* mind and body control. Duncan names several EEG cloning programs:

TEMPEST or Van Eck phreaking

DREAM (Data Repository Establishment and Management) simulations

MIND (Magnetic Integrated Neuron Duplicator)47

TAMI (Thought Amplifier and Mind Interface, 1976)

---

45 Found at Cheryl Welsh's *mindjustice.org/Russian.pdf.* Lopatin was a member of the Russian Duma and ambassador to Japan. Tsygankov graduated from the Odessa Electrical Engineering Institute of Communication, specializing in radio engineering, and collaborated with neurophysiologist P.K Anokhin, whose books were once in UC Davis' Blaisdell Medical Library.

46 Duncan claims to have been tortured and programmed by the very technologies he worked on for the Department of Defense, U.S. Navy, and NATO. His background has yet to be confirmed, though wiping backgrounds is not unheard of in intelligence circles. His two print-on-demand books are *Project Soul Catcher: Secrets of Cyber and Cybernetic Warfare Revealed* (2010) and *How to Tame a Demon: A short practical guide to organized intimidation stalking, electronic torture, and mind control* (2014).

47 "John Ginter's Story With the MIND Machine: A San Quentin prisoner's amazing account told to Cheryl Welsh on September 30th, 1994," mindjustice.org/ginter.htm.

SATAN (Silent Assassination Through Amplified Neurons)BEAST (Battle Engagement Area Simulator/Tracker) 3D holographic image projections

FOCUS (Flexible Optical Control Unit Simulator) projects images onto the retina[48]

The coupling is accomplished by manipulating plasma—yet another perk for maintaining an ionized atmosphere. In 2007, *Live Science* exposed how electrified gas or plasma "antennas" are utilized for stealth targeting *in any battlespace*:

> These antennas only work when energized, effectively vanishing when turned off, with the plasma reverting back to normal gas. This is key for stealth on the battlefield—metal antennas can scatter incoming radar signals, giving away their presence.[49]

Gas plasma antennas "can rapidly adjust which frequencies they broadcast and pick up by changing how much energy the plasma is given," while a "smart" gas plasma antenna "can steer a beam of radio waves 360 degrees to scan a region and then find and lock onto transmitting antennas."[50]

According to TI Carolyn Williams Palit, when gas plasma transmitters are beamed through roofs and walls to remote-view and listen to thoughts, a plasma beam will attach to the ears, eyes, head, genitals, etc. by means of magnetic polarity like strings of pearls exposing what you are doing in real-time and communicating with satellite computers. Satellites do the tracking, but it is plasma-cloaked aircraft like jets, stealth fighters, and helicopters that do the real attacks, thanks to the conductive metal particulates in our ionized atmosphere.[51]

Gas plasma orbs are often mistaken for ball lightning because they are made of high-density plasma. Epidemiologist Rosalie Bertell (1929–2012) pointed to "microwave radiation within a plasma shell":

> A phenomenon associated with lightning is ball or globe lightning. This is a glowing, floating, stable ball of light, occurring at times of intense electrical activity in the atmosphere. On contact, these balls release large amounts of energy. Ball lightning occurs near to the ground during thunderstorms and may be red, orange or yellow. It is accompanied by a hissing sound and has

---

48 Mark Bond, "Electronic Mind Control." *Newsfinder*, last updated July 21, 2002.

49 Charles Q. Choi, "Radio Antenna Made of Plasma." *Live Science*, December 5, 2007.

50 Ibid.

51 Carolyn Williams Palit, "Gas Plasma transmitters generated from chemtrails," May 28, 2005, portland.indymedia.org/en/2005/05/318265.shtml.

a distinct odour. The causes of ball lightning are unknown but speculated causes include air or gas behaving abnormally; high-density plasma; an air vortex containing luminous gases (a vortex is a whirling phenomenon like a miniature whirlwind); and *microwave radiation within a plasma shell.*

Scientists have theorized that a microwave generator could be used to fire a plasmoid, a blob of plasma not unlike ball lightning, into the path of an incoming missile, its warhead, or an aircraft. The theory is that as a missile passes through the fireball, its electronics and navigational systems are disabled. Electromagnetic energy also interferes with the isotopes of a nuclear warhead, effectively disarming the weapon.[52] [Emphasis added.]

If "microwave radiation within a plasma shell" can couple with navigational systems and disable them, might it not do the same with a human brain?

Non-pulsing plasma orbs are not alien visitations. They are used as transceiver weapons that can piggyback voices and holographic images into and out of brains. First, barium gas clouds are vaporized into geometric plasma orbs (plasmoids, microplasma), then affixed to targets' heads. Palit goes into more detail:

They are also photographing gas plasma generation due to the heating of chemtrails by electromagnetics. The technical names for vertical and horizontal plasma columns are columnar focal lenses and horizontal drift plasma antennas. Various sizes of gas plasma orbs are associated with this technology. These orbs can be used as transmitters and receivers because they have great refractory and optical properties. They also are capable of transmitting digital or sound. Barium, in fact, is very refractive—more refractive than glass.[53]

On December 13, 2010, *New Scientist* announced how the solid-state Plasma-Silicon Antenna (PSiAn) revolutionizes high-speed wireless communication, miniature radar, and directed energy weapons:

PSiAn consists of thousands of diodes on a silicon chip. When activated, each diode generates a cloud of electrons—the plasma—about 0.1 millimetres across. At a high enough electron density, each cloud reflects high-frequency radio waves like a mirror. By selectively activating diodes, the shape of the reflecting area can be changed to focus and steer a beam of radio waves. This "beam-forming" capability makes the antennas crucial to ultrafast wireless applications [like Wi-Gig], because they can focus a [narrow] stream of high-frequency radio waves that would quickly dissipate using normal antennas.[54]

---

52 Rosalie Bertell, *Planet Earth: The Latest Weapon of War*. Women's Press Ltd., 2000.

53 Carolyn Williams Palit, "What Chemtrails Really Are." *Rense.com*, November 9, 2007.

54 David Hambling, "Wireless at the speed of plasma." *New Scientist*, 13 December 2010.

Once that "stream of high-frequency radio waves" locks onto neighborhood phased arrays, brains could be made to think at other frequencies. Any cell tower's gigahertz waveform (one billion Hertz) is capable of penetrating seven inches into the skull, but only if modulated with an ELF waveform with which the brain resonates. Investigative journalist Jim Stone:

> If you transmit an ELF waveform by itself, it will need a many miles long antenna to pick it up properly. It will not interface with the body very well because your body is not a big enough antenna for the ELF wave to efficiently "short out" in. You would need a really intense signal to do anything. But if you can take that ELF frequency and use it to modulate a super high frequency carrier, that carrier will drop all of its energy directly into your skull and let the modulation of that energy do the rest.[55]

The gas plasma antenna is a more versatile all-in-one dual-use weapon than the solid-state PSiAn because it can morph into a transmitter or receiver when necessary. Whereas the solid-state PSiAn can only work at high frequencies (1–100GHz), ionized gas plasma antennas can operate anywhere in the radio frequency spectrum (lower than 300GHz), which of course includes the ELFs that are the carriers of our thoughts.

Brain-to-brain coupling is called *techlepathy*, the communication of information directly from one mind to another with the assistance of technology—in the name of "sharing a social world,"[56] of course. Perpetrators coupled with computers are able to "mind-meld" with targeted individuals—hooking up their own brain with the target's brain in such a way that the target's thoughts run through the perpetrator's brain to be recorded on computers.[57] This kind of linkage between neural systems is part of the virtual reality (VR) bridge to the cybernetic storage of memories for later download into other brains whose memories have been wiped—possibly under the technology known as electronic dissolution of memory (EDOM), a generator-transmitter able to interrupt and "wipe" memories by increasing the brain's production of acetylcholine. This is generally how a target's experience of "lost time" is remotely generated.

Memory wiping now comes under *optogenetics*, the branch of science that couples brains with computers, transfers life memories and calls it immortality.

Neuroscientists have long sought the location of these memory traces, also

55 Jim Stone, "Mind control via electronic manipulation," June 14, 2012.

56 Uri Hasson, et al. "Brain-to-brain coupling: a mechanism for creating and sharing a social world." *Trends in Cognitive Science*, February 2012.

57 "Cybergods," *Boycott Brazil* (www.brazilboycott.org). Released on the Internet in February 1996 in English, Swedish, Japanese, Spanish, German, and Danish. See both mptilton.tripod.com/cybergods-29.html and www.lambros.name/CYBERGODS.pdf

called engrams. In the pair of studies, [Susumu] Tonegawa and colleagues at MIT's Picower Institute for Learning and Memory showed that they could identify the cells that make up part of an engram for a specific memory and reactivate it using a technology called optogenetics.[58]

OS/EH (GANGSTALKING / GASLIGHTING)

The 2001 Space Preservation Act (HR2977) mentioned in Chapter 10 specified that DEWs are capable of targeting individuals and populations from space. Target a pilot and the aircraft can be made to crash; target a young man's brain and he can be made to shoot up a theater of people; target a city population and violence can erupt; target a whistleblower and he or she has a sudden heart attack or brain aneurysm.

Gangstalking or *gaslighting*, also known as *organized stalking / electronic harassment (OS/EH)*, is a primary covert no-touch torture "field experiment" and moneymaker utilizing the multipurpose Smart Grid. *Wikipedia* calls gaslighting "a form of mental abuse in which false information is presented with the intent of making a victim doubt his or her own memory, perception and sanity."[59] Break it down into gang and stalking and you get the picture: a low-life, outlaw, or intel network working together via cell phones, laptops, vehicles, and DEWs (a.k.a. enhanced radiation weapons or ERWs) to see what it takes to either drive the target crazy or to commit a violent act. Former FBI agent Bob Levin says OS/EH is CIA "fox and hound exercises" whereby random targets are chosen for agent surveillance practice and research into the four principles of the Torture Paradigm:

(1) Cause self-inflicted harm;

(2) Cause sensory disorientation;

(3) Attack individual fears;

(4) Attack cultural identity.[60]

---

58 Anne Trafton, "Neuroscientists plant false memories in the brain." *MIT News*, July 25, 2013. Films like *Total Recall* (1990, 2012), *Dark City* (1998), *Paycheck* (2003), *Eternal Sunshine of the Spotless Mind* (2004), *The Butterfly Effect* (2004), etc., allude to such a future.

59 *Wikipedia* says that the term arose from the 1938 British play *Gas Light*, renamed *Angel Street* in the U.S.

60 Deborah Dupre, "Ex-FBI agent: 'Gangstalking' term self-harm for Gov. no-touch torture eugenics." *Examiner.com*, February 23, 2011. Levin's "Blackfile Intelligence Reports" can be found at www.scribd.com/doc/133239477/Blackfile-Intelligence-Report-0172-Electronic-Cointelpro-and-Surveillance. As footnoted in Chapter 10, Evelyn Waugh's 1957 novel *The Ordeal of Gilbert Pinfold* depicts Waugh's own targeting with a LIDA machine.

Invisible 24/7 no-touch torture is more common than one would think. Agents tweak the nociceptor response to various levels of pain, objectively observe and record target reactions, derive sexual perks from drugged or gassed victims. *Can they make the targets kill themselves or others? Can they force the target to seek psychological or medical help that will end in involuntary outpatient or inpatient commitment?* are the questions these torturers ponder.

Lobbyists, dissidents, activists, investigative journalists and researchers, politicians, whistleblowers, "former" intelligence and military agents, Special Forces or commissioned officers, New Age channelers, UFO contactees, Hollywood and rock stars are often targeted for political agendas, but the difference is their targeting generally ends when they "see the light" and behave accordingly. But targeted individuals (TIs) stuck in "lifelong case files" or those utilized for agents practicing how to align the four cortical brainwave metrics for mind-melds to be seamlessly transferred to computers are in for years of torture followed by death. "Throwaway" or marginal populations are used first— immigrants, ethnic neighborhoods, women living alone, the rurally isolated, prisoners, homeless, elderly, poor children, the mentally ill, Native Americans, etc.—but apparently there are never enough victims.

Twenty years ago, the extraordinary study called "Cybergods" described an already Orwellian Sweden:

> What the Public Inquiry really means is. . .the life of an individual is also stored on disc and can be called up at will: these discs are acquired by the Public Records Office (*Riksarkivet*) after death. This procedure was drawn up in 1986 after a meeting attended by Minister of Justice Sten Wickbom. Also present were Mats Borjesson from the secret police SAPO, the general director of the Public Records Office, and a further 30 or so researchers, bureaucrats and politicians . . .[61]

TIs seeking medical help are often sent on to a psychiatrist and "detained" for psychiatric evaluation, at which point anti-psychotic or anti-depressant drugs are introduced so that the TI's mind is further owned by the very system that covers for OS/EH and writes DSM-V diagnoses to obscure classified nonlethal operations.[62] Functional neuroimaging[63] ("brain fingerprinting") via NIRSI (near infrared spectroscopic imaging), PET (positron emission tomography), EEG, CT scan, MEG, or fMRI may then be ordered to map and measure (MASINT) the TI's brain and organ activity for their "lifelong case file." The TI

---

61 "Cybergods," 1996.

62 "United States Secret Government At War With the Mentally Ill," *hightechharassment.com/*. Also see Gail Kansky, "President's Message." *The National Forum* [CFIDS/ME, FMS, GWI, MCS and Related Illnesses], Spring 2013 regarding the latest "Somatic Symptom Disorder."

63 A part of *remote neural monitoring*, the at-a-distance monitoring of a brain and its thoughts.

might also be driven to commit an act requiring arrest or detainment, at which point a polygraph will provide additional brainmapping before court-ordered confinement in a mental institution.

As I indicated above, *remote neural monitoring (RNM)* appears to be reality TV for insiders of the Deep State. Swedish TI Magnus Olsson was a successful businessman who had studied economics at the Cesar Ritz in Switzerland, the American University of Paris, and Harvard University in Boston, Massachusetts (1988–1992). In 2005, he entered a Swedish hospital where he was sedated, implanted, and transformed into a victim of non-consensual experimentation. Despite around-the-clock torture, he continues to speak out to educate the public about RNM:

> RNM has a set of certain programs functioning at different levels, like the signals intelligence system which uses electromagnetic frequencies (EMF) to stimulate the brain for RNM and the electronic brain link (EBL). The EMF Brain Stimulation [EBS] system has been designed as radiation intelligence which means receiving information from. . .electromagnetic waves in the environment. However, it is not related to radioactivity or nuclear detonation. [Non-ionized radiation, not ionized radiation.] The recording machines in the signals intelligence system have electronic equipment that investigate electrical activity in humans from a distance. This computer-generated brain mapping can constantly monitor all electrical activities in the brain. The recording aid system decodes individual brain maps for security purposes.
>
> For purposes of electronic evaluation, electrical activity in the speech centre of the brain can be translated into the subject's verbal thoughts. RNM can send encoded signals to the auditory cortex of the brain directly bypassing the ear. This encoding helps in detecting audio communication. It can also perform electrical mapping of the brain's activity from the visual centre of the brain, which it does by bypassing the eyes and optic nerves, thus projecting images from the subject's brain onto a video monitor. With this visual and audio memory, both can be visualised and analysed. This system can, remotely and non-invasively, detect information by digitally decoding the evoked potentials in 30–50 Hz, 5 milliwatt electromagnetic emissions from the brain. The nerves produce a shifting electrical pattern with a shifting magnetic flux which then puts on a constant amount of electromagnetic waves. There are spikes and patterns which are called evoked potentials in the electromagnetic emission from the brain. The interesting part about this is that the entire exercise is carried out without any physical contact with the subject . . .[64]

Thus when fifty-four-year-old Susan Burns, top assistant to the president of the Woodrow Wilson Center, attempted to destroy the Gauguin painting *Two*

---

64 Magnus Olsson, www.mindcontrol.se/?page_id=7488.

Tahitian Women at the National Gallery of Art, saying, "I am from the American CIA and I have a radio in my head," she may not have been delusional; she may have been electronically controlled, her brain "plugged in" to transmissions she was receiving.[65] *Auditory voices modified by a transducer into ELF audiograms and then superimposed on a pulse-modulated microwave beam can be transmitted directly into a target brain (microwave hearing*[66]*) and make the target feel crazy.*[67]

Signs of targeting going on in the 'hood may be unusual night traffic, trespassers, aerial activity, phone-computer-radio-TV interference, odd lights, high-frequency crackling, and electrical appliances "pulsing." Keep an eye out for "national security" contractors paid not in cash but in drugs, material goods, and home improvements, not in cash.

It's *1984* in spades.

TWO TI CASE STUDIES

Would of really told the whole story if they had included the metals,
nanotech, and pseudo life that has infiltrated and completes the circuits.

— Email from "fred," a targeted individual (TI),
in reaction to Jesse Ventura's December 17, 2012
"Brain Invaders" segment of *Conspiracy Theory*

I have personally known several targeted individuals,[68] two of whom are sketched below. Neither TI appears to have been randomly chosen.

During the Clinton years, attorney *Karen Dobson* was wrongly persecuted by the NSA, CIA, and FBI because they thought she knew something about Bill Clinton shifting the CIA drug shipment "protection racket" from Mena, Arkansas to the Oregon Coast once he was elected President in 1992. Basically cocaine was being flown from Colombia to coastal Oregon, then loaded onto trucks for East Coast distribution.

---

65 Paul Bedard, "Gauguin Painting Attacker Aided Lee Hamilton, Jan Harman." *Washington Whispers*, April 6, 2011.

66 Andrew Trotman, "Sky Deutschland to broadcast adverts directly into train passengers' heads." *The Telegraph*, 3 July 2013. "The voice comes from a Sky-branded transmitter made by Audiva that is attached to the train window."

67 See R.J. MacGregor, "A Direct Mechanism for the Direct Influence of Microwave Radiation on Neuroelectric Function." RAND Corporation, June 1970. ". . . power densities of close to 1,000,000mw/cm² can produce auditory hallucinations in a field that would be averaged as low intensity and non-thermal."

68 I recommend readers read "Organized Stalking: Information for people for whom this crime is a new issue" at electronicmindcontrolandharassment.com/wp-content/uploads/2016/01/organized-stalking-information-for-people-for-whom-this-crime-is-a-new-issue.pdf.

No matter where she fled, Dobson was targeted and tortured with DEWs. When she fled to Mexico, she wrote:

> The reason I am not nearly so ill as I was made to be in the US is only the lack of fiber optics in Mexico, and their ability to use a dedicated satellite for pulsing me every 30 to 90 seconds as if I were a radio tower. This is probably what grounds the right foot and makes holes and burns in the bottom and sides of it. . .Through MCI and the privatization of Mexico's power and utility companies (bought up by US interests, of course), Mexico—whose citizens by and large have no phones, pagers, beepers, much less cell phones—will welcome fiber optics throughout their states by the end of the year 2000. This is not because it is needed or economically profitable; I think, rather, it is for use in the fashion that it has been used on me.[69]

So arrogant were her federal case officers that they even met with her at a Denny's restaurant (CIA-owned) and admitted exactly what was being done to her and that there was nothing she could do about it. VLF (3–30 kHz) produced 24/7 nausea, dizziness, sleep deprivation, and general aches, but didn't kill her. Masers (MW + laser), however, burned her mouth, skin, eyes, and internal organs. Because the eyes are electromagnetic and mostly water, they are the hardest hit by EM weapons, the current entering the left eye and exiting the right foot. VLF was used on her acupuncture meridians, including her solar plexus. To counter the masers, in November 1996 she bought Body Armor (probably with RAM or radar-absorbent material)[70] from a "former" CIA agent in St. Louis. She listed the following arsenal she thought they were using in 1999:

- Fiber optics are perfect for DEWs, which can then be run through electricity and phone cables
- HAARP-based DEWs
- Doppler for VHF weapons; used to amplify ELF and ULF
- Huge antenna systems
- Ion laser (cooling tubes) used to ventilate the equipment usually connected to an outside vent that emits steam
- Infrared light for night viewing connected to a big screen telescope
- Hand-held equipment in laptop black cases for rural (no power or phone cables), airplanes, and rest areas

---

69 Karen Dobson, J.D. *Revealed: Total Immersion Microwave Harassment*, unpublished manuscript, ˜2003. I have been unable to contact Karen since spring 2011.

70 Body Armor emits EM pulses that alter perpetrator readings.

- MW gun that looks like a black hand-held hairdryer, its oval end emitting red energy

- Small MW weapons the size of cell phones or beepers

Her warning about fiber optics was deadly serious: *"AVOID FIBER OPTIC CABLE, A TOOL TO BE USED FOR VIGILANCE, EAVESDROPPING, AND HARASSMENT."*

Before sleep, Karen removed all things metal and slept with strong magnets and magnetic rocks around her head, washing her eyes several times a night with water treated with colloidal silver. In the mornings, she walked barefoot on the Earth. She avoided TV, microwaved foods, and synthetic bedding. All computers had to be analog, not digital. The following metals were helpful *to a point* in breaking up the MW used as carriers: aluminum foil, metal screens around hands and feet and temporal lobes above the ears, metal pot scrubbers on hands and fingers, and solar aluminized plastic sun shields. However, when the MW was torqued up, aluminum became an oven.

The last time I spoke with Karen (spring 2011), she had just returned from Mexico and warned me that calling her would mean that the NSA would be picking up first. I have been unable to reach her since.

Judah Botzer, a massage therapist and teacher of Hebrew living in Tres Piedras, New Mexico, served as a paratrooper in Israel's elite 890th Regiment in the 1980s. His targeting began in New Mexico on October 28, 2010 after he spoke out at local public meetings against the U.S. Air Force Low Altitude Tactical Navigation (LATN) plan to expand Cannon Air Force Base and the Piñón Canyon Maneuver Site (PCMS) outside Trinidad, New Mexico to seven million acres of southern Colorado and northern New Mexico under the Joint Forces Future Combat Systems.[71]

Future Combat Systems is wired into fourteen weapons, drones, robots, sensors, and hybrid electric combat vehicles—all plugged into the wireless network, supercomputers, and AI—perfect for operating in big-city canyons and looking through walls with variance-based radio tomographic imaging. The drones weigh twenty-nine pounds and fly like tiny helicopters peering with infrared eyes. The small, unmanned ground vehicle (SUGV) iRobot runs on rubber tracks, weighs thirty pounds, and has a long flexible neck with camera and sensors on top.

Thanks to his military past, Botzer surmised that USAF Special Ops high-tech electronic warfare (EW) weaponry was connected to the increasing chemicals in the no longer azure sky over Taos and believed that his targeting was specifically related to having spoken out against Raytheon Corporation, one-time owner of HAARP and other DEW patents.

---

71 "USAF proposes Low Altitude Tactical Training Area." Cannon AFB press release, September 2, 2010.

On that October day, three low-lying chemtrail columns were lined up like *vigas* over his isolated house. Whatever was in the aerosols made him double over and throw up a slightly luminescent bile, then blood. A light silvery dust coated his car and house. The air smelled foul and his lungs seized. He recorded each dive-bomb assault from 20,000 feet to about 1,000 feet and counted more than a thousand assaults from dawn to dusk and dusk to dawn. Military helicopters streaked over his house at two hundred feet.

He began writing articles for the local *Taos News*. The rain sample he sent to a lab revealed aluminum 125X over the EPA safety standard. In one article, he referenced Nazi gas chambers that "used common air as a weapon" in an attempt to alert citizens that a repeat of history might be imminent ("Bad Air in Tres Piedras," January 14, 2011). In the March 3–9, 2011 *Taos News*, he wrote "Revolution, now," meaning a revolution to take back the skies. A few months later, he referenced an even more distant event:

> And when all the land was famished, the people cried to Pharaoh for bread, and Pharaoh said unto the people, "Go unto Monsanto. What he sayeth to you, do." And the famine was over all the face of the earth. And Monsanto opened all the storehouses and sold it unto the famished. . .("Chemtrails: What in the World?" May 27, 2011)

In "Wave of the Future" (December 8, 2011), he stated that his poisoning was ongoing (disoriented, kidneys throbbing, barely able to urinate). He talked about specialists refusing to treat him, cell phone harassment, police showing up, roadblocks, being on the run, etc.

During my 2014 book tour, I visited Taos and met briefly with Judah outside an espresso café just off the Plaza. As he passed documents to me, a man heckled him from a few feet away, obviously privy to his targeting and unconcerned about any fallout from his off-the-cuff aggression. (The arrogance of the few federal agents I have encountered is extraordinary.)

An era of Nazi mental rape and extreme human rights abuse is upon us. DEWs assure plausible deniability in a jurisprudence system foundering on a bygone age. RFID readers track our driver's licenses and subcutaneous nanosensor microchips.[72] Discs like stadium loudspeakers direct bi-phase polarity frequencies from microwave towers. The Homeland Smart Grid is locked in place for the Space Fence Transhumanist Age speeding humanity toward an "enhanced" future.

---

72 Katherine Albrecht and Liz McIntyre, *Spychips: How major corporations and government plan to track your every move with RFID*. Plume, 2006.

Conclusion

Cura te ipsum! (Cure yourself!)

The Hopi prophecy of "cobwebs crisscrossing the sky" has arrived. Once considered the beacon of freedom and human rights, America is now bound up in conductive metal nanoparticles and a radio frequency 200 million times what it was for millennia. Our transceiver bodies and brains are now pulsing with wireless devices, mobile phones, laptops, tablets, power lines, televisions, towers—all resonating with the Space Fence infrastructure. Without our consent, *frequency following responses (FFRs)* generated by continuous low-frequency tones, *repetitive electrical impulse noise (REIN)*, and *flicker rates* (light transmitting data faster than the eye can see) are profoundly influencing our minds and bodies.

This era of invisible assault challenges us to peer into the invisible and take back the responsibility we once expected of government and civil society.

In this book, I have attempted to put the major puzzle pieces together of the vast, futuristic machine known as the Space Surveillance System (SSS) Space Fence now under U.S. Air Force jurisdiction. Just as the acronym HAARP serves to describe a vast system of ionospheric heaters, radar, etc., so the term "Space Fence" entails much more than what is described in Air Force press releases. The Space Fence spans low-earth orbit (LEO) all the way around the globe and includes a multitude of ground-based military installations, radomes, towers, and phased-array antennas like HAARP. The next layer is an atmosphere suffused with nano-sized conductive metals, sensors, microprocessors, and genetically engineered species, all of which we inhale, ingest, and wear.

Could any of this technology be used for the good? Tesla dreamed of dispensing with transmission wires and helping people to access the free æther energy everywhere. Certainly, HAARP-like technology could be used to minimize tornadoes and hurricanes instead of building them to wreak havoc and produce profits for the few. Drought and hunger could well be challenges of the past with food production enhanced by longer growing seasons and ample rainfall.

But a final goodbye to the out-of-control military-industrial-intelligence complex with its "Death Ray" mass mind control agendas would require an overnight shift in moral development, wouldn't it? New laws (even with teeth) would not be enough to right the wronging of more than a century. The "Star Wars" Space Fence might save us from a cosmic catastrophe, but it seems just as likely that it will land us in one.

Now that geoengineering of the atmosphere is being normalized, I see no quick and easy solution to our enslavement within a vast planetary wireless / chemical Transhumanist contrivance, our bodies and brains now walking electrified towers, antennas, and Petri dishes for which only the 0.01% have access to the antidotes.

Study of our real condition is essential, even in the wake of the seventy-year media blitz that has weakened our minds and will. No authority figure or religious/spiritual leader is coming to our rescue, nor can we enroll in a university that will teach us how to survive the future now laid out for us. We must discipline ourselves to develop the *discernment* for what is true and what is false in the bits and pieces of the broad picture begun in this book, striving always for accurate contextualizing as futuristic revelations come hot and heavy. We must understand the invasive nature of Transhumanist technologies, right down into our cells and DNA, all the while striving to live by virtue of an *inner life,* not merely by our outward life that is being constantly buffeted by messages of fear. Think, analyze, and contextualize so your emotions do not carry you—the *real* you—away. The eye at the center of any raging storm is calm.

This war for the birthright of human beings to evolve in their own time and way is long and will occasion many battles that may look small and inconsequential, having only to do with the daily grind, but do not be deceived: when battling for the future of life on Earth, *no battle is small.* Your incarnation here and now is not accidental. Each of us who loves this planet and the human birthright of an evolving consciousness must set the example for the generations coming after by keeping our house in order for a long, sustained struggle. If I might paraphrase Ephesians 6:12, this struggle is not against flesh and blood but against spiritual forces. In the days of CERN, this may be truer than ever.

I dedicate the rest of this Conclusion to remedies and wise specialists. More guidance can be found in the Conclusion "Look Up!" of my previous book, *Chemtrails, HAARP, and the Full Spectrum Dominance of Planet Earth.* I will be writing one more book in this series regarding how Transhumanism and its synthetic biology (like Morgellons) depend upon this ionized atmosphere-Space Fence contrivance.

Gird up your loins. Ours is a time for courage, not distraction, not avoidance. Reexamine your lifestyle, laziness, entitlement, self-esteem, fears, depression, and dreams. (Living in dangerous times can be good for self-development.) Watch children playing and cling to your courage to serve the human future of individual choice, not a "hive mind."

This is our watch. Let us meet its challenges with our finest qualities while caring for the body's immune system and the soul's immune system known as consciousness.

The following is by no means a complete list of possible remedies, nor am I recommending that anyone follow these suggestions. I have followed a *yin/yang* macrobiotic diet[1] for almost half a century with great success. Though I do not include fine, simple, low-tech approaches like turmeric, oregano oil, colloidal silver, or even more sophisticated approaches to health like the wave genetics of Peter Gariaev, it does not mean I do not recommend experimenting with such approaches and always, always sharing the results.

Take notes, experiment safely, share with others what you have found. No two psychosomatic bodies are the same, but Nature has certain laws (like *yin/yang*) that seem to hold fast through millennia, if you know how to apply them to your body's chemical laboratory. A low-tech approach to protection and healing *must* be experiential, as many medical experts have been programmed (and paid) to push Big Pharma "solutions." My "NO" list is small but power-packed.

Stay alkaline! Your body is your chemistry lab! *Cura te ipsum!* (Cure yourself!)

THE NO LIST (EXCEPTING SPECIAL OCCASIONS)
NO Electronic cigarettes
Read "Electronic Cigarettes Contain Higher Levels of Toxic Metal Nanoparticles Than Tobacco" by Sayer Ji, *GreenMedInfo*, March 26, 2013.
NO Microwave ovens
Read "Why Did the Russians Ban an Appliance Found in 90% of American Homes?" by Dr. Mercola, RealPharmacy.com, no date.
Read "The Hidden Hazards of Microwave Cooking" by George J. Georgiou, Ph.D., www.aaimedicine.com/jaaim/apr06/hazards.php?printable.
NO Psychiatrists
NO Hospitals
NO Sugar
NO Animal products
NO Vaccinations or shots

FIBERS DROPPING FROM THE SKY
Independent scientist Clifford Carnicom (www.carnicominstitute.org) provides a thorough guide to collection of the genetically engineered fibers dropping from the skies, including plate ionization, electrostatic, HEPA, and high-grade furnace filters. Methods of analysis encompass solubility, pH, precipitation, chromatography, electrode, electrolysis, flame, and spectroscopy. At "Microscope's Power to Examine Your Own Samples" (June 24, 2008),

---

1 Independent scientist Leuren Moret also recommends a macrobiotic regimen in "Leuren Moret: Fukushima radiation is intentional extermination and ecocide that HAARP Tesla technology can reverse," Alfred Lambremont Webre, June 9, 2014.

you will find instructions for examining what is dropping at magnification, and if you wish to conduct electrostatic precipitation studies, he recommends that you build your own Van de Graaf generator ("Biological Operations Confirmed," February 25, 2001) and your own magnetometer ("Introducing A Magnetometer," October 31, 2002).

ELFS

You can even build an ELF sensor. Go to "ELF Sensor by Steve Rouch," www.ghostweb.com/elf.html. When Carnicom utilized the ELF-LF circuit (carnicominstitute.org, "A Question of Alfvén?" November 12, 2002), he consistently picked up 75–100 kHz. Were they Alfvén or whistler waves?

LUNGS

Aerial spraying helps the viral envelope fuse with lung cells, permitting easier penetration and infection.

Email from Mary in London, winter 2014–2015: "Still dropping those ice bacteria here, can feel it going into my chest and cough I've had for 7 months now. Fortunately, it does not get infected because of the zapping, barley grass, castor oil, etc. The castor oil helps make the debris slide out the body, the zapper must break up the little critters—the Bob Beck one is the best, really, better than the Hulda Clarke one. The guy that sells them here is *www. goodvitality.com/becks-protocol-on-line-training-1018-p.asp*, and the barley grass stimulates the immune system. I must say I have to constantly up the ante, but given that I am in London, I am keeping on top of it."

IRANIAN PHYSICIST M.T. KESHE

In "Penny Kelly, 'New Worlds of Energy,' Part 3 of 3, INACS & IONS-Austin, 2013-03-19," Institute for Neuroscience And Consciousness Studies, March 27, 2013 www.youtube.com/watch?v=96bfHhFq6jM), the intriguing work of former University of Michigan biophysicist William C. Levengood is cameoed. At 14:25, a participant wonders if plasma and consciousness might be a solution to chemtrails. Penny then refers to the Iranian physicist Mehran Tavakoli Keshe's new technology that removes radioactivity and discusses how he was kidnapped by CIS (Canada) in 2011 after offering to help Japan in the wake of the Fukushima meltdown. In 2016, *Veterans Today*[2] resurrected the old nightmare of the Belgian Royal Family and its involvement with the Red Circle pedophile ring strangely revolving around attempts to control Keshe.

---

2 Gordon Duff, "VT Exclusive: Largest Pedophile Ring in History, 70,000 Members, Heads of State, the Rats Scramble." *Veterans Today*, August 24, 2016.

Now a year later, a criminal trial is underway in Kortrijk, Belgium charging Keshe and his wife Carolina De Roose with fraud and illegal medical practice.

ECOWARRIORS AND REICH LOW-TECH

Ecowarriors may not be able to get to the satellites or HAARP installations, but they have been known to go for the towers, one by one, the spirit of Edward Abbey's *Monkey-Wrench Gang* guiding them. While an Australian drove a tank through Sydney on Bastille Day, managing to knock down six cell phone towers and an electrical substation, ecowarriors operate more subtly with backpacks and belts loaded with tools, Gauss meters, using tow trucks and ropes. They are known for planting little Reich cupcakes called *holy hand grenades (HHGs)* under towers so as to interfere with transmissions. Unfortunately, a telecom truck or white van tends to arrive immediately due to instruments able to detect the interference along with exactly where the low-tech ether-generating HHGs are buried.

Block infrared (IR) = hide your heat signature

Space blanket of Mylar foil

Wool blanket

Insulated jacket, pants, hat

Trees or brush

EM PROTECTION

www.consumeraffairs.com/news/new-maternity-wear-may-protect-fetuses-from-em-radiation-071516.html

Shielded Faraday Tents hollandshielding.com/158-Faraday%20tents

Thanks to Magda Martine: "Total Shield cutcat.com/item/Total_Shield/570 is what I bought two or three years ago. It is a great EMF block for 20,000 ft surround. It detects, interrupts and eliminates grid lines, geomagnetic, electromagnetic standing waves, ELF, and other harmful waves by broadcasting these disturbances through a Tesla Coil of 180 degrees. A second generator generates 7.83 earth resonant frequency. I bought mine at Cutting Edge Catalog in Santa Fe, NM. Not too pricey and worth the protection. Namaste."

Read "How to hide from a magnetic field" physicsworld.com/cws/article/news/2012/mar/22/how-to-hide-from-a-magnetic-field

EMF RF Shielding & Conductive Paints www.lessemf.com/paint.html

RF Analyzer www.amazon.com/Hf35c-Rf-Analyze-800mhz-Detection/dp/B007L09DLY

ThorShield against EM weapons www.thorshield.com

Smart Meter Guard smartmeterguard.com

Peat fibers, www.wandil.de/index.php/en/.

"Welcome to Refugium": electroplague.com

DEEP TISSUE HEALING, VITALITY
scalarwavelasers.com

bio-mats.com

thequantlet.com

Spooky2 Rife machine, spooky2.com; read "Rife Frequency Healing Machine: Scam or Real?" by Pao Chang, energyfanatics.com/2014/03/04/rife-frequency-healing-machine-scam-or-real/#

Z-App (Rife frequency generator), play.google.com/store/apps/details?id=com.zappkit.zappid&hl=nl

528 Hz — Read 528records.com/pages/528hz-sound-miraculously-cleaned-oil-polluted-water-gulf-mexico-according-new-study-canadian-r

www.biogeometry.ca/home

How to make Essiac tea naturalsociety.com/essiac-tea-4-secret-ingredients-cured-thousands-cancer-patients/#ixzz2l7SGfUHn

Fisher Wallace Stimulator for sleep: YouTube "Camille using the Fisher Wallace Stimulator® - Week 3," Fisher Wallace Laboratories, May 10, 2013.

"Targeted treatment of cancer with radiofrequency electromagnetic fields amplitude-modulated at tumor-specific frequencies," Jacqueline Zimmerman et al., *Chinese Journal of Cancer*, November 2013, www.ncbi.nlm.nih.gov/pmc/articles/PMC3845545/.

SKIN HEALING
Amish black salve recipe, foodsandhealthylife.com/amish-black-salve-the-recipe-for-the-most-powerful-healing-ointment-2/

PlasmaDerm: "Plasma makes wounds heal quicker," PhysOrg, June 8, 2015, phys.org/news/2015-06-plasma-wounds-quicker.html.

CAVEAT: "CIA's Venture Capital Arm Is Funding Skin Care Products That Collect DNA," *The Intercept*, April 8, 2016.

TOOTH DECAY

Electrically Accelerated and Enhanced Remineralisation (EAER):
"No more fillings as dentists reveal new tooth decay treatment," *The Guardian*, 16 June 2014, www.theguardian.com/society/2014/jun/16/fillings-dentists-tooth-decay-treatment.

HEAVY METAL DETOX

Barbara M.V. Scott, www.soulmedicinejourney.com: "The tea for aluminum removal is horsetail female plant, stinging nettle, yarrow, mint for flavor and maybe some raspberry leaves, and elderflower blossoms. Tons of silica in horsetail..."

Heavy metal hair analysis, optimalhealthnetwork.com

"What Hair Analysis Testing Reveals: 10 Things Your Hair Can Tell You," butternutrition.com/what-hair-analysis-testing-reveals/

INFECTIONS

Natural antibiotic — www.healthylifeidea.com/powerful-natural-antibiotic-ever-literally-kills-infections-body/

Homemade penicillin www.realfarmacy.com/homemade-penicillin/

LABS, TEST KITS

www.healthrangerstore.com/products/cwc-labs-heavy-metals-analysis?variant=20627172865

ANTIRAD-Plus filter, www.pureeffectfilters.com/cartridges/antirad-plus.html#.VTgcIrWMv-c.facebook

ESMO Technologies Magnetic Interference Cloud, www.esmotech.com/micloud

Read "Billboard in Peru Produces Water out of Thin Air," isupportorganic.blogspot.com/2013/10/billboard-in-peru-produces-water-out-of.html#bPRvCxiDLwBTscd1.99

basiclab.com

www.pacelabs.com/ — Via *Feed Your Brain Magazine*/Christopher Everard

Alberta Hair Analysis Project — Please send any hair analysis results with only your geographical location to: thegardenofresistance@gmail.com. The locations and the results will be marked on a virtual map hosted on a website for the public to browse in the near future.

Hill Laboratories, 1 Clyde Street, Private Bag 3205, Hamilton 3240, New Zealand

www.mccampbell.com

BRAINWAVES

Frequency list www.electroherbalism.com/Bioelectronics/
FrequenciesandAnecdotes/BrainwaveFrequencyList.htm

How to construct a Thought Screen Helmet www.stopabductions.com/main.
htm

Fasting: "Neuroscientist shows what fasting does to your brain and why Big
Pharma won't study it" by Arjun Walia, Collective Evolution, December 11,
2015, www.collective-evolution.com/2015/12/11/neuroscientist-shows-what-
fasting-does-to-your-brain-why-big-pharma-wont-study-it/.

DESTROY NANOTECHNOLOGY, PATHOGENS, IMPLANTS

Tony Pantalleresco — ionic anti-nano footbath construction,
augmentinforce.50webs.com/Anti%20Nano.htm#Anti%20Nano

YouTube "AntiNano Device," HerbsPlusBeadWorks, September 17, 2015.

Negative Magnetism www.morgellonscure.com

The Zapper www.drclark.net/en/the-zapper and www.yourhealthbydesign.
com/z4ex-extreme-3-frequency-zapper/.

Bioresonance drsircus.com/medicine/
deta-elis-star-trek-medicine-bioresonance/

Read "Electrical and Frequency Effects on Pathogens,"
www.electroherbalism.com/Bioelectronics/IntroductiontoBioelectronics/
ElectricalandFrequencyEffectsonPathogens.htm

Ormusite — Susan Ganière: "ORMUS or ORME minerals are superconducting
elements, mostly the platinum group elements on the periodic chart that can
act as a conduit or antenna for very beneficial high-frequency scalar energy that
flows between other dimensions of spacetime." www.freshandalive.com/fresh-
and-alive-content/products-ormusite-scalar-disk-6inch.htm#.V_V9KjKZPEZ

Fulvic Acid thegoodlyco.com/product/humalife-fulvic-acid-concentrate/

YouTube "Morgellons Cure Discovered! – The Cure Is Obvious (832)343-5425,"
Michael Chapala, April 7, 2013. Also see his "Morgellons Cure Discovered! –
Negative EMP (832)343-5425," July 23, 2015.

"Electric fields remove nanoparticles from blood with ease," PhysOrg,
November 23, 2015.

"Infrared Saunas for Treating Lyme Disease," photonlightenergycenter.
com/infrared-saunas-lyme-disease/?utm_campaign=shareaholic&utm_
medium=facebook&utm_source=socialnetwork

CHRONIC PAIN
Transcutaneous electrical nerve stimulation (TENS or TNS), en.wikipedia.org/wiki/Transcutaneous_electrical_nerve_stimulation

ALZHEIMER'S
Read "New Alzheimer's treatment fully restores memory," *Science Alert*, March 18, 2015.

HOME
Green Bank, West Virginia. YouTube "The Town Where High Tech Meets a 1950s Lifestyle," *The Atlantic*, December 8, 2014, www.theatlantic.com/video/index/383477/the-west-virginia-town-that-banned-cell-phones/?utm_source=SFFB

The Grow Dome Project on Facebook.

YouTube "Orgone Energy — A breakthrough that has already happened - GLOBAL BEM conference Nov 2012," Georg Ritschl, March 27, 2013; also read "Operation Paradise: Environmental Healing with Orgone" by Georg Ritschl, orgoniseafrica.com/book/operationparadiseorgonite.pdf.

James DeMeo's Research Website, Orgone Biophysical Research Lab (OBRL), Saharasia, www.orgonelab.org.

Electricity-generating solar cells - www.solarwindow.com

Read "This Tower Pulls Drinking Water Out of Thin Air" by Tuan C. Nguyen, Smithsonian.com, April 8, 2014.

Read "Year-Round Survival: $300 Underground Greenhouse Could Change How we Eat," tv.naturalsociety.com/year-round-survival-underground-greenhouse-eat/.

DroneDefender, www.minds.com/blog/view/538487741193658368?utm_source=fb&utm_medium=fb&utm_campaign=marchagainstmonsanto

Susan Ganière: "Ken Rohla has tested orgonite from many of the major vendors online and most is weak at best. Welz's orgonite is properly constructed and works well, but it is fairly expensive and it does not include ORMUS technology."

YouTube "Cloud Buster Technology – Cloud dissipation," andrewg85, September 30, 2009.

YouTube "How to Use Orgonite To Protect Your Home," Radiomysterium, August 9, 2014.

"Visible effects of orgone on chemtrails," www.quebecorgone.com/orgonite-chemtrails.

QuWave Defender, quwave.net/defender.html.

YouTube "Portable Environmental Aerosol Spectrometer 11-E," GRIMMAerosolTechnik, June 24, 2014.

YouTube "Ver.2 UNDO GMO AT HOME (Heirloom Electrostatics) -The Ebner Effect," synespro, April 5, 2014.

RayGuard for electrosmog protection. YouTube "334 - Dr. Doepp: Introduction 1 of 3 to Electrosmog Protection," Creatrix13, May 27, 2009.

Read "Big Steps in Building: Change Our Wiring to 12 Volt DC" by Lloyd Alter, September 4, 2007, www.treehugger.com/sustainable-product-design/big-steps-in-building-change-our-wiring-to-12-volt-dc.html.

Read "Open Source Farming: A Renaissance Man Tackles the Food Crisis" by Dahr Jamail, August 10, 2014, www.truth-out.org/news/item/25438-open-source-farming-a-renaissance-man-tackles-the-food-crisis.

Tesla Psi-Generator, mindtechenterprises.com/index.html

IMMUNE SYSTEM

Macrobiotic diet — visit Denny Waxman's Facebook site or www.dennywaxman.com.

Read "Miso Protects Against Radiation, Cancer and Hypertension" by Margie King, August 20, 2013, www.greenmedinfo.com/blog/miso-protects-against-radiation-cancer-and-hypertension?page=1

Read "Soon Censored? Koren Scientists Successfully Kill Cancer With Magnets" by Lisa Garber, October 10, 2012, naturalsociety.com/scientists-successfully-kill-cancer-magnets/.

Read "Scientist Demonstrates How Cancer Can Be Destroyed By Frequencies" by Sean Adl-Tabatabal, January 24, 2016, yournewswire.com/scientists-demonstrate-how-cancer-can-be-destroyed-by-frequencies/

YouTube "Shattering cancer with resonant frequencies: Anthony Holland at TEDxSkidmoreCollege," TEDx Talks, December 22, 2013.

YouTube "Cancer is an Electrical Phenomenon," ElectricHealing's channel, July 13, 2010.

YouTube "Burzynski: Cancer Is Serious Business | Full Documentary," Burzynski Movie, November 4, 2012.

Read "Amazing Food Science Discovery: Edible Plants 'Talk' to Animal Cells, Promote Healing" by Sayer Ji, www.greenmedinfo.com/blog/amazing-food-science-discovery-edible-plants-talk-animal-cells-promote-healing#

BioVin, www.newhope.com/managing-your-business/ no-mere-antioxidant-cyvex-launches-biovinr-advanced-red-wine-extract.

Read "Electromagnetic Frequencies Prevent Germination" by Catherine J. Frompovich, January 14, 2016, www.wakingtimes.com/2016/01/14/ electromagnetic-frequencies-prevent-germination/

Polycontrast Interference Photography (PIP), www.electrocrystal.com/pip.html

YouTube "Dr. Jerry Tennant – pH and Voltage – "Healing is Voltage," HealthAbundance411, April 30, 2013.

YouTube "Sodium Bicarbonate (Baking Soda) and pH Medicine," Dr. Sircus, February 28, 2013.

Read "Your body is electrical and runs on electrons – not sugar, protein or fat!" Dr. Robert O. Young, blog.phoreveryoung.com/2015/08/19/ your-body-is-electric-and-runs-on-electrons-not-sugar-protein-or-fat/

Read "How the Western diet has derailed our evolution" by Moises Velasquez-Manoff, November 12, 2015, nautil.us/issue/30/identity/ how-the-western-diet-has-derailed-our-evolution

TRUSTED HELPERS (MANY MORE!)
Hulda Regehr Clark, N.D. (1928–2009)

YouTube "Dr Peter Gøtzsche exposes big pharma as organized crime," John McDougall, April 1, 2015.

Manfred Doepp, Ph.D.

Konstantin Meyl, Ph.D. — Tesla kit

Hildegarde Staninger, M.D.

Dietrich Klinghardt, M.D.

BLESSINGS ON YOUR HEADS! ONWARD!

GLOSSARY

ACID pH <7

ACOUSTIC GRAVITY WAVES Infrasound low frequency (20–100 Hz) modulated by ultra-low infrasonic waves (0.1–15 Hz). Many of the groans and hums heard around the world are acoustic gravity waves.

ALBEDO Latin, "whiteness." Reflected sunlight; the diffuse reflectivity or reflecting power of a surface

ALKALINE pH >7

AM Amplitude modulation

ANTHROPOGENIC Man-made

ATMOSPHERE Composed of the *troposphere* (6–20km), stratosphere (20–50km), mesosphere (50–80km), thermosphere (80–690km)

ATMOSPHERIC RIVER Bands of enhanced water vapor in the atmosphere several thousand kilometers long and a few hundred kilometers wide. One atmospheric river can carry a greater flux of water than the Amazon River. Three to five atmospheric rivers in a hemisphere at any given time.

BOLOMETER Measures incident electromagnetic radiation by heating a material with a temperature-dependent electrical resistance; from the Greek *bole* (βολή) for something thrown, as a ray of light

CATION An ion or group of ions having a positive charge

C4 Command, control, communications, cyberwarfare

CIA Central Intelligence Agency

CME Coronal mass ejection

DB Decibel; a measurement used in acoustics and electronics, such as gains of amplifiers, attenuation of signals, and signal-to-noise ratios

DIA Defense Intelligence Agency

DOPPLER NEXRAD (NEXt generation weather RADar) is a Doppler radar system.

ELECTROJET A charging electric current 90–150 km (56–93 miles) in the ionosphere; the *equatorial electrojet* (magnetic equator) and *Auroral electrojets* (Arctic and Antarctic Circles) are the ionospheric fluctuations.

ELF Extremely or extra low frequency, 3–300 Hz

EHF Extremely high frequency. The advantage of EHF is that transmissions travel in a straight line (a beam).

EPA Environmental Protection Agency

ERP Effective radiated power of directional radio frequency

FAA Federal Aviation Administration

FREQUENCY The rate at which a particle moves back and forth in space. Electrons behave collectively in waves; ions move more slowly and make waves with a smaller frequency.

FM Frequency modulation

FSB Frequency-selective bolometer; low noise sub-millimeter to mid-infrared sensor

GAUSS (G) Measurement of magnetic flux density or magnetic induction

GE/GM/GMO Genetically engineered / genetically modified organism

G-FORCE Gravitational force exerted on pilots who jink at close to 500 mph

GIGAHERTZ GHz, one billion (1,000,000,000) cycles per second (Hz)

HARMONICS The mathematics that connect wavelengths of matter, gravity, and light

GLOSSARY

HERTZ (HZ) Measure of frequency in cycles per second. In the U.S., electrical power is an alternating cycle (AC) of current and voltage with a frequency of 60 cycles per second regardless of the amount of power used:

kHz (kilohertz, 10^3 Hz)
MHz (megahertz, 10^6 Hz)
GHz (gigahertz, 10^9 Hz)
THz (terahertz, 10^{12} Hz)

HUMINT Human intelligence

HYDROSCOPY Optical device used for viewing objects far below the surface of water

HYGROSCOPY A substance's ability to attract and hold water molecules from the surrounding environment.

HYGROSCOPE instrument showing changes in humidity.

IMINT Imagery intelligence (NGA)

IONIZATION The process by which ions are formed, (1) passing radiation through matter, and (2) heating matter to high temperatures to give the atoms energy and force the electrons to leave the atom to become free electrons (negative charge). (Whatever is left of the atom becomes a positive ion.)

IONOSPHERE From the top of the mesosphere to the top of the thermosphere.

INTERFEROMETRY HAARP is a scalar interferometer. See Myron W. Evans, P.K. Anastasovski, T.E. Bearden et al., "On Whittaker's Representation of the Electromagnetic Entity in Vacuo, Part V: The Production of Transverse Fields and Energy by Scalar Interferometry" (*Journal of New Energy*, 4(3), Winter, 1999, 76–78); and www.padrak.com/ine/JNEV4N3.html.

JET STREAMS Fast-flowing air currents in the upper atmosphere. *Polar jets* 9–12 km (5.5–7.4 miles) above sea level, or *subtropical jets* 10–16 km (6.25–10 miles), one of each in the Northern and the Southern Hemispheres. Jet streams are caused by planetary rotation and atmospheric heating combined.

KILOMETER 0.621371 miles (a little over half a mile)

LASER Light amplification by stimulated emission of radiation

LIDAR Light Detection and Ranging / Laser Imaging Detection and Ranging

MAGNETOSPHERE Overlaps the ionosphere and extends into space to 60,000km (37,280 miles) toward the Sun, and over 300,000km (186,500 miles) away from the Sun (nightward) as the Earth's magnetotail

MAGNETRON A barrier microwave technology that operates non-ionizing radiation at the extremely high-frequency (EHF) end of the spectrum by generating electrons with a cathode-anode cylinder to combine a magnetic field with an electrical field. A large magnetron can generate a consistent EHF beam of microwave pulses equaling 10 million watts (10MW) per pulse.

MASER Microwave Amplification by Stimulated Emission of Radiation

MASINT Measurement and signature intelligence (DIA)

MEGAHERTZ MHz, one million (1,000,000) cycles per second (Hz)

METAMATERIAL (Greek, *meta* — beyond) Engineered to have a property that is not found in nature

MICRON (μM) Micrometer; one millionth of a meter (0.000039 inch). 1 micron is 1/70 the thickness of a human hair.

NANOMETER (NM) One-billionth of a meter (0.000000001 m)

NASA National Aeronautics and Space Administration

NGA National Geospatial-Intelligence Agency

NOAA National Oceanic and Atmospheric Administration

NONIONIZING Nonthermal radiation; insufficient energy to cause ionization *radiation* or heating: electric and magnetic fields, radio waves, microwaves, infrared, ultraviolet, and visible radiation

NRO National Reconnaissance Office (housed at one time in Room 4C1052 of the Pentagon)

ONI Office of Naval Intelligence

ORGONE Wilhelm Reich's term for the life-giving ether

OTH Over-the-horizon radar

PINEAPPLE EXPRESS Atmospheric river affecting the West Coast of North America

PM Particulate matter; particles that are 10 microns or less. Coarse particle = 10,000-2,500 nanometers; fine particles = 2,500-100 nanometers; nanoparticle (ultrafine) = 1-100 nanometers.

PPM Parts per million

PLASMA Fourth state of matter; an ionized (electrically conductive) gas in which atoms have lost electrons and exist in a mixture of free electrons and positive ions. Positively charged particles (electrons and ions) governed by electric and magnetic forces possess collective behavior.

QUORUM SENSING (QS) Cell-cell communication.

RADAR Radio Detection And Ranging

RADON A chemically inert, radioactive gaseous element produced by the decay of radium

RFR Radio frequency radiation

SCALAR Having only magnitude, not direction; longitudinal as opposed to EM waves which are transverse

SDI Strategic Defense Initiative; "Star Wars"

SIGINT Signals intelligence

SLF Super low frequency

SRM Solar Radiation Management

SYNERGY Increased intensity caused by the combination of two or more substances

TECHLEPATHY Communication of information directly from one mind to another (i.e., telepathy) with the assistance of technology

TTA Tesla tech array

WATTS Electrical power. For Direct Current (DC), watts = volts x amps. For Alternating Current (AC), calculating watts is more complicated but if using the Root Mean Squared (RMS) values for voltage and current, watts = volts x amps. Normal US household electricity is 120 volts, so if your appliance draws 2 amps, watts = 120 volts x 2 amps = 240 watts.

MAPS AND SITES

CLIMATE AND SPACE

Billy Hayes: "I remind all that computer weather animations are forecasting tools (CGI = computer graphics imagery), not actual satellite maps. Aviation alerts are for real as the U.S. military conducts airborne radar tests along the Atlantic coast."

"The Hubble Space Telescope Photos or Virtual Reality Deceptions?" Godrules, Feb. 22, 2015, www.youtube.com/watch?v=aAuUXmNObSw

Compare Google's with Jim Lee's map, one misleading, the other striving for reality:

www.google.com/maps/d/viewer?hl=en_US&mid=1ZAjP-EnlrT5AgQNBmipZokb_mtw - HAARP tech locations

Ionospheric heaters + missile defense radars, Star Wars, Space Fence + SuperDARN Radar + High-powered lasers

Go to ClimateViewer.org and click on the sidebar folder labeled "Electromagnetic" climateviewer.org/mobile/index.html?layersOn=star-wars,sky-heaters,superdarn,directed-energy+digisonde + lasers + ELF/VLF

Earthquakes: EarthScope Array Network Facilty (ANF) anf.ucsd.edu/recenteqs/

NexSat: goo.gl/ldwzHa - the only thermal-imaging satellite or radar or cloud mass

climate.cod.edu/hanis/satellite/hemi/index.php?type=polar-wv-1-24&checked&prodDim=100&overDim=100 - NEXLAB water vapor for polar sector

tropic.ssec.wisc.edu/real-time/mimic-tpw/main.html5.php?&basin=global - SSMI/SSMIS/TMI-derived Total Precipitable Water - Global

spaceweather.tv

sdo.gsfc.nasa.gov - Solar Dynamics Observatory

earth.nullschool.net

SkyGrabber.com

www.antipodesmap.com - other side of the Earth

flightaware.com/live/

earthquake.usgs.gov/earthquakes/map/

earthquake.alaska.edu/earthquakes

www.swpc.noaa.gov/products/solar-cycle-progression

droughtmonitor.unl.edu

www.ssd.noaa.gov/goes/west/

thehum.info - the world hum and database project

www.antennasearch.com - for your neighborhood

enipedia.tudelft.nl/maps/PowerPlants.html - power plants in your area

mapsat.com

allskycam.com

spaceweather.com/flybys/index.php?PHPSESSID=rmkao5begup6glpji1grsv2no5 - check spy satellites overhead

www.weatheronline.co.uk

www.heartmath.org/research/global-coherence/gcms-live-data/ - magnetometers

www.intellicast.com/Local/WxMap.aspx?latitude=39&longitude=-97&zoomLevel=4&opacity=1&basemap=0014&layers=0039

www.eol.ucar.edu/all-field-projects-and-deployments - Earth Observing Laboratory projects over the years

www.carnicominstitute.org/javascript/climate_model/climate_model_videos.html - "A Geoengineering & Climate Change Model," Videos & Tutorials, Clifford E. Carnicom, January 26, 2015.

Clifford: "This is a model to estimate climate change from applying changes in greenhouse gas concentrations relative to the current rates of increase. It also provides for introducing various aerosols and into the atmosphere. It also includes the simulation of random events upon climate change. It will also estimate mortality impacts upon the population. The model is based upon several factors, such as a differential analysis of the specific heat of a mixture of gases, the specific heats and albedos of various aerosols, and mortality projections."

SAFETY

www.whocalledme.com - Find out who called you

map.norsecorp.com/#/ - indicates cyber attacks, by and to whom, type of attacks

www.naturalnews.com/054354_heavy_metals_testing_Health_Ranger_laboratory_CWC_Labs.html?utm_source=facebook.com&utm_medium=social&utm_campaign=Postcron.com - Heavy metals testing

www.optimalhealthnetwork.com/Hair-Tissue-Mineral-Analysis-s/1050.htm

www.youtube.com/user/MfromCanada2 - Morgellons Coverup YouTubes by Marcia Pavlis

fukushimainform.ca

www.electroherbalism.com/Bioelectronics/FrequenciesandAnecdotes/BrainwaveFrequencyList.htm - brainwave frequencies

espace.cern.ch/be-dep/BEDepartmentalDocuments/BE/LHC_Schedule_2016.pdf - CERN schedule

www-elsa.physik.uni-bonn.de/accelerator_list.html - particle accelerators around the world

RECOMMENDED READING

No Natural Weather: Introduction to Geoengineering 101 by WeatherWar101 (Author), Sofia Smallstorm (Foreword). Kindle Edition, 2014.

Chemtrails Exposed: A New Manhattan Project by Peter A. Kirby. Kindle and paperback, 2012.

INDEX

ABOUT THE AUTHOR

Elana Freeland is a writer, ghostwriter, lecturer, storyteller, and teacher who researches and writes on Deep State issues, including the stories of survivors of MK-ULTRA, ritual abuse, and invasive electromagnetic weapons (*Nexus*, October 2014). She is best known for *Chemtrails, HAARP, and the Full Spectrum Dominance of Planet Earth* (Feral House, 2014).

She has also self-published the *Sub Rosa America* series of four books, a fictional American (occult) history of the Deep State that has dominated and destroyed the United States since President John F. Kennedy's public assassination. Most recently, she wrote a story called "What Would Solon Have Done?" for the 2017 book *If I Were King: Advice for President Trump*, edited by Harry Blazer and Introduction by Catherine Austin Fitts.

Freeland's undergraduate degree was in creative writing with a second major in biology. Her Master of Arts degree from St. John's College concentrated on historiography. She lives in Olympia, Washington.